Freshwater Fishes of South-eastern Australia

FRESHWATER FISHES
OF SOUTH-EASTERN AUSTRALIA

Edited by R.M. McDowall

REED

First published in 1980 by
A.H. & A.W. Reed Pty Ltd, Sydney

This completely revised edition published in 1996 by
REED BOOKS
a part of Reed Books Australia
Level 9, North Tower
1–5 Railway Street
Chatswood, NSW 2067

© Contributing authors, 1996

All rights reserved. No part of this publication may be reproduced, stored in a retrieval system or transmitted in any form or by means.electronic, mechanical, photocopying, recording or otherwise, without the prior written permission of the publishers.

National Library of Australia
Cataloguing-in-Publication Data

Freshwater fishes of south-eastern Australia
[Rev. ed.]
Bibliography.
Includes index.
ISBN 0 7301 0462 1.

1. Freshwater fishes - Australia - Identification.
I. McDowall, R. M. (Robert Montgomery), 1939- .

597.092994

Edited by Peter Meredith
Designed by Warren Penney
Printed in Hong Kong by South China Printing

Contents

Contributors 6
Preface to the second edition 7

1. Introduction 9
2. Studying fishes 15
   Glossary 19
3. Key to families of fishes found in the fresh waters of south-eastern Australia 25
4. Family Mordaciidae Shortheaded Lampreys 32
5. Family Geotriidae Pouched Lamprey 36
6. Family Anguillidae Freshwater Eels 39
7. Family Clupeidae Herrings 44
8. Family Chanidae Milkfish 48
9. Family Elopidae Tarpon and Oxeye Herring 50
10. Family Galaxiidae Galaxiids 52
11. Family Aplochitonidae Tasmanian Whitebait 78
12. Family Salmonidae Salmons, Trouts And Chars 81
13. Family Retropinnidae Southern Smelts 92
14. Family Prototroctidae Southern Graylings 96
15. Family Cyprinidae Carps, Minnows etc. 99
16. Family Ariidae Salmon or Fork-Tailed Catfishes 107
17. Family Plotosidae Eel-tailed Catfishes 109
18. Family Cobitidae Loaches 114
19. Family Poeciliidae Livebearers 116
20. Family Atherinidae Silversides or Hardyheads 123
21. Family Melanotaeniidae Rainbowfishes 134
22. Family Pseudomugilidae Blue-eyes 141
23. Family Scorpaenidae Scorpionfishes 144
24. Family Chandidae Glassfishes, Chanda Perches 146
25. Family Percichthyidae Australian Freshwater Cods and Basses 150
26. Family Terapontidae Freshwater Grunters or Perches 164
27. Family Nannopercidae Pygmy Perches 168
28. Family Cichlidae Cichlids 176
29. Family Apogonidae Cardinalfishes and Mouthbrooders 181
30. Family Percidae Freshwater Perches 183
31. Family Gadopsidae Freshwater Blackfishes 186
32. Family Mugilidae Grey Mullets 191
33. Family Bovichtidae Congolli 198
34. Family Gobiidae (subfamilies Eleotridinae and Butinae) Gudgeons 200
35. Family Gobiidae (subfamilies Gobiinae and Gobiinellinae) Gobies 220

References 229
Index 241

Contributors

This book results from collaboration between fisheries scientists and specialists, most of whom are professionally active in research on Australian freshwater fishes. This collaboration made it possible to produce a book of authoritative quality that would not have been possible with single authorship. No one author could marshall the information available to the many workers in the field, much of it not formally published.

The editor of this book, Dr Robert M. McDowall, is a scientist with the National Institute of Water and Atmospheric Research in Christchurch, New Zealand. He has worked widely on the freshwater fishes of cool southern lands, particularly New Zealand, but also extensively on selected Australian fish families, particularly as a result of a period of study at the Australian Museum, Sydney. The contributing authors, listed below, are active in fisheries research in a variety of museums, universities, and State and Commonwealth fisheries research organisations, mostly in eastern Australia.

G.R. Allen, Western Australian Museum
A.P. Andrews, Tasmanian Museum
A. H. Arthington, Griffith University
J.P. Beumer, Queensland Department of Primary Industries
I.C. Briggs, formerly Australian Museum
A.R. Brumley, East Gippsland Community College of TAFE
J. Burchmore, New South Wales Fisheries
P.L. Cadwallader, Victoria Department of Conservation and Natural Resources
L.E.L.M. Crowley, Macquarie University
P.E. Davies, Freshwater Systems
T.L.O. Davis, Commonwealth Scientific and Industrial Research Organisation
W. Fulton, Tasmanian Inland Fisheries Commission
J.H. Harris, New South Wales Fisheries
D.F. Hoese, Australian Museum
P.A. Humphries, Murray-Darling Freshwater Research Centre
W. Ivantsoff, Macquarie University
P.D. Jackson, Queensland Department of Primary Industries
J.D. Koehn, Arthur Rylah Institute
R.H. Kuiter, Aquatic Photographics
H.K. Larson, Museum and Art Gallery of the Northern Territory
M. Lintermans, ACT Parks and Conservation Service
L.C. Llewellyn, NSW National Parks and Wildlife Service
J.R. Merrick, Macquarie University
P. Parker, formerly New South Wales Fisheries
D.A. Pollard, New South Wales Fisheries
I.C. Potter, Murdoch University
M.A. Rimmer, Queensland Department of Primary Industries
S.J. Rowland, New South Wales Fisheries
A.C. Sanger, Tasmanian Inland Fisheries Commission
J.M. Thomson, University of Queensland

Preface to Second Edition

Since publication of the first edition of this book about 15 years ago, knowledge of the freshwater fishes of south-eastern Australia has grown rapidly, more rapidly than at any other time in Australia's history. There have been major revisions of large families like Atherinidae and Melanotaeniidae; new species in diverse families have been described from the area; knowledge of the life histories, habitats and diets of many species has increased; the Australian Society for Fish Biology has established a listing of threatened and endangered species; several States have initiated species recovery plans to help ensure the survival of species in decline; and there has been serious and intense interest in the freshwater fishes of Australia and Papua New Guinea by amateur naturalists and aquarists through the Australia New Guinea Fishes Association (ANGFA). ANGFA has been responsible for publication of both periodic newsletters as well as the journal *Fishes of Sahul*, in which can be found a mass of information on the fishes covered by this book. Other books have been written on the fauna, some addressing the fauna of various States, others covering the entire freshwater fish fauna of Australia.

All these factors have led to substantial changes in this new edition, both in the information provided about species described in the first edition, as well as the addition of new species. Some of them are described as new in the past 15 years, but some are added as exotic fishes that have become established in south-eastern Australian waters in that time.

Several people have contributed to the information published here, in addition to those listed as authors. These include K. Bishop, P. Brown, R. Faragher, D. Jerry, M. Mallen-Cooper, S. Saddlier, P. Unmack, and R. Watts. Rudie Kuiter and Neil Armstrong generously provided a large number of colour slides from which to choose those printed here.

1
Introduction

The continent of Australia covers a vast area and is, overall, one of the driest parts of the world; large areas of central and western Australia are desert or semi-desert and lack flowing waters at most, if not all, times of the year. Major river systems are restricted to northern coasts, the south-west, and eastern coasts and mountain ranges from northern Queensland to Victoria. Tasmania, except for the east coast, is well watered. A dominating feature of eastern Australia is the huge, inland Murray-Darling River system, one of the great rivers of the world, which drains southern Queensland, the whole of New South Wales and Victoria west and north of the Great Dividing Range, as well as the south-eastern corner of South Australia.

This book provides an account of the freshwater fishes of much of this eastern region of Australia (see p. 10), including those rivers that drain the mountains of the Great Dividing Range and flow east or south towards the coast, and extending from the south-eastern corner of Queensland south and west to eastern South Australia (to about Spencer Gulf, west of the mouth of the Murray); the entire Murray-Darling system, which drains westwards from the Great Dividing Range and then south-west to the southern coast of Australia; and Tasmania. It has been compiled to enable naturalists, fisheries biologists, aquarists, anglers, enthusiastic amateur observers and others to identify the freshwater fishes of this area, to obtain a brief outline of what is known about the natural history of these fishes, and also to locate up-to-date and authoritative sources of more detailed and/or technical published information about them.

Australia's freshwater fish fauna is quite small, totalling about 180 indigenous species; to this, human introductions have added about 25 more species, making for a fauna that is much less diverse than is found in comparable areas in Asia, Africa or the Americas. This, of course, is due in large measure to the excessive dryness of much of the Australian continent, as mentioned above.

The fauna was very poorly known until relatively recent decades. However, extensive exploratory and taxonomic work has been carried out on the fauna since the early 1970s, as a result of which the fauna has become much better understood. Order has been created from what was formerly a rather chaotic situation: far more species were formerly recognised in some families than are now recognised, while in other families new species have been discovered and described (and quite a few remain to be described). The life histories and ecology of some species have also been described. In addition to the work undertaken by professional fish biologists, considerable advance in knowledge of the fauna can be attributed to Australia's vigorous and committed band of freshwater fish aquarists and amateur naturalists. They have come to recognise that Australia's freshwater fishes, though perhaps not as diverse or grand or distinctive as the faunas of the Amazon or the great Rift lakes of Africa, are nevertheless beautiful and interesting to keep and

INTRODUCTION

rewarding to study. And Australia has given the world of aquarists the rainbowfishes (family Melanotaeniidae), justifiably famous everywhere in the world of tropical fish keeping.

Of the 180 freshwater fish species known in Australia, about 108 are found in the south-east, in the area covered by this guide; of these, 86 are native to Australia, while 22 are introduced. Nearly all of the native species are found only in Australia, many only in the south-east, but some occur elsewhere in Australia, usually either in the south-west, or otherwise further north and west into Queensland. A very few are found beyond Australia, four being present in New Zealand and two of these four also in South America, while a few are found in the islands of the south-western tropical Pacific. In addition to species that are clearly 'freshwater fishes', the fauna includes quite a few that may or must spend a part of their lives in the sea. Some are just freshwater fishes that become estuarine or coastal wanderers, but others have life histories that involve obligatory movements to and from the sea at well defined and regular life history stages—species described as 'diadromous' fishes. Finally, there are many marine fishes that may wander into estuaries, or even further into truly fresh waters. Thus, determining the boundary between 'freshwater fishes' and 'sea fishes' is far from explicit, making for some debate about what species should be included in a book on the 'freshwater fishes'. We have attempted to err on the side of including too many species in the freshwater fauna rather than excluding some. Rather more families of fishes have been included in the Key to Families of fishes, so that at least a family identification can be made, even though no further information may be provided here about such species; help can be found in books on the marine fishes of eastern Australia.

The main objective of this book is to enable identification; because of this, a variety of potential ways of identification has been provided. Keys, a well-known if sometimes daunting tool for identification, have been prepared to enable first the identification of families to which a fish belongs, and then identification to the species. In the accounts of each species, details are provided of the most important characters that assist in identification, especially the diagnostic characters that enable the most similar species to be distinguished. As far as possible, the characters listed can be seen with the naked eye and without the need for dissection of a specimen; however, sometimes this is not easily possible, and always the intention has been to provide enough information for a certain identification to be achieved. Most of the species are illustrated with black-and-white drawings, many also with colour photographs.

In addition to information about identification, we have included a summary of what is known about the natural history—the habits and habitats—of each species. There is a great deal of variation in what is known. Some species have long attracted the attention of naturalists or fisheries biologists—because they are common, or are involved in fisheries, are beautiful and perhaps of high interest to aquarists, or are highly peculiar, or maybe moderately well known. Others are rare or little known, perhaps apparently 'uninteresting' or of no particular 'value', or have only recently been discovered. Virtually nothing is known about them beyond a superficial description of where they live and perhaps some interpretation of how they 'make a living' judged by their characteristics or by species to which they are closely related. Because of this variation in knowledge, the length of accounts of species varies greatly. Overall, we have tried to provide a summary of what is known, each species being treated as equally important, whether rare or common, useful in fisheries or not, beautiful or ordinary, native or introduced.

This book has several sections: Chapter 2 tells how to use this book, how to make identifications using the keys and how to find and distinguish the characters that are used to identify fishes. Information in this chapter should be used in conjunction with the Glossary (p. 19), in which various body parts and structures of fishes, especially those important to identification, are described; some of the technical terms used in the text are also defined. Chapter 2 also deals briefly with methods of handling and preserving fishes, indicates how they should be stored and labelled and provides some information on where to find help with identification, or where further information on Australian freshwater fishes can be found. Chapter 3 consists mainly of a Key to Families of fishes found in this book, including the true freshwater fishes as well as many of the marine fishes occasionally found in fresh waters. Use of the Key to Families enables the observer to determine the family of fishes to which a given fish belongs, and this greatly facilitates identification. For freshwater fish species, the Key to Families then indicates where to turn in the book to obtain the identity of the species using the Key to Species in each family (where that family has more than one species).

Each of the subsequent chapters deals with a particular family of fishes. These chapters vary greatly in length since some families may contain only one Australian species, or only one species altogether, which may be poorly known, whereas other families may contain many species, some of which may be well studied and known. Regardless of how many species are included, a common format is followed for each family.

- There is a brief introduction to the family.
- For each family with more than one species there is a Key to Species that assists with rapid and sure identification of members of that family.

Following the Key to Species in each chapter are individual species accounts. These too follow a consistent format. The section is headed by the accepted common name and the correct scientific name, following the most recent authority. Information is then given under the following subheadings, in a fixed order:

Description A brief summary of the important and diagnostic characteristics of the species (the aim being identification rather than description).
Sexual dimorphism This is a term used when there are easily seen differences between the sexes, and if there are such differences, this section is included; it is important to know when such differences exist, as very often in the past, as a result of sexual dimorphism, the male has been regarded as belonging to a species different from the female, and much confusion has arisen.
Colour Obviously of great importance to identification, colour of the species, and variations, are described. Often colour patterns are more important than the actual colours themselves. Colour of fish may vary, depending on geographical locality, the colour of the water and the lake or stream bed, the vegetation in the water and the amount of light that reaches it. Moreover, in some species, colour differs with age or size or with the stage of sexual maturity; or the male may differ from the female, usually being more brightly coloured than the female.
Size The greatest known size of each species is given, and in many instances the commonly attained size is also included. Weights reached may also be indicated (mostly for large fishes of importance to anglers). Length is taken from the tip of the snout, either to the base of the fork in the tail or—in species with rounded tails—to the most distant point of the tail.
Distribution Distribution of the species within the area covered by this guide is described and summarised in a small map; where the species occurs beyond the region, this is also outlined. For introduced species this section is labelled 'Introduction and Distribution'.
Conservation status A separate section on the conservation status of indigenous species is provided. This section will say whether the species has been listed by the Australian Society for Fish Biology (ASFB) as being of particular conservation concern.

The ASFB has assigned Australian freshwater fishes to a series of categories of conservation concern as follows:
Extinct Taxa that are no longer found in the wild or in a domesticated state;
Endangered Taxa that have suffered a population decline over all or most of their range,

whether the causes of this decline are known or not, and which are in danger of extinction in the near future (special measures required if the taxa are to continue to survive);

Vulnerable Taxa not presently endangered but which are at risk by having small populations and/or populations that are declining at a rate that would render them endangered in the near future (special measures required if the taxa are to continue to survive);

Potentially threatened Taxa that could become vulnerable or endangered in the near future because they have a relatively large population in a restricted area; or they have small populations in a few areas; or they have been heavily depleted and are continuing to decline; or they are dependent on a specific habitat for survival (require monitoring);

Indeterminate Taxa that are likely to fall into the Endangered, Vulnerable or Potentially Threatened category but for which insufficient data are available to make an assessment (require investigation);

Restricted Taxa that are not presently in danger but which occur in restricted areas, or which have suffered a long-term reduction in distribution and/or abundance and are now uncommon;

Uncertain status Taxa whose taxonomy, distribution and/or abundance are uncertain but which are suspected of being restricted.

The ASFB has taken particular interest in the conservation status of Australian freshwater fishes, has a special committee that addresses this problem and maintains a continually updated register that lists the status of threatened and vulnerable species (Jackson, 1993). There is also a detailed account of the conservation status of the fauna (Wager and Jackson, 1993) that provides a rationale for the categories assigned to species of conservation concern. The ASFB categories are included where appropriate in this section for each species' account.

Natural history This section covers details of the habits and habitats of each species, e.g. what sort of conditions the species usually lives in, where, when and how it breeds, how the young behave, what the fish feed on, what other fish feed on it. This section is thus a mélange of information that varies greatly in extent and coverage, depending on how much is known. Often, very little is known.

Utility Freshwater fishes have a variety of uses to humans—as angling fishes, as food from either wild stocks or fish farms, as bait fishes for anglers or commercial fishermen, as forage fishes for larger predatory species that may or may not be useful for humans, as biological controls for troublesome insects or damaging aquatic weeds, or as decorative and interesting aquarium and pond fishes. Some species may be harmful to the environment within which they live, or they may affect the occurrence or survival of other fish that share their habitats—especially where introduced into new fish communities either from beyond, or other parts of, Australia. Details of this sort are discussed in this section. Notes on catching fishes of importance to anglers may also be included.

Similar species Although we believe that the keys and descriptions, together with the illustrations, should be enough for identification, confusion is bound to arise between species that look similar even though they may not be closely related. This section lists those species that we think are likely to cause confusion, making for easier cross-checking of descriptions and illustrations.

Other names This brief section has two parts:
1. Common names used vary widely, sometimes from place to place; some authorities even suggest two or more common names; whatever the case, the one used here may not be the one that some readers are familiar with. Only common names in fairly wide usage are listed. Australian fish biologists at present have a real problem with the use of common names for fish. Different common names are used by different people; sometimes books offer duplicate names, which only adds to the confusion, and they are always changing. There are even instances where the same name is used for different species. This is a problem that needs attention, perhaps best from the Australian Society for Fish Biology.
2. Scientific names that have been used widely in recent years but which are no longer regarded as the correct names are given in this section. Changes in scientific names are unfortunate as they can create much confusion; even though their use is strictly controlled by the International Commission for Zoological Nomenclature (which issues the Code for Zoological Nomenclature), some changes do need to be made. Examples of changes that have caused particular trouble or consternation include the following: The Murray cod, for many years known to science as *Maccullochella macquariensis*, is correctly known as *Maccullochella peeli*—the name *Maccullochella macquariensis* is the proper name for the trout cod, formerly called *Maccullochella mitchelli*. The common jollytail was known for many years as *Galaxias attenuatus* but must now be known as

INTRODUCTION

Galaxias maculatus. A recent change made in North America has significance to Australia: the rainbow trout, known for more than a century as *Salmo gairdnerii*, is now called *Oncorhynchus mykiss*. The reasons for this last change are complex but relate to the fact that there is an old name used for the fish in Siberia (*mykiss*) and to opinions that the rainbow belongs with the Pacific salmons (genus *Oncorhynchus*) rather than with the Atlantic salmon and brown trout of Europe and eastern North America.

Literature At the end of each species account is a list of a few publications to which readers may refer to obtain additional information about the various fishes. These references include only the author and date; the complete reference, needed to obtain the appropriate publication from a library, is given in the References section on p. 229. Several sources recur frequently in this section. One of these is Merrick and Schmida's (1984) book on the entire Australian freshwater fish fauna, which brings together in one place more information than anywhere else available about the fauna. Much of that information is drawn from other literature but much also derives directly from the authors' experience with these fishes. Another major source is Koehn and O'Connor's (1990) very comprehensive summation of the literature on the freshwater fishes of Victoria. The information there is widely scattered through their report but can nevertheless be brought together by species with a little effort. Several other significant books have been published on the fauna, beginning with Lake (1971, 1978), and later McDowall (1980), Cadwallader and Backhouse (1983) and Allen (1989). In general these are not cited as they are essentially summary compilations and contain less original information than other sources. Nevertheless, they are listed where the authors involved have special knowledge of a group of fishes or where there is minimal other available information.

2
Studying fishes

The study of fishes, and the use of keys to identify them, depends on some knowledge of fish structure, and of the methods used in measuring and counting the various parts and structures. These structural characters are defined in the Glossary (p.19).

Fish descriptions are based on several types of characters. General descriptive characters, such as 'fish elongate and slender', are used and are helpful, but more precise body proportions, and counts of serially-repeated body parts such as fin rays and vertebrae, are also important and widely used. *Body proportions* are obtained from the measurement of body parts and are usually given as proportions or percentages of other, larger body proportions, e.g. head length as a percentage of standard length, or eye diameter as a percentage of head length (Fig. 2.1). *Fin spines* and *fin rays* may be counted; in all fishes all the spines are counted, but in some fishes not all the rays are counted, specifically the very tiny, unsegmented and unbranched rays at the front of the fins, known as *procurrent rays* (Fig. 2.1). Only the segmented rays in the dorsal and anal fins are counted. Often the last ray in the dorsal and anal fins appears to be double, as it is divided to the base, but it is counted as one ray. The standard procedure in the study of the caudal (tail) fin is to take a *principal ray count*, which includes all the branched rays in the tail plus the two longest, unbranched rays, one above and one below the branched rays; usually these two unbranched rays almost reach the tip of the fin. Spine counts are given in Roman numerals (e.g. VIII) and ray counts in Arabic numerals (e.g. 13). Where both spines and rays are present in a fin, the count is given in the form VIII, 13. *Vertebral counts*, although difficult

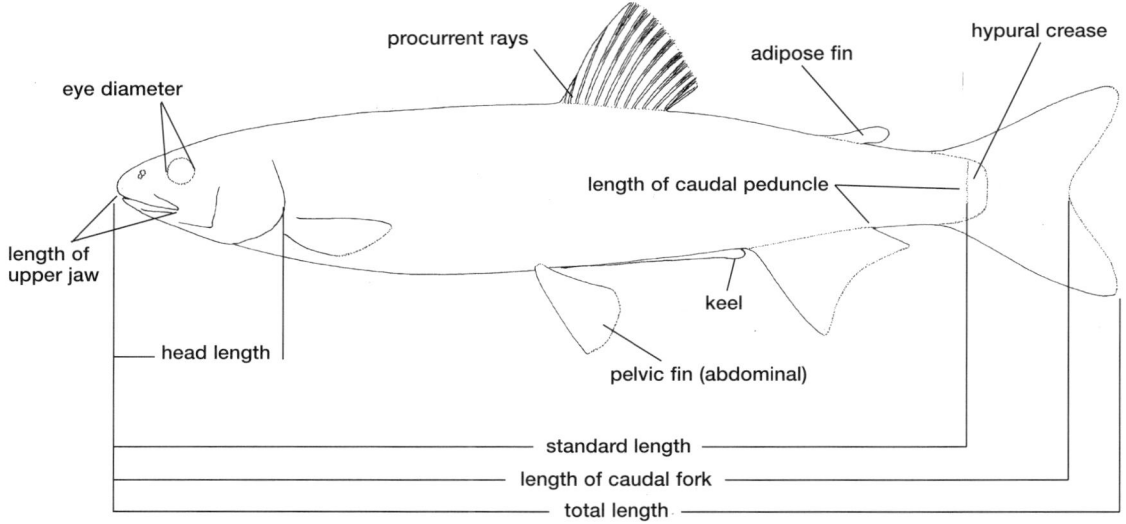

2.1 A typical soft-rayed fish showing the external structures and dimensions measured in the study of fishes (see also Fig 2.2). (R.M. McDowall)

to obtain, are sometimes very important in fish identification and can be obtained by dissection, from radiographs (X-rays), or by a method of clearing the body tissues and staining the bones with alizarin dye (Taylor, 1967). *Scale counts* along the lateral line are obtained by counting the number of pored scales in the series from the upper angle of the opercular opening above the pectoral fin to the base of the tail (Fig. 2.2). Some fishes have no lateral line and in these the number of scales along the side (where a lateral line would be) is counted. *Gill rakers* are counted in the 1st gill arch, usually on the left side, either the total number of rakers or the number in the lower limb of the arch. In some fishes the number and arrangement of *open pores* on the head are important in identification.

Scientific and common names Biologists use scientific names for referring to fishes for which most people have common names, but they can be a real source of confusion for those not used to them. Each species has a two-part scientific name consisting of a *generic* name, which always has a capital letter, and a *specific* name, which never has a capital (though sometimes a capital was used in earlier times). Both the generic and specific names are printed in italics, and both are used together to refer to any animal species (not just fishes!). A genus may contain one or many species, and the inclusion of several species in one genus implies close relationship between these species. Scientific names have the *form* of Latin, even though the *words* from which the name is derived may not be Latin in origin (*Maccullochella* is a typical example, derived from the name of early Australian ichthyologist A.R. McCulloch). Even though they may be hard to remember and pronounce, scientific names are important as they are usually more stable and consistent than common names. They also have universal application to species that are very widespread and which may have different common names in different places.

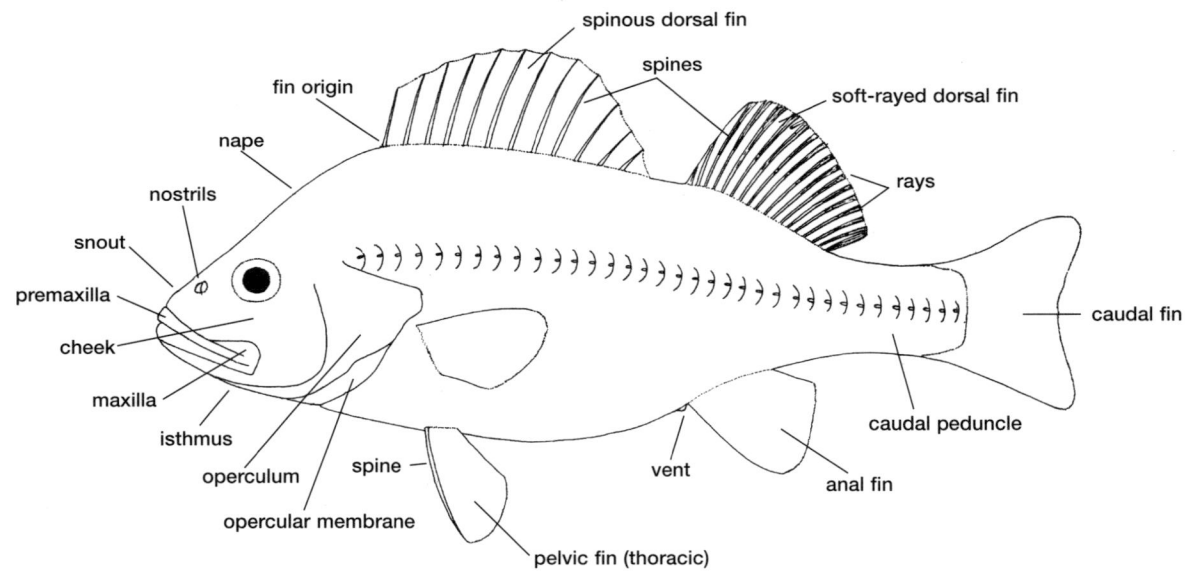

2.2 A typical spiny-rayed fish, showing the external structures and dimensions measured in the study of fishes. (see also Fig. 2.1). (R.M. McDowall)

Preserving fishes Much can be learnt about fishes by observing them in their natural habitats. It may take time and patience to remain still while fishes overcome their natural shyness and begin to move around in the open, although some fishes seem to have little fear of movement by objects on the bank. Some species live in the open and are easily seen, but many are nocturnal, coming out only at night, or live in such turbulent waters that they are not visible from the surface. In Australia many fresh waters are turbid, so that visibility is extremely poor. For a variety of reasons, therefore, to get to know the species and to study some of their characteristics, it is necessary to catch them, either so that they can be kept alive in captivity or so that a collection of preserved fishes can be established for reference. In most instances a permit is required to collect fishes. Fish caught for retention alive are best placed in an Esky, particularly cold-water species from more southern and higher-altitude locations. To enhance survival, fish are best put in a large plastic bag with a few centimetres of water, and the bag should be inflated, sealed and put in an Esky. Aeration of the water with a portable, battery-operated aerator, or with oxygen from a pressure bottle, is worthwhile. Fish caught for preservation are best anaesthetised as they are caught, because otherwise they tend to die from asphyxiation with their mouths gaping widely and their shape distorted. Fish can be anaesthetised easily and safely by placing a small amount of a narcotic such as benzocaine in the water in which they are placed. Benzocaine can be obtained from drug companies and is best carried in the field as a solution in ethyl alcohol; it does not dissolve easily in water.

Unless fish are wanted for some special purpose, they are best preserved in 5–10% formalin. Formalin can be obtained from pharmacies, drug and chemical companies and often from stock and station agents. When preserving larger fish—over about 300 mm—it is best to cut a small slit in the side of the belly (on the right side) to allow the formalin to flow into the body cavity; or it can be injected there with a large syringe. Formalin is a poisonous and very unpleasant liquid. If it should splash into the eyes, they should immediately be flushed with large quantities of fresh, cold water. If in doubt see a doctor. Containers used to carry formalin should be robust and have reliable seals on their lids.

Fish are much easier to study if they are preserved straight, so it is best not to stuff them into small jars or plastic bags. Instead, it is preferable to cover the fish in 2–3 cm of strong (20–30%) formalin solution in shallow plastic (photographic) trays for about thirty minutes before bottling them. Otherwise the anaesthetised fish can be put carefully into jars that are amply large and the jars laid on their sides so that the fish lie straight. Alcohols, such as ethyl or methyl alcohol, should be used as a preservative only when there is no alternative, except when fish are preserved before extraction of their otoliths for aging studies (otoliths tend to become opaque and unreadable after formalin preservation).

After fish have been stored for about a week or more in formalin, they should be soaked for 2–3 days in several changes of fresh water and then transferred to either ethyl alcohol (70%) or isopropyl alcohol (40%). The latter is less favoured now than formerly as it seems that after long storage fish in it become soft and mushy. Long-term storage in formalin is not recommended as it becomes acidic and eventually dissolves the fish bones; acidity can be counteracted by buffering the formalin with borax. In addition, working with fish stored in formalin is very unpleasant and could be damaging to health. The colour of fish always fades rapidly after preservation in any fixative; some colours cannot be retained by storage in any known fixative, but fading occurs more slowly if fish are stored in the dark, or at least away from strong sunlight (which accelerates fading). Colour photos of freshly caught fish are the best means of recording colouration.

Fish specimens are of little use without proper labels, which are best made of stout, durable, water-resistant card (very hard to obtain), calico, or synthetic tracing parchments, or the more modern polypropylene 'papers', which are almost indestructible. Information on labels should be written with either soft, black lead pencil or waterproof ink. Ballpoint pen ink fades rapidly in liquids, especially alcohols, and is therefore quite useless. Labels are best put into the containers with the fish. If they are attached to the outside of the jar, they may come off (especially if the jar becomes wet); but more important, they are not easy to transfer if fish are shifted from one container to another. Labels should include, as a minimum, information on the locality and the date of capture. Additional details can include any of the following: collector's name; more detail about locality of capture, such as map coordinates (or latitude and longitude); information about the habitat in which the fish were caught, with notes on bottom type, flow characteristics, water temperature and colour; altitude; general surroundings; other fish species caught at the same locality. However, there is a limit to what can be crammed onto a small label. Where all these details cannot be fitted on a label, they can be recorded in a field notebook. Field notebooks are most valuable when they are used on a consistent and methodical basis, with notes on all collections made over a period of time. Some workers give a serial collection number to each sample, and the notes made on that sample and the habitats from which it came are written in the field notebook under the appropriate number. This makes for easy reference from a sample to a notebook. However, even though samples are numbered, this should not be a replacement for at least minimal information on a label included with the fish. It is all too easy for samples and field notebooks to become separated and the notebooks lost. For durability a field notebook should have a hard cover and preferably be made for use in damp conditions; as with labels, notebooks with polypropylene 'papers' are available and have much advantage.

Identifying fishes This book has been written with the primary purpose of accurate and rapid identification of fishes found in the fresh waters of south-eastern Australia. To help in making identifications there is a Key to Families of fishes (p.25) that makes it possible to determine the family to which a fish belongs. A simple outline drawing of a representative of each family helps to confirm that the correct family has been found, and the page with the account of each family and of the species within it is noted alongside the family name. In some families there is only one species, so obtaining the correct family leads straight to the right species, but where a family contains more than one species, each family introduction is followed by a key for identifying the species.

Although keys are a well-known means of identifying fishes, many amateur naturalists may not be familiar with them and may find them hard to use at first. A key can be considered essentially as a series of pathways with forks in them; as long as one starts at the beginning, and as long as the directions given at each fork in the pathway are followed carefully, the right destination should be reached. The keys used here consist of pairs of alternative statements, only one of which can be correct.

Thus, when using a key, start at the first pair of statements, read the first alternative, and if it does not apply to the fish, try the second alternative (which should!). Each statement in the pair will end either in the name of a family (or species), or with a number. If the family key ends with a name, turn to the page indicated. If it ends with a number, scan down the left margin of the page for a further pair of alternative statements given under that number. For instance, the Key to Families (p.25) begins with:

```
1   No jaws or paired fins. . . . . . . . . . . . . . . . . . . . 2
    Jaws and paired fins present, sometimes
    no pelvic fins . . . . . . . . . . . . . . . . . . . . . . . . . 5
```

Depending on which of these two statements is correct for a fish being identified (and only one of them will be), scan down the left margin for either 2 or 5. If your fish has no jaws or paired fins, look for 2, where you will find that the key offers:

```
2   Eyes present. . . . . . . . . . . . . . . . . . . . . . . . . . 3
    Eyes absent . . . . . . . . . . . . . . . . . . . . . . . . . . 4
```

Again, only one of these statements can be correct for your fish, so choose one; continue to follow the numbers until you reach a family (or species) name. If there is some doubt about which pair of an alternative applies, follow each through to an identification and check your fish against the descriptive information and the illustrations. Once you have found a species name for your fish, compare it with the illustrations, check the characters given in the description and see if the species has been found previously in the general area from which your fish came. This latter character is, of course, only a guide, as continually fish are being found in new areas, though not often far from known distributions.

If the keys prove too difficult at first, illustrations provide an easy though somewhat less accurate means of identification. Any identification based on illustrations should always be checked carefully against the text. Usually the species identified should occur within or near the area given under 'Distribution'—although we do not yet know precisely the distributional ranges of some species. Knowledge of distribution can be a help, but it can also be misleading unless care is taken.

Information given in the descriptions should be enough to enable identification, but if more is needed it can be found in the literature listed at the end of each species description. Many of these publications are readily available, if not in public libraries, certainly in the libraries of museums, universities, scientific societies, and other institutions. Readers should realise, however, that very little is known about a great many of Australia's freshwater fishes and that much of what is known has been summarised in this book. Furthermore, there is little doubt that some new species remain to be discovered, named, and described.

When difficulties arise, or should anything new or strange turn up, help can be found in the museums of south-eastern Australia and from naturalists, many of whom have special interests and expertise in fishes. The authors of chapters in this book are such experts. Fish collections are maintained in the Queensland Museum in Brisbane, the Australian Museum in Sydney, the National Museum of Victoria in Melbourne, the South Australian Museum in Adelaide, the Queen Victoria Museum in Launceston and the Tasmanian Museum in Hobart. Research is going on into the habits and habitats of freshwater fish in various research organisations, including: the Fisheries Branch of the Queensland Department of Primary Industries, Brisbane; the Fisheries Research Institute, of the New South Wales Department of Agriculture in Cronulla, with a field station at Narrandera in inland New South Wales; the Arthur Rylah Institute for Environmental Research, of the Victoria Department of Conservation, Forests and Lands in Heidelberg, with stations also at Shepparton and the Snobs Creek Hatchery at Alexandra in inland Victoria; the Tasmanian Inland Fisheries Commission in Hobart. There are also fish and fisheries workers at most universities. Many of the scientists in these various institutions are willing to assist serious amateur naturalists, but remember that they are mostly very busy people.

Glossary

abdomen (*adj.* abdominal)—that part of the body containing the digestive and reproductive organs; the lower part of the body in front of the vent (*q.v.*).

adipose eyelid—thick fleshy tissue surrounding and partly covering the eyes (Fig. 2.3).

adipose fin—a small dorsal fin without rays, usually rather thick and fleshy, and the hindmost of the dorsal fins (Fig. 2.1).

aestivation (*v.* aestivate)—a period of dormancy or inactivity during warm, dry periods.

ammocoete—the distinctive, eyeless, larval form of the lamprey.

amphidromous—describes fish that migrate between the sea and fresh water at a regular life-history stage but not directly to spawn (*cf.* anadromous, catadromous, diadromous).

anadromous—describes fish that migrate from the sea into fresh water, as adults, to spawn (*cf.* diadromous, amphidromous, catadromous)

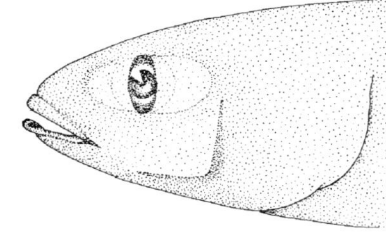

2.3 Adipose eyelid, as in the grey mullet, *Mugil cephalus*. (R.M. McDowall)

GLOSSARY

anal fin—the unpaired fin situated ventrally just behind the vent (*q.v.*) (Fig. 2.2).
anterior—forwards, on or near the head (*cf.* posterior).
axillary process—a small, fleshy or scale-like projection sometimes at the bases of the pectoral and/or pelvic fins (Fig. 2.4).

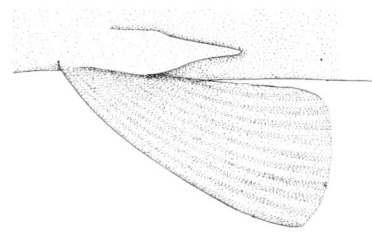

2.4 Axillary process, here at pelvic fin origin, as in many trouts and salmons. (R.M. McDowall)

barbel—slender, finger-like structures around the mouth or on the snout, with sensory function, usually taste (Fig. 2.5).

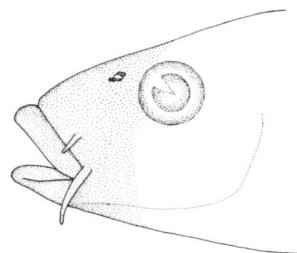

2.5 Barbels around mouth, as in the common carp, *Cyprinus carpio*. (R.M. McDowall)

benthic—bottom living.
billabong—an isolated river pool or backwater.
bottom fauna—animals that live on or among the rocks and sands at the bottoms of streams and lakes.
breast—ventral surface of body below pectoral fins.
canine teeth—jaw teeth conspicuously larger than others nearby (Fig. 2.6).

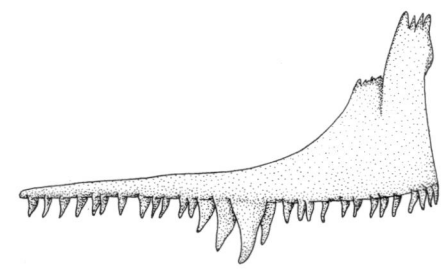

2.6 Enlarged canine teeth in the jaws, as in some species of *Galaxias*. (R.M. McDowall)

carnivore (*adj.* carnivorous)—feeding on other animals (*cf.* herbivore, omnivore).
catadromous—describes fish that migrate from fresh water as adults to spawn at sea (*cf.* anadromous, amphidromous, diadromous).
caudal peduncle—the posterior of the body, behind the vent, usually slender, measured from the posterior of the anal fin base to the end of the vertebral column, as indicated by the hypural crease (*q.v.*) (Fig. 2.2).
cheek—that part of the head below the eyes (Fig. 2.2).
chevron-shaped—v-shaped.
ciliated scale—a scale with a single row of spines along the outer margin (Fig. 2.7).
circulus (*pl.* circuli)—concentric rings found on a fish scale.
cloaca—combined external openings of urinogenital and alimentary systems.
compressed—flattened from side to side (*cf.* depressed).
concave—hollowed out, curved inwards.
cryptic—applied to fishes that live among sheltering and concealing cover or which have protective colouration, or both.
ctenoid scale—a scale with a patch of small spines on the outer surface (Fig. 2.7) (*cf.* ciliated, cycloid).
cusp—a projection on a tooth, often sharp.
cycloid scale—a scale lacking all spines, the outer margin more or less smooth (Fig. 2.7) (*cf.* ctenoid, ciliated).
deciduous—easily detached (scales).
demersal—living on the bottom or sinking to the bottom (of eggs).
depressed—dorso-ventrally flattened (*cf.* compressed).
detritus—organic material derived from decomposing animals and plants.
diadromous—describes fishes that migrate between fresh and salt water at a regular life-history phase, in either direction, but not necessarily to spawn (*cf.* anadromous, amphidromous, catadromous).
dimorphism—existing in two forms, usually refers to differences between the sexes in body shape and/or colour, i.e. sexual dimorphism.
diurnal—active during the day (*cf.* nocturnal).
dorsal—of the upper surface, back (*cf.* ventral).
dorsal fin—any fin on the back, unpaired, and up to three in number (Fig. 2.1, 2.2).
elver—a young eel (Fig. 6.6).

GLOSSARY

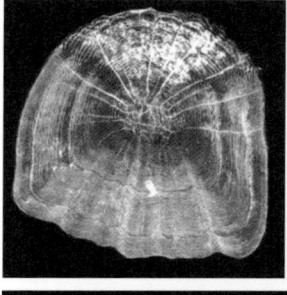 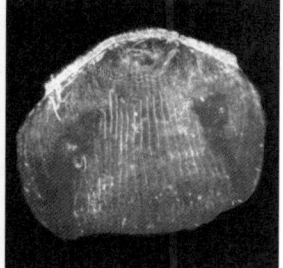

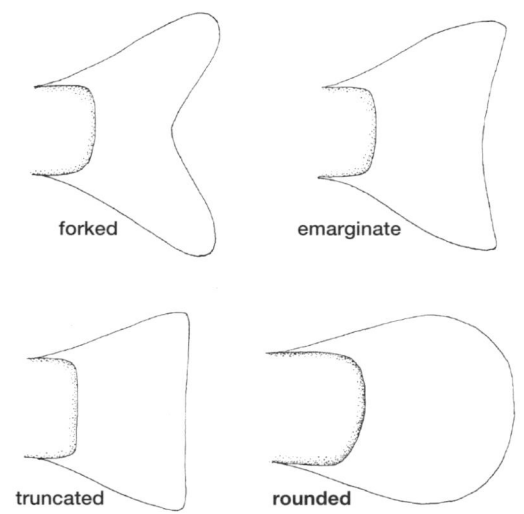

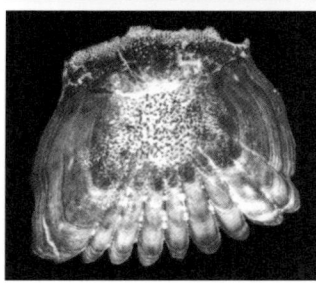

2.7 Types of fish scales. TOP LEFT: *cycloid*—note absence of spines. TOP RIGHT: *ciliated*—note row of spines along outer margin. LEFT: *ctenoid*—note spines in patch on surface. (J.A. Bahler)

2.8 Shapes of caudal (tail) fin in fishes. (R.M. McDowall)

emarginate—slightly concave or notched (Fig. 2.8).
endemic—native and restricted to a given area.
euryhaline—able to live in a wide range of water salinities.
eye diameter—the horizontal measurement across the eye (Fig. 2.1).
falcate—scythe-shaped.
family—one of the categories in animal classification, the name ending in *-idae*; contains one or more genera (*sing*. genus, *q.v.*).
fauna—the animals of a given area or locality.
fin ray—a slender bone supporting a fin, divided into two halves that are side by side, segmented along its length and often branched one or many times (Fig. 2.9).
fin spine—a slender bone supporting a fin, not divided into two halves, not segmented and not branched, sometimes sharp and stout (Fig. 2.10).
forage fish—a species that provides food for another, larger fish species.
frenum—a fold of skin.
fusiform—spindle-shaped, tapering to both ends.

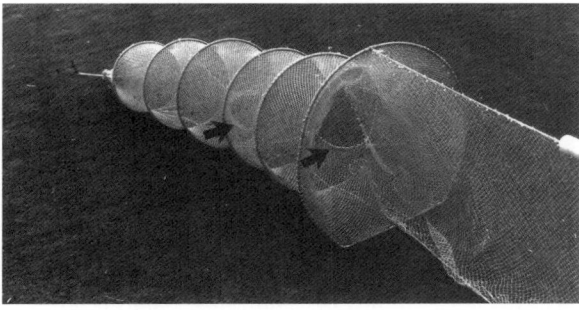

2.11 Fyke net as used for eel and other sorts of fishing in fresh water. (R.M. McDowall)

fyke—a trap net, usually made of knotted mesh, supported on hoops (Fig. 2.11).
genital papilla—a small fleshy protuberance just behind the vent, on which the openings of the

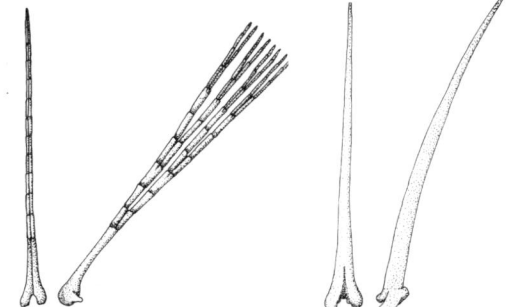

Fig 2.9 Fin ray—note paired, segmented and branched structure. (R.M. McDowall)

Fig 2.10 Fin spine—note unpaired, unsegmented and unbranched structure. (R.M. McDowall)

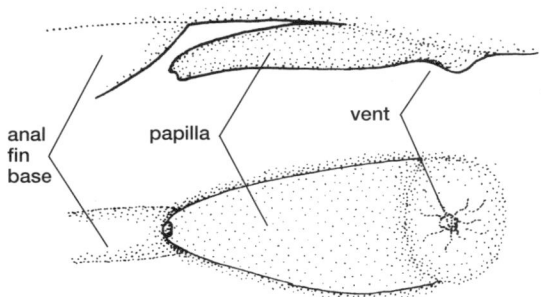

2.12 Genital papilla, as in *Gobiomorphus*. TOP: lateral view. BOTTOM: ventral view. (R.M. McDowall)

GLOSSARY

reproductive organs occur (Fig. 2.12).

genus (*pl.* genera)—one of the categories in animal classification that contains one or more species (*q.v.*). The species in a genus are regarded as more closely related to one another than to excluded species.

gill raker—a bony or horny tooth- or filament-like structure on the anterior (internal) edges of the gill arches, used for filtering food organisms or for protection of the gills. Vary from few to many and from short and stout to long and slender (Fig. 2.13).

glass eel—the young eel, on migration from the

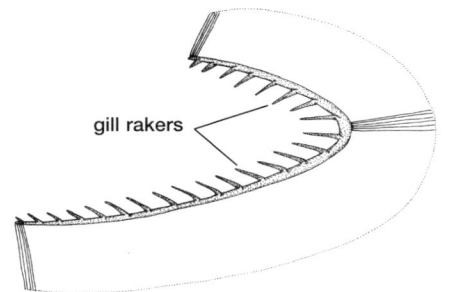

2.13 Gill rakers, as in rainbow trout, *Oncorhynchus mykiss*. (R.M. McDowall)

sea, and lacking all colouration (Fig. 6.5).

gonad—organ containing reproductive products: ovaries in female, testes in male.

gonopodium—a specialised portion of the anal fin in the males of some fish species, used to transfer reproductive products to the female.

gravid—ready to spawn, abdomen swollen with gonads.

gregarious—tending to live in groups rather than solitary.

head length—distance from the tip of the snout to the posterior edge of the opercular membrane (Fig. 2.1).

herbivore (*adj.* herbivorous)—feeds on plants.

heterocercal—refers to a tail with unequal lobes, the upper lobe larger than the lower.

hypural crease—a crease that forms at the base of the tail when the tail is pushed to one side. Defines the end of the standard length (*q.v.*).

indigenous—native, although not necessarily restricted, to an area.

interorbital—the area on top of the head between the eyes.

isthmus—the area below the head between the gill openings (Fig. 2.2).

jugular—refers to the isthmus (*q.v.*); below the head between the gill openings. Pelvic fins are sometimes jugular.

keel—a ridge along the ventral surface of the abdomen, either bony or horny.

kype—the strongly upturned tip of the lower jaw in male trouts and salmons (Fig. 2.14).

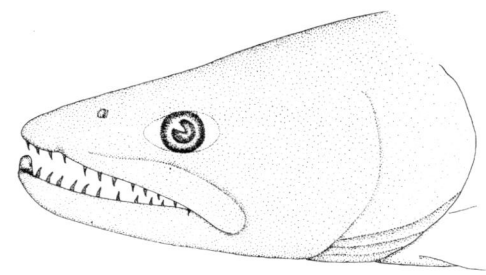

2.14 Kype, as in brown trout, *Salmo trutta*. (R.M. McDowall)

lanceolate—tapering to a point, lance-shaped.

lappet—flat or finger-like structures surrounding the oral disc in some lampreys.

larva (*pl.* larvae, *adj.* larval)—the youngest life-history stage, usually with structure and shape that differ obviously from those of the adult.

lateral—referring to the sides.

lateral line—a row of pored scales in the skin along the sides, the pores opening to a sensory canal along the sides (Fig. 2.2).

laterally compressed—flattened from side to side.

laterosensory pores—small pores on the head that are part of the lateral line system (Fig. 2.15).

leptocephalus—characteristic larval form found in eels and some other fishes; very strongly compressed, willow-leaf shaped, with a small head (Fig. 6.4).

macrophthalmia—the early adult stage of the lamprey, following the ammocoete (*q.v.*); migrates down to sea in some species.

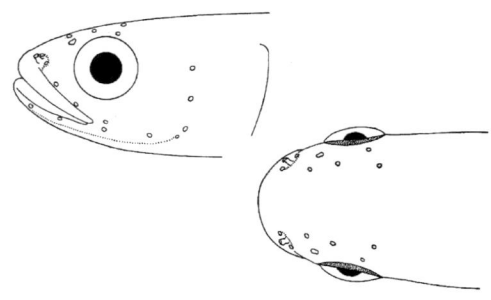

2.15 Open laterosensory pores on head, as in *Galaxias* species. (R.M. McDowall)

maxilla (*pl.* maxillae)—an upper jawbone forming the lateral margin of the upper jaw, carrying teeth in some primitive fishes (Fig. 2.2, 2.17).
meristic—describes structures often associated with body segments that may be counted, such as vertebrae and fin rays.
metamorphosis—transformation from larval form to juvenile or adult form.
microphagous—feeding in microscopic organisms.
milt—the male reproductive organs (testes, *q.v.*), and their reproductive products.
molar—describes teeth used for grinding food.
monotypic—describes a genus with only a single species.
nape—the back of the head, behind the eyes (Fig. 2.2).
nomenclature—the naming of animals and plants.
noxious—injurious or harmful.
nuptial tubercles—small fleshy, horny, or bony protuberances that develop on the head (often), trunk and fins (less often), of breeding fishes, more frequently in males than females, often

2.16 Anal fin of male *Retropinna*, showing nuptial tubercles along fin rays. (J.A. Bahler)

called pearl organs (Fig. 2.16).
omnivore (*adj.* omnivorous)—a fish that eats a wide variety of animal and plant foods (*cf.* carnivore, herbivore).
operculum—a large bony plate that covers the gills (Fig. 2.2).
opercular membrane—a fleshy flap that covers and allows closure of the gill openings (Fig. 2.2).
oral disc—the area surrounding the mouth opening of the lamprey.
origin—applied to fins, being the anteriormost point of attachment to the body.
ova (*sing.* ovum)—eggs.
ovaries—female reproductive organs containing ova (*q.v.*).

oviduct—a tube along which ova pass when spawned, leading from the ovaries to the exterior in most fish species.
palate—the roof of the mouth.
palatine—a paired bone forming the margin of the roof of the mouth towards the front. Palatine teeth are carried on this bone (Fig. 2.17).

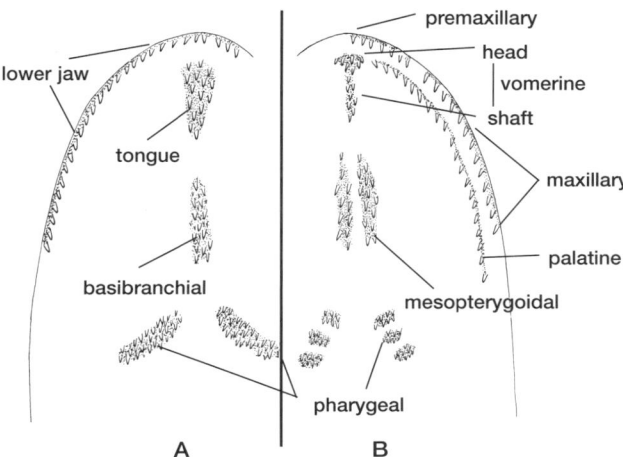

2.17 Location of teeth in the mouth of a hypothetical fish: **A** lower jaw and floor of mouth; **B** upper jaw and roof of mouth. (R. M. McDowall)

papilla—a small, fleshy protuberance.
parr—a distinctive juvenile stage in trouts and salmons, carrying parr marks, dark, vertical bands along the trunk (Fig. 12.3).
pearl organs—*see* nuptial tubercles.
pectoral fins—paired fins that are lateral, just behind or below gill openings (Fig. 2.2).
pelagic—living in open waters.
pelvic fins—paired fins positioned on the ventral surface between the head and vent, either jugular (*q.v.*), thoracic (*q.v.*), or abdominal (*q.v.*); often referred to as ventral fins; sometimes absent (Fig. 2.1, 2.2).
pharyngeal teeth—teeth in the back of the throat.
piscivore (*adj.* piscivorous)—feeds on fish.
plankton (*adj.* planktonic)—minute organisms living in the open waters of the sea and lakes.
premaxilla (*pl.* premaxillae)—paired bones that form some of the front of the upper jaw in a few fishes, more often the entire front; the main tooth-bearing bone in the upper jaw in most fishes (Fig. 2.1, 2.17).
preopercular margin—the margin of the preopercular bone that lies in front of the operculum (*q.v.*), just behind the eyes (Fig. 2.18).

GLOSSARY

preopercular ridge—an elevated ridge on the preopercular margin (Fig. 2.18).

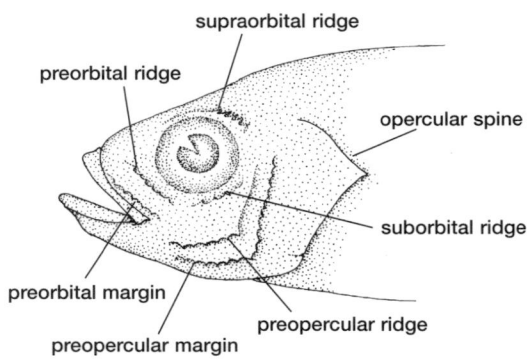

2.18 Bony and serrate ridges on head of fishes. (R.M. McDowall)

preoperculum—the anterior bone in the opercular series, just behind the eyes.

preorbital—a bone lying in front of the eyes (Fig. 2.18).

procurrent rays—small fin rays that are divided but unbranched and unsegmented, occurring at the front of some fins in more primitive fishes and in the tail in most fishes (Fig. 2.1).

protrusible—a condition of the jaws in which the jaws project forwards as a tube when the mouth is open.

pterygiophore—slender bones in the body musculature that support fin rays and spines.

pungent—refers to a spine that projects through the skin.

pyloric caecum (*pl*. caeca)—few to many blind tubes attached to the stomach (Fig. 2.19).

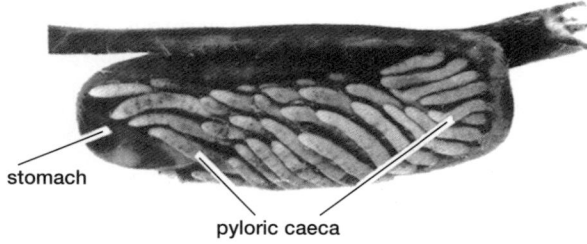

2.19 Pyloric caeca on stomach, as in rainbow trout, *Oncorhynchus mykiss*. (J.A. Bahler)

ray—see fin ray.

redd—a trough, excavated in the gravel of a river bed by trouts and salmons, in which the eggs are laid and in which they develop.

ripe—describes fish ready to spawn, when milt and ova can be expressed with gentle pressure on the abdomen.

scute—hard, bony structures, usually modified scales, that sometimes form ridges on fish, especially along the back and belly.

serrate—notched at the edges, saw-like.

sexual dimorphism—*see* dimorphism.

snout—that part of the head in front of the eyes (Fig. 2.2).

spathulate—spoon-shaped.

species—groups of actually or potentially interbreeding populations that are reproductively isolated from other groups of populations. Species are usually but not always morphologically distinct.

spent—describes fish that have recently spawned.

spine—see fin spine.

spring tide—very high tide that coincides with or follows shortly after the full and/or new moons.

standard length—distance from tip of snout to hypural crease (*q.v.*) (Fig. 2.1).

subequal—nearly, but not quite equal.

suborbital—a bone beneath, sometimes slightly behind, the eyes (Fig. 2.18).

synonym—one of several names for a single species. Only one name, the senior synonym, is valid.

taxonomy—the science of classification of animals and plants.

testes—the male reproductive organs.

thoracic—the area of the abdomen below the pectoral fin bases, just behind the head. Pelvic fins may be thoracic (Fig. 2.2).

tricuspid teeth—teeth with three cusps (*q.v.*).

truncate—literally 'cut off'; often refers to caudal fin with posterior margin more or less straight (Fig. 2.8).

tuberculate—covered with tubercles, either soft or hardened projections on the surface of a fish's skin or scales.

urinogenital papilla—a protuberance on which the urinary and genital organs open.

vent—the external openings of the alimentary canal, anus.

ventral—refers to the lower surface of the body, belly (*cf*. ventral fin).

ventral fin—see pelvic fin.

vermiculation—irregular, wavy colour patterns, like worm tracks.

vomer—a bone at the front of the roof of the mouth, behind the premaxillae (*q.v.*), sometimes toothed (*cf*. vomerine teeth Fig. 2.17).

zooplankton—small animals that live in the surface waters of the sea and lakes.

3
Key to families of fishes
found in the fresh waters of south-eastern Australia

As an initial step in identification, a key to the families of fishes found in the fresh waters of south-eastern Australia follows. Advice on how to use this and other keys in the guide can be found in Chapter 2. The key in this chapter includes not only those families containing species that *regularly* and *normally* occur in fresh water (i.e. 'freshwater fishes'), and which are discussed in detail in the main body of the text, but also those families that contain one or more species that are *occasional invaders* of fresh waters. Such species are not dealt with further.

No doubt some families that include sporadic invaders of fresh water will have been missed from the key—no claim is made for comprehensive coverage of these occasional invaders—but the species that are covered should be sufficient to be generally useful.

Although the key is a key to families, it is not meant to enable the identification to family level of all the *marine* fishes in those families, but only the family identification of those species that do enter fresh waters. Thus, for example, the key characters for the family Lutjanidae, a primarily marine family, allows us to distinguish the mangrove-jack, *Lutjanus argentimaculatus*, which enters estuaries and fresh waters, but it may not be successful in obtaining the family Lutjanidae for other lutjanids that do not enter fresh waters.

Preparation of a key to families is quite difficult, and it has become necessary to use some quite technical characters. These have been minimised as far as possible; those that remain are necessary, and their meaning can be determined by reference to the Glossary, beginning on p. 19.

Key to families of fishes

1. No jaws or paired fins 2
   Jaws and paired fins present (sometimes no pelvic fins) 5

2. Eyes present 3
   Eyes absent 4

3. Two large tricuspid teeth on oral disc above mouth, teeth on disc few, sharp and in radial rows; disc not surrounded by fringed lappets; eyes dorsolateral . . . . **Mordaciidae** (*adults*) p. 32

   One tooth on oral disc above mouth with 2 large, spathulate cusps laterally and 2 smaller, sharp, central cusps; teeth on disc numerous, blunt and arranged in irregular spirals; disc surrounded by fringed lappets; eyes distinctly lateral . . . . **Geotriidae** (*adults*) p. 36

4. Snout to anterior margin of vent 83% or more of total length. . . . . **Mordaciidae** (*ammocoete*) p. 32
   Snout to anterior margin of vent 80% or less

of total length .... **Geotriidae** (*ammocoete*) p. 36

5 Five pairs of external gill openings; no operculum (gill cover) 6
   One pair of external gill openings; an operculum (gill cover) present 8

6 Body strongly dorsoventrally flattened 7
   Body not dorsoventrally flattened...........
   .... **Carcharinidae** (marine)

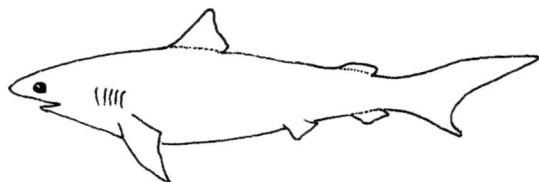

7 Snout produced as an elongate, flattened, strongly-toothed saw; a typical shark-like heterocercal tail without spines ....
   .... **Pristidae** (marine)

   Snout not elongate and without teeth; tail elongate and whip-like with serrated spines....
   .... **Dasyatidae** (marine)

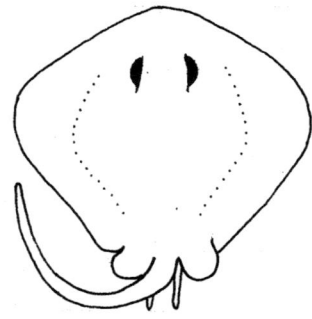

8 Trunk and tail very elongate and encased in an armour of bony rings; mouth extremely small, at the end of a long, tubular snout, usually toothless .... **Syngnathidae** (marine)

   Trunk and tail not encased in an armour of bony rings; mouth small to large, but not at tip of a long tubular snout, usually toothed ... 9

9 Gill openings narrow lateral, vertical slits just in front of pectoral fin bases; no pelvic fins .. 10
   Gill openings more extensive and usually extending below head; pelvic fins usually present (may be very reduced) 11

10 Dorsal, caudal and anal fins continuous and inseparable; fish very elongate and slender, tubular; teeth not fused to form a beak ....
   .... **Anguillidae** p. 39

   Dorsal, caudal and anal fins quite separate; fish stocky; teeth fused to form a beak ....
   .... **Tetraodontidae** (marine)

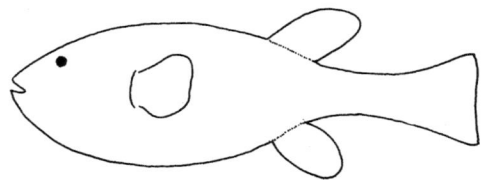

11 Adipose fin present (may be very small) 12
   No adipose fin 16

12 Stout spines present in leading edges of dorsal and pectoral fins; barbels present around mouth .... **Ariidae** p. 107

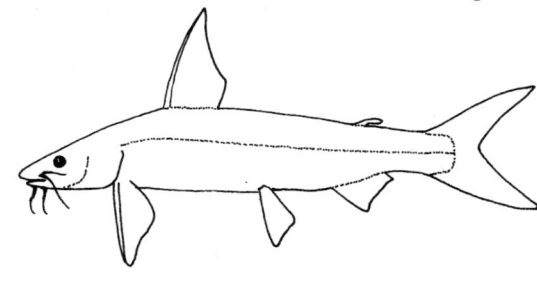

   Fins entirely soft-rayed; no barbels 13

13 Dorsal fin origin in front of pelvic fin bases; axillary processes at bases of pelvic fins ......
   .... **Salmonidae** p. 81

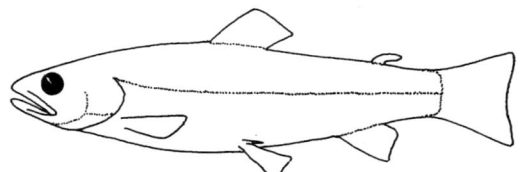

   Dorsal fin origin behind pelvic fin bases; no axillary processes at pelvic fin bases 14

14 Scales lacking; lateral line present (indistinct) . . . . **Aplochitonidae** p. 78

Scales present; lateral line lacking 15

15 Dorsal fin base wholly in front of anal fin origin and just behind pelvic fin bases; horny shelf surrounds lower jaw, which is shorter than upper. . . . . **Prototroctidae** p. 96

Dorsal fin partly or largely above anal fin, well behind pelvic fin bases; no horny shelf around lower jaw; jaws about equal in length . . . . **Retropinnidae** p. 92

16 Ventral surface of abdomen serrate owing to the presence of scaly skutes . . . . . . **Clupeidae** p. 44

Ventral surface of abdomen not serrate 17

17 Jaws toothless. 18
Jaws toothed (teeth may be very small) 19

18 No spines in fins; eyes with an adipose eyelid. **Chanidae** p. 48

Stout spines in dorsal and pectoral fins; no adipose eyelid . . . . **Cyprinidae** p. 99

19 Last ray of dorsal fin an elongate filament reaching base of tail; adipose eyelid present . . . . **Elopidae** p. 50

Last ray of dorsal fin not elongate; adipose eyelid rarely present. 20

20 Long barbels around mouth 21
No barbels around mouth (occasionally 1 short one beneath chin). 22

21 Second dorsal fin, tail and anal fins continuous . . . . **Plotosidae** p. 109

Second dorsal fin, tail and anal fins separate . . . . **Cobitidae** p. 114

22 Head with a very elongate and slender beak . 23
Head without an elongate, slender beak 24

23 Beak formed from an elongation of both upper and lower jaws; jaws about equal in length. . . . . **Belonidae** (marine)

FAMILY KEY

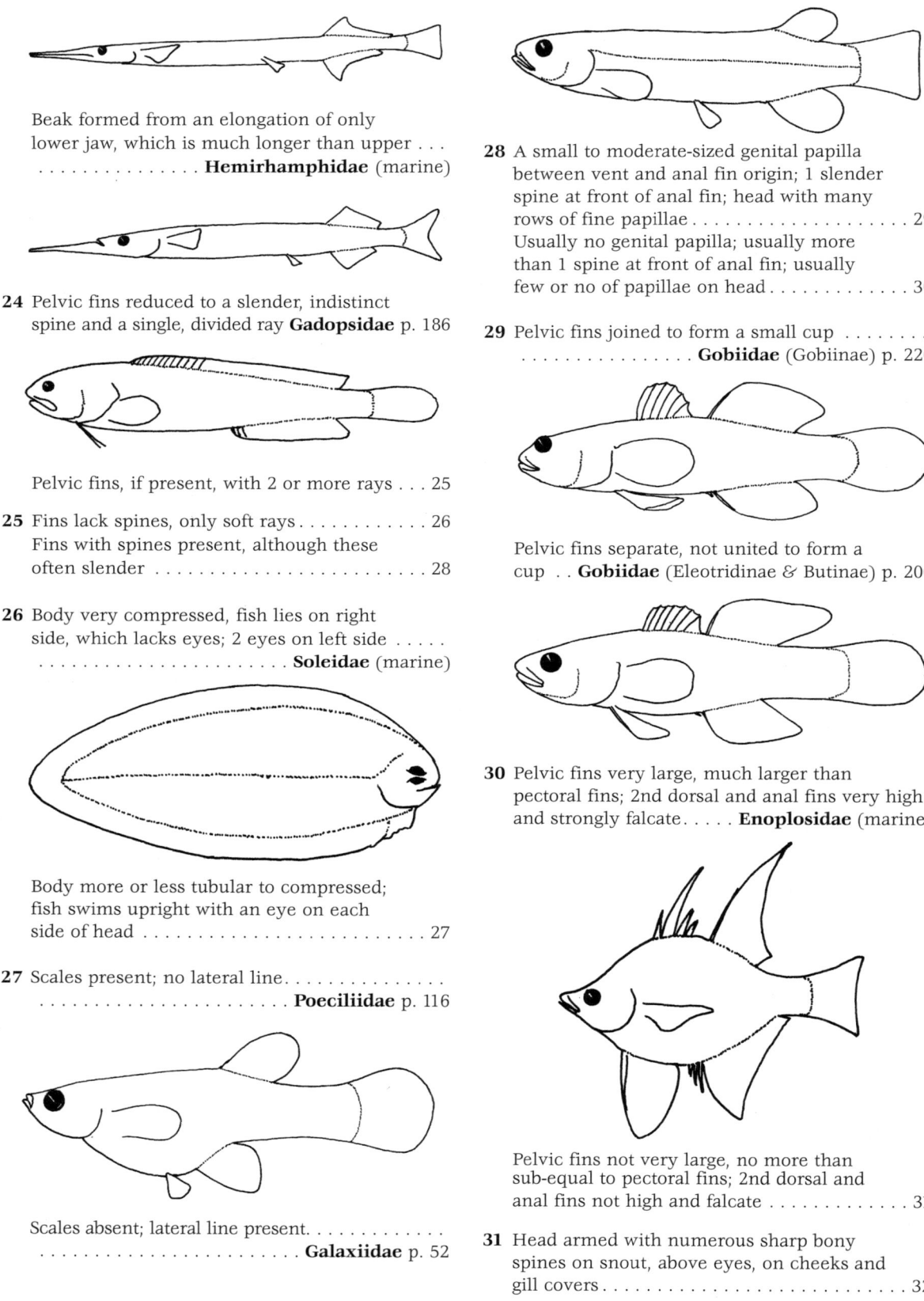

Beak formed from an elongation of only lower jaw, which is much longer than upper .... **Hemirhamphidae** (marine)

24 Pelvic fins reduced to a slender, indistinct spine and a single, divided ray **Gadopsidae** p. 186

Pelvic fins, if present, with 2 or more rays ... 25

25 Fins lack spines, only soft rays 26
Fins with spines present, although these often slender 28

26 Body very compressed, fish lies on right side, which lacks eyes; 2 eyes on left side .... **Soleidae** (marine)

Body more or less tubular to compressed; fish swims upright with an eye on each side of head 27

27 Scales present; no lateral line .... **Poeciliidae** p. 116

Scales absent; lateral line present .... **Galaxiidae** p. 52

28 A small to moderate-sized genital papilla between vent and anal fin origin; 1 slender spine at front of anal fin; head with many rows of fine papillae 29
Usually no genital papilla; usually more than 1 spine at front of anal fin; usually few or no of papillae on head 30

29 Pelvic fins joined to form a small cup .... **Gobiidae** (Gobiinae) p. 220

Pelvic fins separate, not united to form a cup .. **Gobiidae** (Eleotridinae & Butinae) p. 200

30 Pelvic fins very large, much larger than pectoral fins; 2nd dorsal and anal fins very high and strongly falcate ..... **Enoplosidae** (marine)

Pelvic fins not very large, no more than sub-equal to pectoral fins; 2nd dorsal and anal fins not high and falcate 31

31 Head armed with numerous sharp bony spines on snout, above eyes, on cheeks and gill covers 32

FAMILY KEY

Head not armed with numerous sharp, bony spines 33

32 Elongate, dorsoventrally flattened, head much broader than deep; spiny 1st dorsal with 8 spines, separate from 2nd dorsal .... **Platycephalidae** (marine)

Stocky and slightly compressed; depth of head subequal to width; spinous dorsal with 15–16 spines, joined to 2nd dorsal .... **Scorpaenidae** p. 144

33 Both dorsal and pelvic fins originate well forward of pectoral fin bases .... **Bovichtidae** p. 198

Dorsal and/or pelvic fin origins level with or behind pectoral fin bases 34

34 Inside of mouth bright orange; a small barbel below chin; 2 spines in anal fin .... **Sciaenidae** (marine)

Inside of mouth not bright orange; no chin barbel; mostly 1 or 3 spines in anal fin (occasionally 2) 35

35 Spinous and soft-rayed dorsal fins quite separate, with a distinct gap 36

Spinous and soft-rayed dorsal fins continuous although there may be a deep but incomplete notch 40

36 Lateral line present; scales usually ctenoid and very thick; mouth large, reaching below eyes . . 37
Lateral line absent; scales cycloid and often very thin; mouth small, not reaching below eyes . . . 38

37 Spinous dorsal short, 6 spines, with no black patch posteriorly; head very large, more than $1/3$ standard length .... **Apogonidae** p. 181

Spinous dorsal longer, 13 or more spines, with a distinct black patch posteriorly; head not so large, less than $1/3$ standard length .... **Percidae** p. 183

38 Soft dorsal and anal fins long-based, dorsal with 9 or more rays, anal with 18 or more rays .... **Melanotaeniidae** p. 134

Soft dorsal and anal fins short-based, dorsal often with less than 9 rays, anal with less than 18 rays 39

39 Spines in vertical fins quite stout and rigid; more than 1 anal spine; preorbital bone serrate in adults; gill rakers very long and slender, very numerous, more than 50 ..... **Mugilidae** p. 191

FAMILY KEY

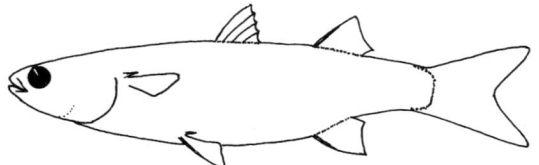

Spines in vertical fins slender and flexible, 1 anal spine; preorbital bone smooth; gill rakers usually short and stubby but sometimes long and slender, less than 50. 40

40 Pelvic fins not attached to abdomen by a membrane along innermost fin ray; scales present between pelvic fin bases and vent. . . . . **Atherinidae** p. 123

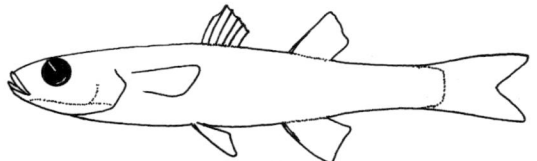

Pelvic fins attached to abdomen by a membrane along innermost fin ray; no scales present between pelvic fin bases and vent. . . . . **Pseudomugilidae** p. 141

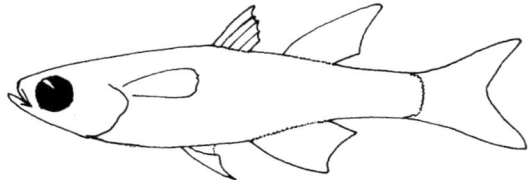

41 Total spines and rays in anal fin more than 35; 4 short spines at front of dorsal fin grading rather abruptly into much longer spines and rays; pelvic fins very reduced in adults, with 1 spine and 2–3 rays. . . . . . . **Monodactylidae** (marine)

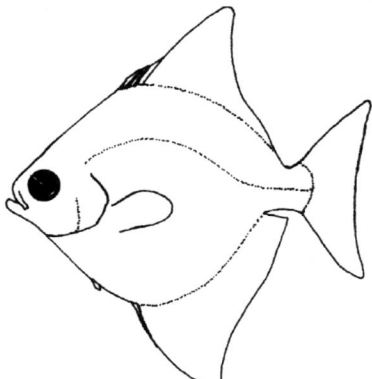

Total spines and rays in anal fin less than 30; sometimes 2–3 shorter spines at front of dorsal, longest spine subequal to longest ray; pelvic fins well developed, usually I, 5. 42

42 Gill membranes joined to each other across isthmus, free from isthmus; scales very small, 80 or more along lateral line. . . . . **Scatophagidae** (marine)

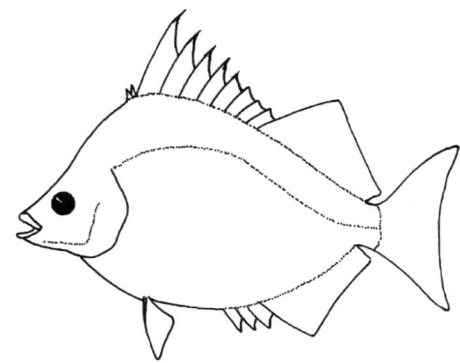

Gill membranes free from each other and isthmus; scales larger, less than 80 along lateral line. 43

43 Dorsal and anal fin bases sheathed with large scales. 44
Dorsal and anal fin bases not sheathed with large scales (though small scales may spread onto fin bases). 45

44 Scales cycloid, large, less than 30 along side, lateral line reduced to a few, anterior, pored scales; preopercular margin strongly serrate and with a parallel serrate ridge; mouth somewhat protrusible. . . . . **Chandidae** p. 146

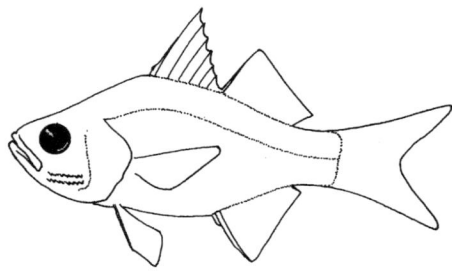

Scales usually ctenoid, moderate, more than 40 long lateral line, which is continuous and distinct; preopercular margin not strongly serrate and with no parallel serrate ridge; mouth highly protrusible . . . **Gerridae** (marine)

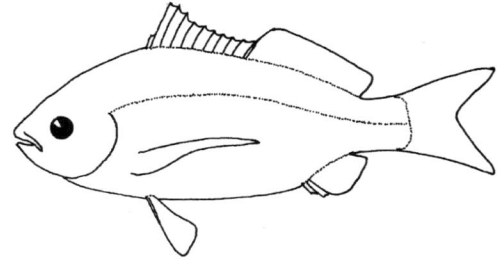

45 Teeth in jaw in a row of stout, sharp fangs forming the outer margin of a pavement of cobble-like teeth inside; pectoral fins large, usually reaching to about vent . . . . **Sparidae** (marine)

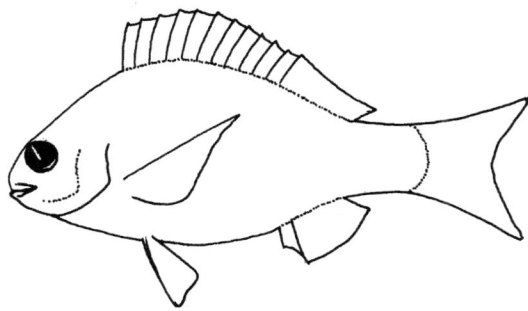

Teeth in jaws lacking a pavement of cobble-like teeth; pectoral fins smaller, not reaching vent 46

46 Lateral line discontinuous, incomplete or absent; scales large, less than 35 along side . . 47
Lateral line continuous and complete; scales smaller, more than 40 along side 48

47 Dorsal fin long-based, often extending back as a long, tapering to rounded point above tail; no distinct notch between spinous and soft-rayed sections . . . . **Cichlidae** p. 176

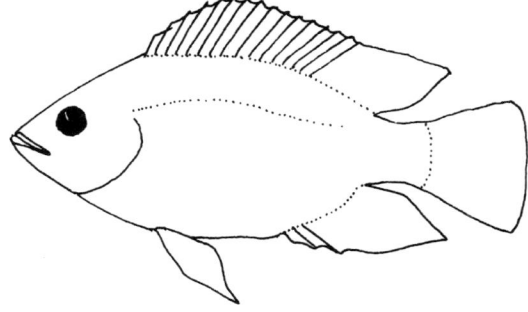

Dorsal fin shorter-based, never extending back over tail; a distinct notch between spinous and soft-rayed sections . . **Nannopercidae** p. 168

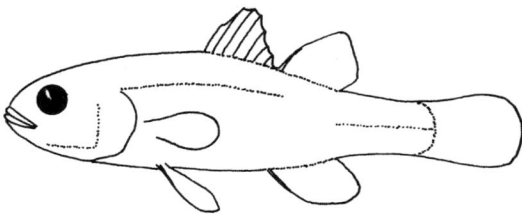

48 Teeth in jaws usually in a few rows, conical, sharp and stout; axillary process present at bases of pelvic fins . . . . **Lutjanidae** (marine)

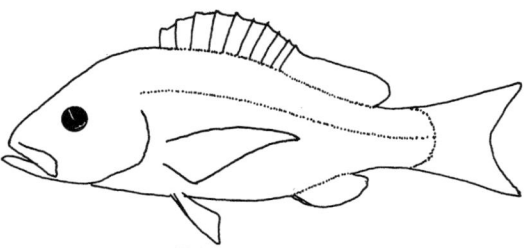

Teeth in jaws small, numerous in several rows; no axillary process at bases of pelvic fins 49

49 Pelvic fin origin below or in front of pectoral fin bases; mouth large, usually reaching to hind margin of eye; tail usually rounded . . . . **Percichthyidae** p. 150

Pelvic fin origin distinctly behind pectoral fin bases; mouth smaller, usually not reaching middle of eye; tail forked or emarginate. . . . . **Terapontidae** p. 164

4
Family Mordaciidae
Shortheaded Lampreys

I.C. POTTER

The lampreys (Petromyzontiformes), together with the hagfishes (Myxiniformes), are the sole surviving representatives of the Agnatha, or jawless vertebrates. The absence of jaws, and also of paired fins, distinguishes both of these groups from the cartilaginous and bony jawed fishes with which, for convenience, they are sometimes collectively grouped under the broad title of fishes.

Lampreys have a larval juvenile stage (ammocoete) that in all species is worm-like and relatively sedentary in habit. The ammocoete lives in the soft substrates of streams and rivers, feeding on algae, detritus and micro-organisms that it filters from the water above its burrow. After a number of years, the ammocoete undergoes a radical metamorphosis into a young adult, which has a tooth-bearing suctorial disc (Figs. 4.3, 4.5).

The young adult of the anadromous (sea migratory), parasitic species of lamprey eventually emerges from its burrow and migrates downstream to the sea, where it uses its suctorial disc to attach itself to host fish. A tongue-like, toothed piston in the centre of the disc is then rocked backwards and forwards, creating a hole in the side of the host from which blood and/or muscle tissue can be extracted. During this marine feeding phase, the lamprey rapidly grows far longer than its ammocoete. The adult eventually stops feeding and re-enters rivers, migrating upstream to spawn and die.

Although all lamprey species enter metamorphosis and develop teeth and a suctorial disc, some do not feed after completing larval life. In contrast to anadromous and parasitic species, nonparasitic species rapidly develop gonads during metamorphosis; they become sexually mature within a year of entering metamorphosis. The adult leaves the substrate at the approach of maturity and, after a short upstream migration, spawns and dies. Nonparasitic species thus complete their life cycles within their natal streams or river systems. Since their adults do not feed, they are similar in size to their larger ammocoetes and are thus much smaller than the adults of anadromous, parasitic species.

The morphology, and especially the arrangement of the teeth, of each nonparasitic species bears a striking resemblance to that of a particular parasitic species. It has thus been proposed that each nonparasitic species has evolved from a form very similar to a particular contemporary parasitic species, possibly in response to factors such as land-locking. The combination of each presumed ancestral parasitic species and its derivative nonparasitic species is referred to as a paired species.

Lampreys have an antitropical distribution. The 36 species of holarctic (northern, cool-temperate to sub-Arctic) lamprey are placed in the family Petromyzontidae, whereas the four species of Southern Hemisphere lampreys are separated into two families, Mordaciidae and Geotriidae. The Mordaciidae contain two anadromous, parasitic species, of which only the shortheaded lamprey (*Mordacia mordax*) occurs in Australia (the other species occurs in southern South America). Since the third *Mordacia* species, the Australian nonparasitic lamprey (*M. praecox*), is very similar to the shortheaded lamprey, it is assumed to have evolved from a form very similar to that species. The Geotriidae is a monotypic family, comprising just the pouched lamprey (*Geotria australis*), an anadromous and parasitic species (see Chapter 5).

Key to shortheaded lampreys

The two species of *Mordacia* are very similar morphologically, though the fully grown adult shortheaded lamprey normally occurs in fresh water at sizes greater than 300 mm, whereas the nonparasitic lamprey is rarely more than 170 mm long. The ammocoetes of the two species are indistinguishable.

SHORTHEADED LAMPREY *Mordacia mordax* Richardson

4.1 Shortheaded lamprey *Mordacia mordax*, adult. (I. C. Potter)

Description Ammocoete eyeless and worm-like (Fig. 4.2), with a prominent hood-like structure that overhangs oral aperture, which does not contain teeth. Seven small, vertically-arranged oval gill apertures on each side of branchial region. Two dorsal fins low, not extending far above body surface. The ventrally-located cloaca lies below a point nearly halfway forward along 2nd dorsal fin (Fig. 4.2).

Dorsal and anal fins in adults relatively far larger than in ammocoetes; eyes present, prominent dorsolaterally (Fig. 4.1). Adults, just prior to and immediately after completion of marine phase, have a well-developed suctorial disc, with radially

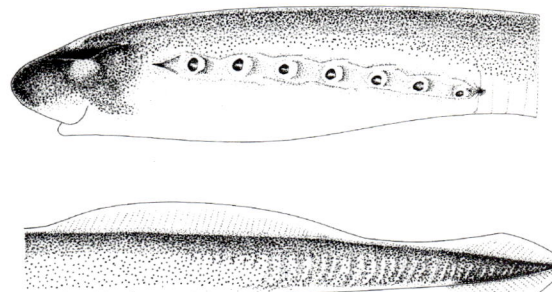

4.2 Shortheaded lamprey *Mordacia mordax*, head (above) and (tail) below of ammocoete larva. (F.J. Neira)

arranged tooth plates and a central, multicuspid, tongue-like piston (Fig. 4.3). In sexually mature adults, the radial tooth plates have disappeared, leaving only inner and outer cusps, and in males the oral disc has greatly increased in size.

Colour Ammocoete an overall brownish colour, darker on back. Adults, prior to and just after marine feeding phase, bluish grey dorsally and silvery ventrally. Colour becomes duller and greyer during spawning run.

Size Length at which ammocoetes enter metamorphosis varies among rivers, typically occurring, for example, at 110–140 mm in the Moruya River in New South Wales and at 100–115 mm in the Wye and North Esk Rivers in Tasmania. Adults typically return to rivers from the sea at 300–440 mm.

Distribution Adults and/or ammocoetes have been found in mainland rivers from the Hawkesbury River in New South Wales and southwards and westwards to the Gawler River in

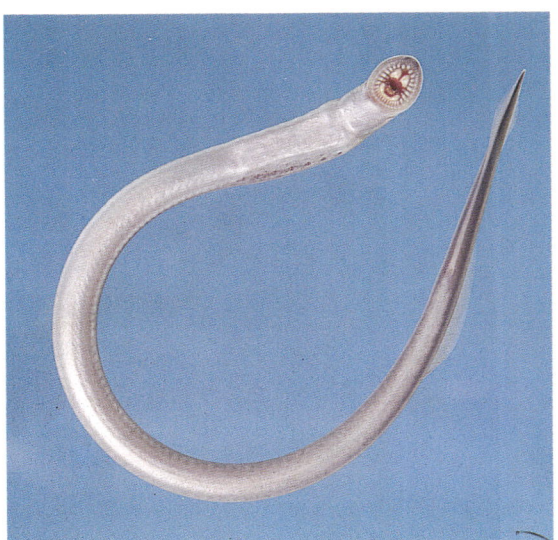

Shortheaded lamprey *Mordacia mordax*, ammocoete. (D. Rodgers)

South Australia; also present in many rivers in Tasmania. The extent of distribution within rivers has still to be determined.
Conservation status Is moderately abundant in some rivers.

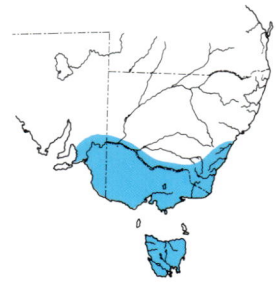

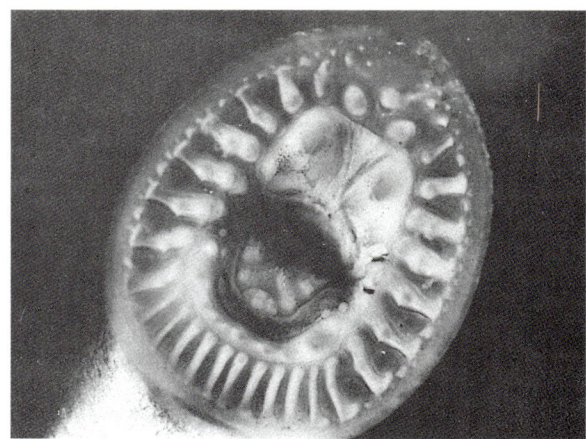

Natural history Metamorphosis typically begins in New South Wales rivers when ammocoetes are on average 3 years old; it starts in late February/early March, about two months later than in Tasmania. Although metamorphosis is completed in 5–6 months, the subsequent downstream migration of many of the young adults in the former region can be delayed by 3–4 months in years of low freshwater discharge.

Young adults can start feeding in estuarine waters when these are large enough to contain suitable hosts, as in the Gippsland Lakes, in Victoria, where shortheaded lampreys prey on brown trout (*Salmo trutta*), black bream (*Acanthopagrus butcheri*) and yelloweye mullet (*Aldrichetta forsteri*). Fishermen have reported that this species also attacks barracouta (*Leionura atun*) in the ocean just outside the Gippsland Lakes.

Samples collected at dams indicate that shortheaded lampreys typically start their spawning migration into Victorian rivers in late winter and spring and into Tasmanian rivers in spring. They are then 300–400 mm long. Adults are unusual in that they burrow into the substrate during this migration. Although breeding has not been observed, the time of appearance of small, new 0+ recruits indicates that the spawning run lasts for about a year. During this time, dentition changes from a condition similar to that of the young adult (Fig. 4.3), with the radially-arranged

4.3 Oral disc of adult shortheaded lamprey *Mordacia mordax*. (I. C. Potter)

tooth plates intact, to one similar to that in the mature adult of the nonparasitic lamprey, in which all but the inner and outer large cusps of those radial tooth plates have been lost (Fig. 4.5).

Utility Limited, though adults are occasionally caught in the Dandenong Creek and Yarra River, in Victoria, for sale and consumption. Ammocoetes can be maintained easily in aerated aquaria containing a muddy substrate and can be supplied with food such as suspensions of yeast. Adults caught on their spawning run can be held in large, well-aerated aquaria containing a sandy substrate into which they can burrow.

Similar species Comparable life stages of other lampreys; resemblance to anguillid eels is only superficial.

Other names Common: none. Scientific: none.

Literature Hughes & Potter, 1968; Johnston *et al.*, 1987; Koehn & O'Connor, 1990; Neira *et al.*, 1988; Potter, 1970, 1980; Potter & Hilliard, 1987; Potter & Strahan, 1968; Potter *et al.*, 1968, 1986.

NONPARASITIC LAMPREY *Mordacia praecox* Potter

4.4 Nonparasitic lamprey *Mordacia praecox*. (I.C. Potter)

Description Ammocoete morphologically indistinguishable from that of shortheaded lamprey; adults also indistinguishable except that those of nonparasitic lamprey rarely exceed 170 mm, while those of shortheaded lamprey are rarely less than 300 mm long at return to rivers from the sea.

Colour Colour of ammocoete the same as for shortheaded lamprey (see p. 33). At maturity, dorsal surface in both sexes bluish black; ventral surface greyish in males but appears yellowish in females owing to the eggs being visible through

the thin abdominal wall (Fig. 4.4).
Size Ammocoetes and adults can reach 172 mm.
Distribution Has been found in the Moruya and Tuross Rivers in southern New South Wales; probably occurs also in Victorian rivers, but difficulties in distinguishing its ammocoetes from those of

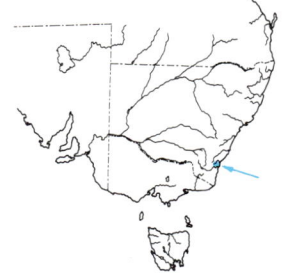

4.5 Oral disc of adult nonparasitic lamprey *Mordacia praecox*. (I. C. Potter)

shortheaded lampreys, together with the restricted period during which its clearly identifiable mature adults are present, makes elucidation of the species' full distribution difficult.
Conservation status Difficult to determine for the above reasons; is for this reason listed as 'potentially threatened' by the Australian Society for Fish Biology.
Natural history In the Moruya River, ammo-

Nonparasitic lamprey, *Mordacia praecox*, mature female (above) and male (below). (I.C. Potter)

coetes typically metamorphose at lengths 10–20 mm greater than those of shortheaded lampreys; it is assumed that, as with 'paired species' of *Lampetra* in Europe, the 'derivative' nonparasitic lamprey has a longer larval life than the 'ancestral' parasitic shortheaded lamprey. Enters metamorphosis in the Moruya River in late spring (November) and spawns late in the following winter (August). During sexual maturation, the radial tooth plates break down, leaving only the cusps at their inner and outer edges (Fig. 4.5). This parallels the situation found in shortheaded lamprey.
Utility None.
Similar species Other lamprey species at similar life stages; anguillid eels, though the similarity is only superficial.
Other names Common: none. Scientific: none.
Literature Hughes & Potter, 1968; Potter, 1970, 1980; Potter *et al*, 1968, 1986.

5
Family Geotriidae
Pouched lamprey

I.C. POTTER

This family contains only a single species, widely distributed in south-western and south-eastern Australia, Tasmania, New Zealand and south-eastern and south-western South America (Patagonia). (See Chapter 4 for a general discussion of lamprey biology.)

POUCHED LAMPREY *Geotria australis* Grey

5.1 Pouched lamprey *Geotria australis*—(above) unpouched adult soon after entering fresh water; (below) pouched male (note reduced length and enlarged snout. (I.C. Potter)

Description Ammocoete similar to that of short-headed lamprey, except that cloaca lies below origin of 2nd dorsal fin (Fig. 5.2). Adults of pouched lamprey have lateral rather than dorsolateral eyes, lateral teeth on oral disc spathulate rather than pointed, and edge of suctorial disc bears fimbriae (Fig. 5.3). During spawning run, males develop a large pouch below head, and suctorial disc becomes greatly enlarged (Figs 5.1, 5.3).

Colour Colour of ammocoete similar to that of *Mordacia* species: sandy-brown, somewhat paler ventrally. Newly metamorphosed young adult is bright silvery with bright blue bands along back; these persist throughout sea life and early return to fresh water, after which adult becomes increasingly drab and eventually a muddy grey-brown.

Distribution Has been recorded in rivers between Lakes Entrance and Gulf St Vincent, in southern Australia, and in many rivers in Tasmania. Also occurs in south-western Australia, New Zealand and rivers of Chile and Argentina. Is arguably one of the most widely dispersed species of freshwater fish. Distribution within rivers still has to be determined.

Size Ammocoetes rarely exceed 120 mm; adults re-enter rivers at 500–700 mm long.

Conservation status Abundance has almost certainly declined following damming and other types of changes to rivers that are likely to affect upstream migration.

Natural history Has not been studied in detail in south-eastern Australia and thus the following account of biology is drawn largely from investigations in rivers of the south-west. Enters metamorphosis in south-western Australia, and also in Tasmania, at about 80-120 mm long and usually when $4^{1/4}$ years old. Metamorphosis begins in mid- to late summer (January–February) and is completed in mid-winter (July), with the downstream migration occurring soon afterwards when

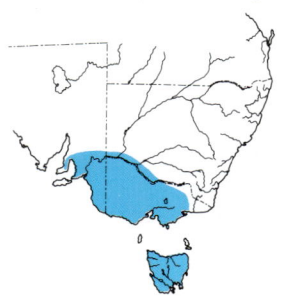

freshwater discharge is seasonally high. There is no information on hosts of pouched lamprey while at sea, though it undoubtedly parasitises marine fish, as discussed for shortheaded lamprey (see p. 34). During the southern summer, albatrosses that breed on South Georgia consume large numbers of adults of South American lamprey populations.

In south-western Australia, the spawning migration into estuaries begins in mid-winter (July). Although migrating adults are regularly caught in front of dams over the next four months as they migrate upstream and have been collected by electrofishing through to February, they have

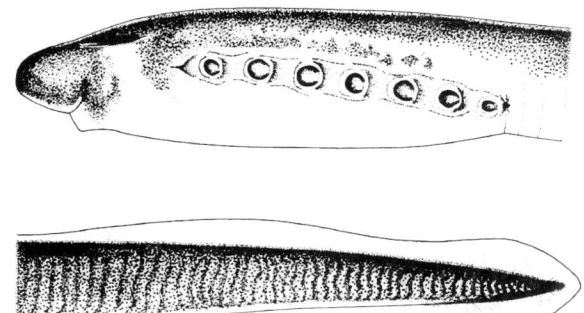

5.2 Pouched lamprey *Geotria australis*—head (top) and tail (below) of ammocoete larva. (F.J. Neira)

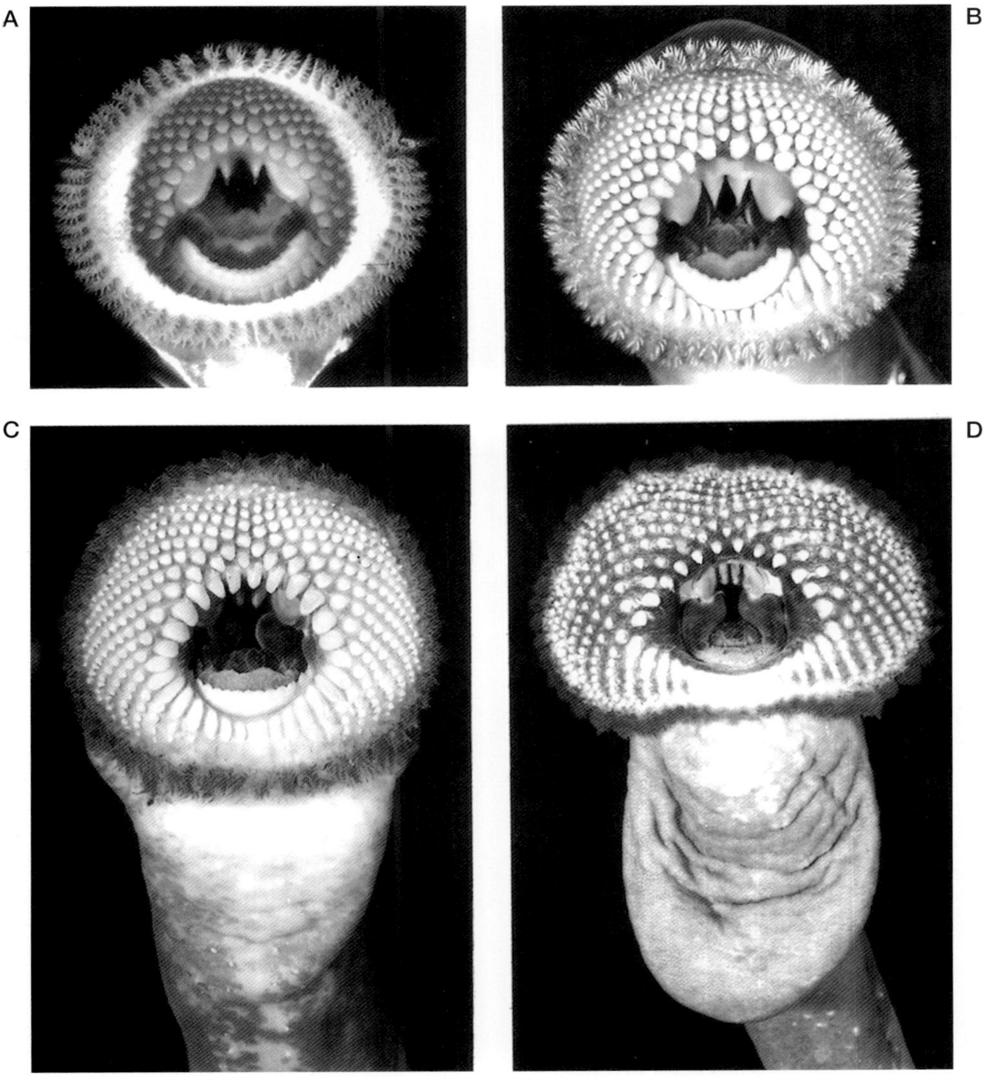

5.3 Pouched lamprey *Geotria australis*—oral discs of: **A** newly metamorphosed young adult (at time of downstream migration); **B** adult at return from sea; **C** nearly mature female; **D** mature male with well-developed pouch

FAMILY GEOTRIIDAE

Pouched lamprey *Geotria australis*, oral disc. (R.M. McDowall)

and extreme modifications to the dentition (Fig. 5.3D).
Utility Taken as food by Maoris in New Zealand, where it is also favoured by immigrants from eastern Europe.
Similar species Similar life stages of shortheaded lampreys; also anguillid eels, though the similarities are superficial.
Other names Common: none. Scientific: *Yarra singularis*.
Literature Bird & Potter, 1981; Hardisty *et al.*, 1986; Johnston *et al.* 1987; Koehn & O'Connor, 1990; Lethbridge & Potter, 1981; Neira *et al.*, 1988; Potter, 1980; Potter & Hilliard, 1986; Potter & Robinson, 1991; Potter & Strahan, 1968; Potter *et al.*, 1980, 1983, 1986.

only occasionally been found in rivers after that. Since the gonads of these adult lampreys are still very immature in February, gonadal development was monitored in adults held in the laboratory under light and temperature regimes that parallel those in the field. Gonad development data indicate that spawning is in October or early November, i.e. 15–16 months after the start of the spawning run. This proposed time of spawning is consistent with the time of appearance of the first, small 0+ recruits. This very protracted migration is accompanied by vast morphological changes that, in the male, include the development of a large pouch

Pouched lamprey *Geotria australis*, head of upstream-migrating adult. (R.M. McDowall)

Pouched lamprey *Geotria australis*, soon after entering fresh water from the sea. (S. Moore)

6
Family Anguillidae
Freshwater eels

J.P. BEUMER

The family Anguillidae contains just one genus, *Anguilla*. Although known mostly from their occurrence and distribution throughout coastal fresh waters, these eels migrate downstream to spawn in the sea when sexually mature. The greater part of their life cycle is spent in fresh water, in both running and still areas.

Freshwater eels occur in wetlands bordering the North Atlantic, the Western Pacific from Japan south through South-East Asia, the Pacific Islands and to New Zealand and Australia, westwards in the Indian Ocean to Madagascar and eastern Africa. Fifteen species of *Anguilla* are currently recognised. Four have been recorded from the coastal wetlands of the Australian continent. Of these, two occur in south-eastern Australia, a third in north Queensland and a fourth species in north-west Australia. The family is characterised by elongate cylindrical form, continuous dorsal-caudal-anal fins, lack of spines in any fins and the absence of pelvic fins.

Key to freshwater eels

Dorsal fin extends little forward of anal fin; vomerine teeth in a broad, short patch;
uniform colour without distinctive markings . . . . **Anguilla australis** p. 39; Fig. 6.1
Dorsal fin extends markedly forward of anal fin; vomerine teeth in a long
narrow band; colour distinctly blotched or mottled . . . . . . **Anguilla reinhardtii** p. 42; Fig. 6.7

SHORTFINNED EEL *Anguilla australis* Richardson

Description Elongate and cylindrical; dorsal, caudal and anal fins continuous, inseparable and without spines; dorsal fin originating only slightly in front of or level with anal fin. No pelvic fins, small ovate pectoral fins just behind narrow, vertical gill openings. Eyes small in immature specimens but increasing by a factor of 2–3 at maturity, when a copper sheen becomes clearly visible. Mouth large, extending to below eyes. Anterior nostrils long and project forwards over upper lip. Scales indistinct (Fig. 6.2) but form a mosaic deeply embedded in thick, fleshy

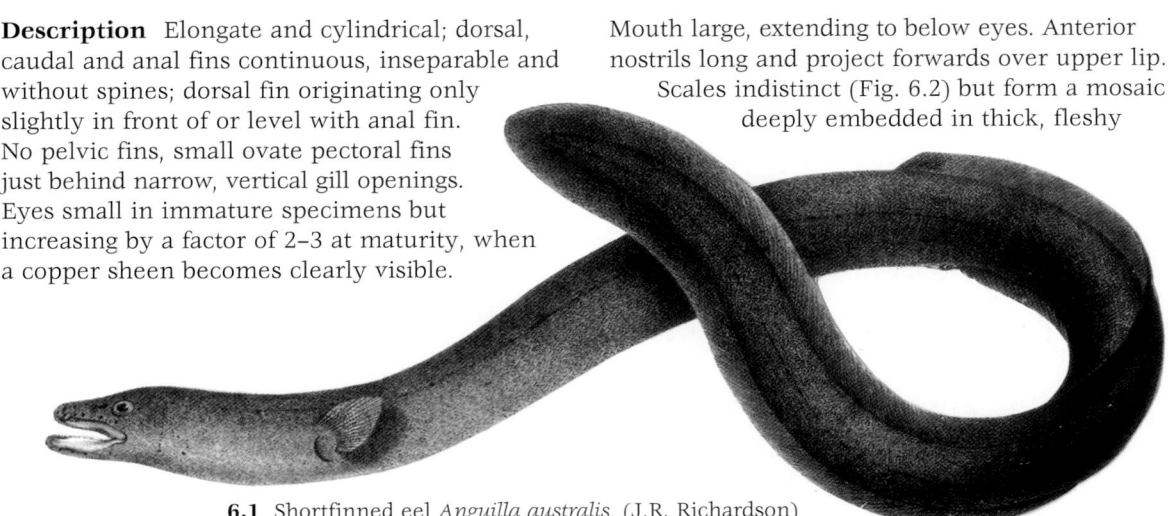

6.1 Shortfinned eel *Anguilla australis*. (J.R. Richardson)

skin; a distinct lateral line. Vomerine teeth form a somewhat club-shaped patch that does not reach back as far as the jaw teeth (Fig. 6.3A); pectoral rays 14–16; vertebrae 109–116.

Colour Back and sides usually a uniform golden olive to olive-green; belly greyish white, sometimes almost a silvery white. Fins dark, like the back. On maturity, dorsal body darkens and silvery belly contrasts markedly.

Size Reaches about 1100 mm in length and 3.2 kg in weight, but usually much smaller; males (500 mm and 250 g) are much shorter than females.

Distribution Present in south-eastern Australia from the Caboolture River in southern Queensland (27°S), south and west to the vicinity of Mount Gambier in South Australia. Occurs primarily in coastal wetlands draining south and east from the mountain ranges, though a few have been reported from the inland Murray-Darling drainage system. Occurs in streams on Flinders and Vansittart Islands in Bass Strait, and is widespread in coastal and lowland rivers in Tasmania.

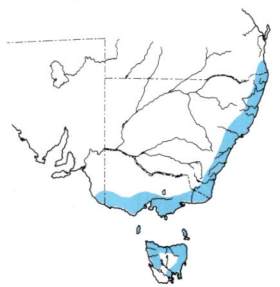

Is also widespread in the western Pacific, occurring on Norfolk and Lord Howe Islands. Not recorded from Fiji and Tahiti. Present in New Zealand and the Chatham Islands.

Conservation Remains abundant across range, though numbers have declined as a consequence of the development of commercial eel fisheries.

Natural history Occurs in a wide variety of wetland habitats, from rivers and creeks to lakes and swamps, but is essentially a still-water species. Migrates to sea to spawn, possibly in or near the Coral Sea. Larval eels, known as *leptocephali* because of their compressed, gum-leaf shape (Fig. 6.4), are carried back from the spawning grounds by the East Australian Current. On nearing the coast, they metamorphose into glass eels, assuming the characteristic eel shape but initially remaining largely unpigmented (Fig. 6.5). These juvenile eels enter fresh water after which development into fully pigmented elvers rapidly occurs, and mass migrations upstream follow. Young eels enter fresh waters during spring and summer, penetrating the upper reaches of rivers, and lakes, lagoons and swamps. Large obstacles such as waterfalls, dams or weirs slow their progress upstream but are not insurmountable provided some water is present along spillway margins. Eels may take 10–20 years to reach maturity and begin their downstream migration to the sea to spawn. Downstream migration is facilitated by floodwaters. Spawning occurs at considerable depths in the ocean. Feeds on a

6.2 Scale of a shortfinned eel *Anguilla australis*. (D.J. Jellyman)

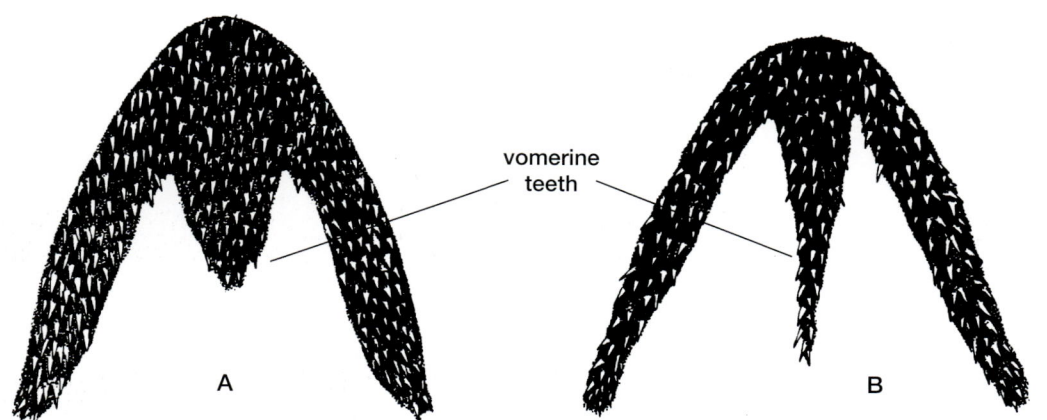

6.3 Vomerine teeth in roof of mouth as in: **A** shortfinned eel *Anguilla australis*; **B** longfinned eel *Anguilla reinhardtii*. (R.M. McDowall)

FRESHWATER EELS

Shortfinned eel *Anguilla australis*. (R.H. Kuiter)

variety of aquatic fauna (insects, crustaceans, molluscs and fish).

Utility Provides good sport for anglers and is enjoyed as a food by some, particularly people from Europe or Asia, where eels are regarded as a delicacy. Easily caught recreationally with any bait on a hook or commercially in baited or unbaited fyke nets or traps, mostly overnight. Forms the basis for a small but lucrative export industry in Victoria (225 tonnes a year) and to a lesser extent in New South Wales and Tasmania. Most is exported frozen, with a small quantity smoked for domestic consumption. Intensive culture of eels, both rearing of glass eels and fattening of small eels taken in the wild, has attracted wide interest but little development. Although it is very easily kept in aquaria, its nocturnal and predatory habits make it rewarding in an aquarium

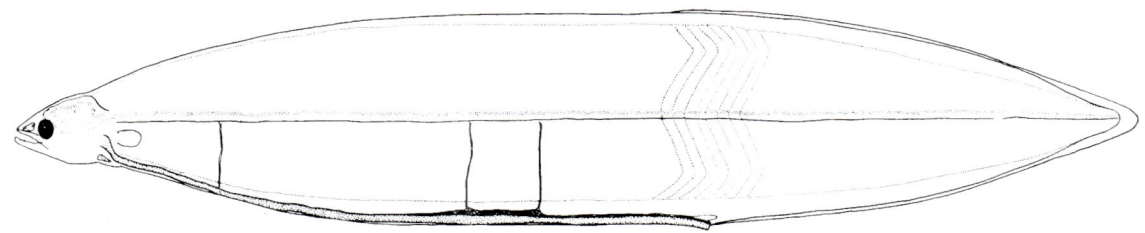

6.4 Leptocephalus of shortfinned eel *Anguilla australis*. (P.H.J. Castle)

6.5 Glass eel of shortfinned eel *Anguilla australis*. (D.J. Jellyman)

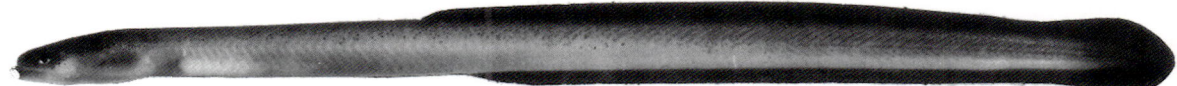

6.6 Elver of shortfinned eel *Anguilla australis*. (D.J. Jellyman)

FAMILY ANGUILLIDAE

Shortfinned eel *Anguilla australis*, head of migrant female. (R.M. McDowall)

only as a single specimen.
Similar species Longfinned eel, perhaps also lampreys, especially when small.
Other names Common: silver eel. Scientific: referred to under subspecies names *Anguilla australis australis* and *A. australis occidentalis;* the first is technically correct but of little significance, the second is quite incorrect and not to be used.
Literature Anderson & Whitley, 1925; Bertin, 1956; Beumer, 1979c; Beumer & Sloane, 1990; Castle, 1963; Ege, 1939; Kershaw, 1911; Koehn & O'Connor, 1990; Lewis, 1942; Schmidt, 1928; Scott, 1953; Whitley, 1956; 1957b.

LONGFINNED EEL
Anguilla reinhardtii Steindachner

Description Closely resembles shortfinned eel but has dorsal fin originating well in front of anal fin; head broader, lips thick and fleshy, and the mouth large, extending back well past eyes. Vomerine teeth in a long, narrow band that extends back about the same distance as teeth in jaws (Fig. 6.3B); 16–20 pectoral rays; 104–110 vertebrae.
Colour Distinctly blotched or mottled dorsally and laterally with olive-green to brownish, paling ventrally; fins dark brownish, pectoral fins tend to be yellowish.
Size Reaches a length of about 1650 mm and a weight of 22 kg; commonly up to about 1000 mm, males being much smaller (650 mm and 600 g) than females.
Distribution Known in Australia from Cape York

in northern Queensland, south to vicinity of Melbourne, on the coastal side of the Great Dividing Range, also northern and eastern Tasmania. Also present in New Caledonia and Lord Howe Island.
Conservation status Remains abundant across range.
Natural history See shortfinned eel; longfinned eels tend to occur more often in rivers than in lakes. Predation on juvenile waterfowl is not uncommon.
Utility Is good angling because of its large size and is easily taken at night on baited hooks. Is of increasing importance commercially for live

FRESHWATER EELS

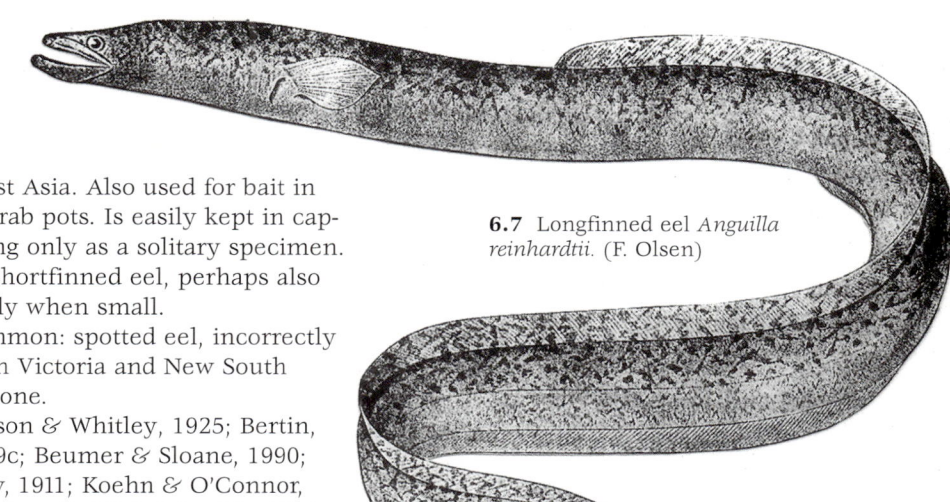

export to South-East Asia. Also used for bait in crayfish and mudcrab pots. Is easily kept in captivity, but rewarding only as a solitary specimen.
Similar species Shortfinned eel, perhaps also lampreys, especially when small.
Other names Common: spotted eel, incorrectly called conger eel in Victoria and New South Wales. Scientific: none.
Literature Anderson & Whitley, 1925; Bertin, 1956; Beumer, 1979c; Beumer & Sloane, 1990; Ege, 1939; Kershaw, 1911; Koehn & O'Connor, 1990; Lewis, 1942; Powell, 1930; Schmidt, 1928; Whitley, 1956; 1957b.

6.7 Longfinned eel *Anguilla reinhardtii*. (F. Olsen)

Longfinned eel *Anguilla reinhardtii*. (R.H. Kuiter)

7
Family Clupeidae
Herrings

I.C. BRIGGS AND R.M. McDOWALL

The family Clupeidae is a large family of commercially very important fishes, including the herrings, sardines and shads, that are sold in fish markets throughout the world and are important in the canning industry. There are quite a few freshwater species as well as some diadromous species that migrate between fresh water and the sea. There are two clupeids in south-eastern Australia, one of which is entirely freshwater-dwelling and the other probably catadromous, migrating out of rivers to spawn at sea. Clupeids can be distinguished by their single, soft-rayed dorsal fin, thin, cycloid scales, absence of lateral line, and a ridged serrate lower margin of the belly caused by stiff, keeled scales.

Key to herrings

Scales along ridge of back in front of dorsal fin spiny and serrate (like belly); last ray of dorsal fin not elongated . . . . **Nematalosa erebi** p. 44; Fig. 7.1
Scales along ridge of back in front of dorsal fin smooth (unlike belly); elongated last ray in dorsal fin . . . . **Potamalosa richmondia** p. 46; Fig. 7.2

BONY BREAM
Nematalosa erebi (Günther)

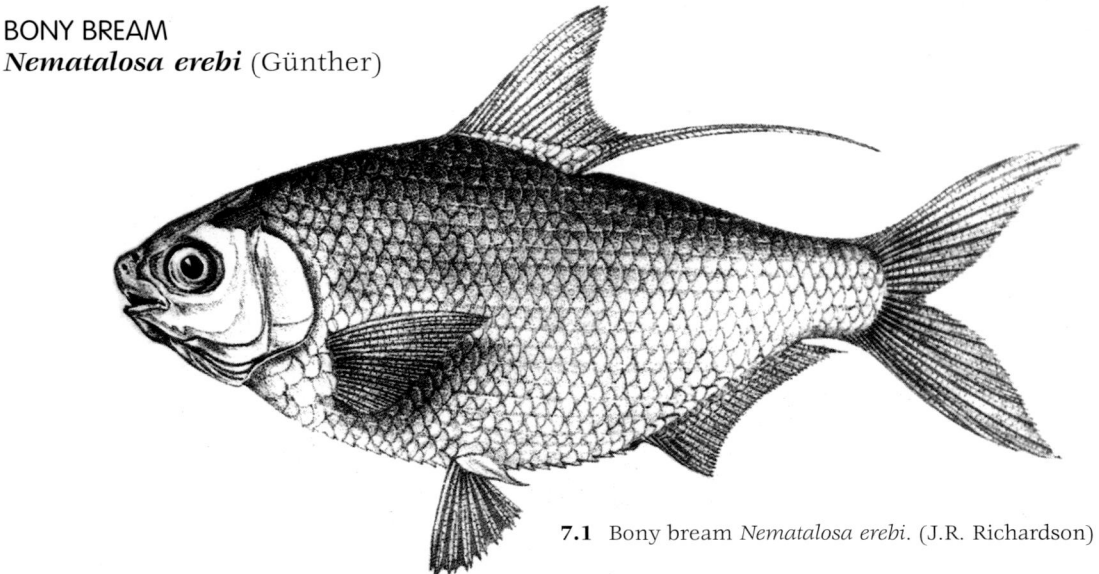

7.1 Bony bream *Nematalosa erebi*. (J.R. Richardson)

Description A deep-bodied, herring-like fish, very compressed; head small, with blunt snout; mouth small, downturned, and bony but toothless, lower jaw with a central notch that fits into a groove in upper jaw when mouth is closed. Scales of moderate size (40–46 along sides), smooth, but forming a sharp serrate ridge along both belly and back in front of the dorsal fin;

Bony bream *Nematalosa erebi*. (I.C. Briggs)

head scaleless. Eyes large and covered by an adipose eyelid. Fins angular; dorsal (14–19 rays) with anterior rays long, tapering to rear, with last ray greatly elongated and in some cases reaching base of tail. Tail large, strongly forked; anal fin long and low (17–27 rays). Pelvic fins abdominal, with a strong axillary spine at base of each. Pectorals small (14–18 rays). Vertebrae 41–45.

Colour Generally greenish on back, bright silvery iridescence on sides, and silvery white on belly; gill covers may be golden to olive. Distinctive reddish colouration may occur, most intense around mouth but spreading elsewhere.

Size Grows to 470 mm long and up to 2 kg weight, but commonly to 120–150 mm.

Distribution Is found throughout the Murray-Darling system at elevations below about 200 m; may abound in tributaries of the Darling, but is less common in the Murray and Murrumbidgee Rivers; also present in inland Lake Eyre drainages. May form huge schools, often in shallow waters. Is not known from the coastal drainages of the south-east. Occurs widely through remainder of Australia and also in Papua New Guinea.

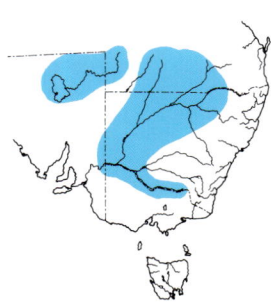

Conservation status Is widespread and remains highly prolific and abundant.

Natural history Is found in most of the main flowing and some standing waters of area, including billabongs, particularly occupying turbid waters; is a robust fish known to tolerate a wide range of temperatures, salinities and other habitat conditions. The life history is poorly understood. Upstream migrations during the day, of adult and subadult fish (60–400 mm long), have been documented at the Torrumbarry fishway on the Murray River. Spawns in spring and early summer, the eggs being very small (less than 1 mm in diameter), and may number several hundred thousand in large females. Probably reaches maturity at end of first year when about 80 mm long. Is an omnivore that feeds particularly on muddy detritus, also algae and aquatic insects and crustaceans. Is highly prolific and may form huge populations. Provides an important food for other, larger fish that share its habitat, such as Murray cod, and also aquatic birds such as pelicans and cormorants.

Utility Is not a recognised quarry of anglers, though it can be caught in large numbers using gill, drum or seine nets. Is well named, providing little flesh but has numerous fine bones. Not regarded as a table fish, it was, however, canned for troops during the World War II.

It is an attractive silvery fish in captivity, but it is very sensitive to handling and not easily kept in aquaria.

Similar species Resembles other small silvery fishes and can be confused with smelt during early life.

Other names Common: various, including pyberry, melon fish, tukari. Scientific: *Fluvialosa richardsoni, Nematalosa richmondia*.

Literature Koehn & O'Connor, 1990; Lake, 1967a; Nelson & Rothman, 1973; Mallen-Cooper, 1994; Merrick & Schmida, 1984.

FRESHWATER HERRING *Potamalosa richmondia* Macleay

7.2 Freshwater herring *Potamalosa richmondia*. (A.R. McCulloch)

Description Slender and compressed, with a longish snout and large eye; mouth small and upturned, with lower jaw protruding. Fins small and rather angular; dorsal high on back, at about level of pelvic fins, rays towards front longest (15–18 rays); tail strongly forked, with pointed lobes. Anal fin small and low, anterior rays longest (15–17 rays). Pelvic fins abdominal, triangular. Pectoral fins low, with 15–19 rays. Scales thin, smooth and quite large (41–48 along side), easily dislodged; head lacks scales. Scales along back in front of dorsal fin smooth, not keeled, but those along belly keeled, making lower margin of belly serrate.

Colour Bright silvery, back somewhat iridescent green, often with a dark band running along upper part of body and occasionally with a few indistinct stripes; eyes silvery. Fins largely colourless or with some slight darkening along fin rays.

Size Known to grow to 320 mm but generally reaches only half this size.

Conservation status Not common, and numbers seem to fluctuate, but is not considered under serious threat.

Distribution Recorded from coastal drainages of

New South Wales and Victoria, especially river systems to the north of about Sydney, and may penetrate substantial distances inland. Does not occur in the western drainages of the Great Dividing Range.

Natural history Is a fast-swimming, schooling fish that prefers clear, moderately fast-flowing streams, though it has been found in a wide variety of habitats, including still waters. Very little is known of the life history, but possibly migrates downstream into estuaries or perhaps the sea during winter to spawn; the young presumably later migrate upstream. Is a carnivore that feeds on insects, prawns and worms.

Utility Is easily taken by angling but is not much fished for. Is reasonable eating, though this is discouraged by the numerous small, intramuscular bones.

Although it is a handsome, silvery fish, the freshwater herring is not regarded as a good

Freshwater herring *Potamalosa richmondia*. (D. Rodgers)

aquarium species, being fragile and subject to scale loss when handled.
Similar species Resembles grayling, large smelt and perhaps mullets.

Other names Common: Nepean herring. Scientific: none.
Literature Bishop & Bell, 1978b; Koehn & O'Connor, 1990; Merrick & Schmida, 1984; Whitley, 1957a.

8
Family Chanidae
Milkfish

H.K. LARSON

The family Chanidae contains only one species found throughout the Indian Ocean and western-central Pacific Ocean in tropical to warm temperate regions. It is characterised by small, cycloid scales, a lateral line present, a single dorsal fin high on the back, the absence of teeth in the jaws, and a huge number of long, slender gill rakers.

MILKFISH *Chanos chanos* (Forsskal)

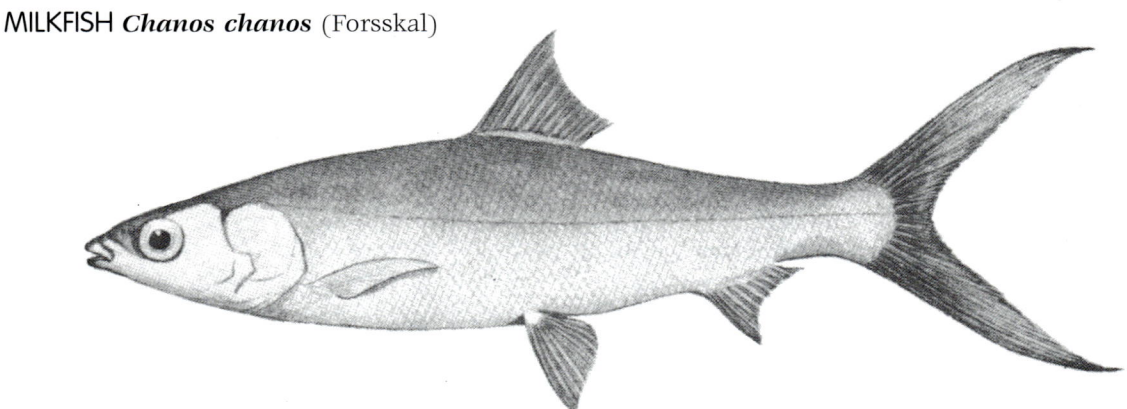

8.1 Milkfish *Chanos chanos*. (G. Coates)

Description Rather slender and compressed, mouth toothless, and small, not reaching back as far as eyes. Single dorsal fin (13–17 rays), high on back, above abdominal pelvic fins; anal fin (9–11 rays) much smaller. Tail deeply forked with long pointed lobes; scaly sheaths along bases of dorsal and anal fins; 16–17 pectoral rays. Scales small and cycloid (75–91 along lateral line), firmly implanted; lateral line easily visible. Several hundred long, slender gill rakers; about 57 vertebrae.
Colour Brilliant silver, greenish to blue-grey on back and silvery-white below. Fish from muddy waters may be darker in colour.
Size Recorded to 1700 mm long and up to 5.5 kg.
Distribution In Australia occurs from Western Australia to northern New South Wales and has been reported from Port Melbourne, Victoria; is essentially a warm-water and marine species but occurs in warm, shallow estuaries and will travel up rivers.

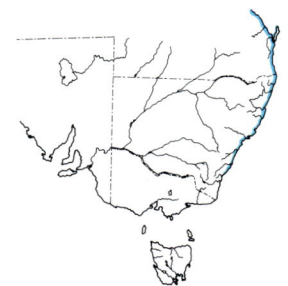

Conservation status Remains abundant through range.
Natural history Little known in natural habitat, most observations being made on pond-reared fish. Adults school over shallow, sandy or muddy coastal areas and may move up into rivers. Reproductive behaviour is not known; eggs very small, a little more than 1 mm in diameter, and have been found in open sea 25 km offshore. A large female carries huge numbers of eggs, counted in the millions—one female 1120 mm long having nearly 6 million. Spawning is probably in shallow bays where water is less saline; larvae move from sea into estuaries, where they feed on

plankton. Adults apparently feed on algae, detritus, and associated animals.

Utility A very important pond-culture and food fish in South-East Asia, particularly the Philippines and Taiwan, less so in the Pacific region. Larvae caught in shallow coastal seas are transferred to rearing ponds, where they are fed on algae and supplementary foods. A marketable size of about 300 g is reached after about 8 months. Is prepared for eating by pickling in brine or smoking, but is not generally eaten by Western people because there are so many fine bones. Is occasionally caught in Australia on hook and line; fights very strongly, leaping from the water high into the air.

Similar species Mullets and some herrings.

Other names Common: bangos. Scientific: none.

Literature Berra, 1981; Jhingran, 1975; Schuster, 1952.

9
Family Elopidae
Tarpon and oxeye herring

D.A. POLLARD

The fishes of this small family occur in tropical and subtropical waters of the Atlantic, Indian and Pacific Oceans. The Atlantic tarpon, *Megalops atlanticus,* which is renowned as a game fish, commonly grows to 40 kg in weight and may reach 100 kg. The Indo-Pacific and northern Australian oxeye herring, *Megalops cyprinoides*, is very similar, though smaller. The family is characterised by an elongate last ray in the dorsal fin, adipose eyelids and conspicuous axillary processes on the pectoral and pelvic fins.

OXEYE HERRING *Megalops cyprinoides* (Broussonet)

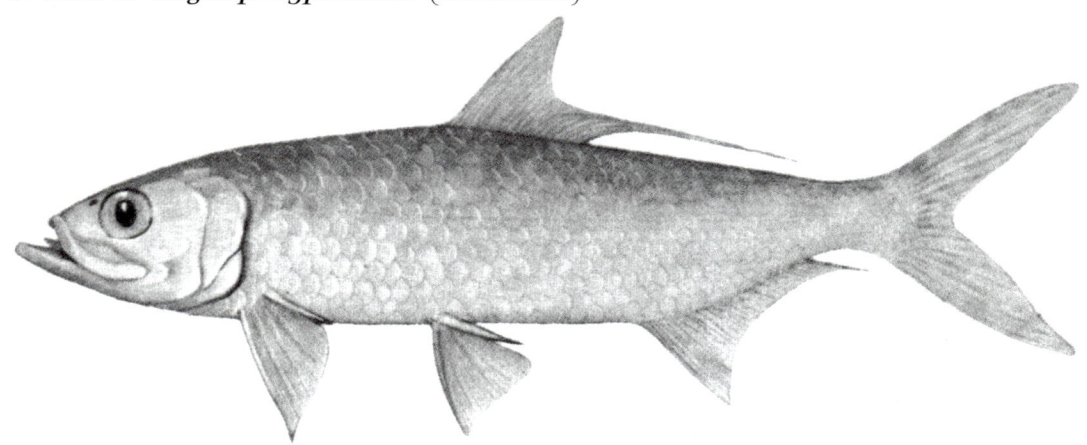

9.1 Oxeye herring *Megalops cyprinoides.* (F. Olsen)

Description Body oblong and laterally compressed; head of moderate size and snout quite sharply pointed; mouth very large, lower jaw protruding strongly. Teeth in jaws small, and a bony plate present below throat. Eyes large, and covered by fleshy adipose eyelids. Scales large, lateral line distinct (36–40 scales). Dorsal fin (17–20 rays) placed high on back, above abdominal pelvic fins, last fin ray very elongated. Tail deeply forked, with pointed tips. Anal fin long and shallow (24–31 rays). Pectoral fins (14–16 rays) and pelvic fins (10–11 rays) with long, fleshy, axillary processes.
Colour Back bluish green, the sides more silvery, the lower surface more whitish, the head darker olive and the tail yellowish; other fins a greenish yellow, the dorsal having a dusky margin.
Size Is reputed to grow to more than 1500 mm.
Distribution Occurs in all tropical Australian seas and adjacent estuaries and coastal fresh waters, extending southwards into northern New South Wales, where it is likely to be found in fresh waters only in the far north (south to at least the Clarence River). Is also

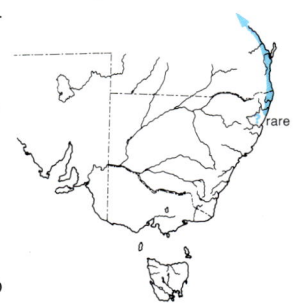

widespread in the tropics, from East Africa, around the northern shores of the Indian Ocean to South-East Asia, and also northwards to Japan and eastwards to Tahiti. Specimens have been taken over 900 km upstream in the Fly River in Papua New Guinea.

Conservation status Abundant throughout most of its range.

Natural history Essentially a marine species as an adult, but smaller specimens (usually between 200 and 500 mm long) inhabit estuaries and are regularly found in fresh water well above tidal influence, as well as in temporarily enclosed or landlocked lagoons and billabongs. Spawning is thought to take place in summer in estuaries and/or shallow inshore waters, the transparent *leptocephalus*-type larvae being of a ribbon-like or willow-leaf shape. In northern Australia, young fingerlings are often found in freshwater creeks after the wet season. Larger juveniles are often found shoaling in estuaries during summer, and subadults return to the sea before they mature. Adults feed mainly on crustaceans and smaller fish, and the young will take insects at the water's surface. The larvae may be plankton feeders. In fresh waters, this species has been recorded from clear to muddy waters at pH 5–9 and temperatures of 23–34 °C. These fish can also withstand low oxygen levels and may be able to breathe air at the water's surface like their close relative *M. atlanticus*.

Utility A very hard-fighting sport fish, although not generally valued as food because its flesh is soft, very bony and relatively flavourless; however, it is raised for food in ponds in India and South-East Asia, where it is also used to control populations of other fish species in brackish aquaculture ponds.

Similar species This fish is rather distinctive, but juveniles may be confused with herrings (family Clupeidae).

Other names Common: tarpon. Scientific: *Clupea cyprinoides, Megalops filamentosus, Megalops setipinnis*.

Literature Coates, 1987; Fortes, 1980, 1985; Grant, 1991; Kaliyamurthy *et al*., 1977; Losse, 1968; Merrick & Schmida, 1984; Pandian, 1969; Rao & Ghosh, 1986; Roberts, 1978; Tsukamoto & Okiyama, 1993; Wade, 1962.

10
Family Galaxiidae
Galaxiids

R.M. McDOWALL AND W. FULTON

This family is one of the largest in the Australian freshwater fish fauna, and certainly the largest in Australia's south-eastern river and lake systems. There are about 20 Australian galaxiids, of which about 17 are found in the area covered by this guide. The family is widespread in southern cool temperate lands, with species present in southern Australia, New Zealand, Patagonian South America, South Africa, as well as some islands, including New Caledonia (in a high elevation lake), Lord Howe, Chathams, Auckland and Campbell Islands and the Falklands; there are about 50 species in all. The Tasmanian fauna is particularly rich, with 16 species recognised in 3 genera.

Galaxiid fishes are generally rather elongate, tubular fish with (except in Tasmania's *Paragalaxias*) the single, soft-rayed dorsal fin set well back on the body, its origin above or a little in front of the level of the anal; there is no adipose fin nor any scales; a lateral line is present, and there are usually 7 pelvic fin rays and 16 principal caudal fin rays.

The common name for the family and its members is a matter of some contention. Various species have been referred to as 'trouts', which, though not totally misleading, causes confusion with introduced salmonids; others are called 'minnows', but these fishes are certainly not minnows in the strict sense (family Cyprinidae). An extreme example, perhaps, is reference to one species as 'trout minnow'! The only distinctive Australian common name is 'jollytail', which seems to be highly suitable for some species. Otherwise, on the whole, it seems best to settle for 'galaxias', modified with some descriptive adjective referring to location, habitat or some such.

Key to galaxiids

1. Dorsal fin forward, above pelvic fin bases 14
   Dorsal fin posterior, distinctly behind pelvic fin bases, more or less above vent 2

2. Dorsal fin origin distinctly behind anal origin; no laterosensory pores below lower jaw;
   less than 16 principal caudal rays, usually 13–14 .... ***Galaxiella pusilla*** p. 70; Fig. 10.15
   Dorsal fin origin usually above or a little in front of anal origin; laterosensory pores
   present beneath lower jaw; usually 16 principal caudal rays 3

3. Pelvic fin rays usually 7, occasionally more, rarely 6 4
   Pelvic fin rays 5-6 .... ***Galaxias parvus*** p. 66; Fig. 10.11

4. Caudal fin distinctly rounded .... ***Galaxias cleaveri*** p. 64; Fig. 10.10
   Caudal fin truncate to forked 5

5. Usually 5–6 pores in series from below eye to anterior nostril............................
   .... ***Galaxias fontanus*** p. 59; Fig. 10.5
   Only 4 pores in series from below eye to anterior nostril 6

6. Gill rakers short and stout; anal fin origin usually below middle of dorsal or further back 7
   Gill rakers long and slender; anal fin origin below dorsal origin or as far back as mid-dorsal ... 9

7   Two pyloric caeca of moderate length; Clarence Lagoon Tasmania
    .... ***Galaxias johnstoni*** p. 57; Fig. 10.3
    0–1 pyloric caeca, occasionally 2 8

8   Head short, bluntly rounded; usually 14–16 pectoral rays; widespread in
    south-eastern mainland Australia .... ***Galaxias olidus*** p. 55; Fig. 10.2
    Head longer, flattened above and tapering to a slender snout; usually 12–13
    pectoral rays; Lake Pedder, Tasmania .... ***Galaxias pedderensis*** p. 58; Fig. 10.4

9   Pyloric caeca long; lower jaw shorter than upper .... ***Galaxias brevipinnis*** p. 53; Fig. 10.1
    Pyloric caeca reduced to vestiges or lacking; jaws about equal 10

10 Caudal fin forked; pectoral and pelvic fins small, reaching less than half distance from
    bases to bases of pelvic and anal fins respectively; slender-bodied 11
    Caudal fin truncate to emarginate; pectoral and pelvic fins long, reaching more than half
    distance from bases to bases of pelvic and anal fins respectively; chunky and deep bodied. . . 12

11 Mouth small, reaching only to front of eyes; in coastal drainages of south-eastern
    coasts and Tasmania .... ***Galaxias maculatus*** p. 67; Fig. 10.12
    Mouth larger, reaching well below eyes, only in the inland Murray-Darling system
    .... ***Galaxias rostratus*** p. 69; Fig. 10.14

12 Dorsal, anal and pelvic fins with grey-black margins 13
    Dorsal, anal and pelvic fins without grey-black margins; Arthurs and Woods Lakes,
    Tasmania .... ***Galaxias tanycephalus*** p. 63; Fig. 10.9

13 Widely spaced round spots along sides, each with a halo; a distinct dark blotch,
    sometimes double, above pectoral fin base; vertebrae 56–62, usually more than 59;
    Victoria and Tasmania .... ***Galaxias truttaceus*** p. 60; Fig. 10.7
    Closely spaced vertical bars along sides, becoming fragmented into vertical series
    of oval blotches; no distinct dark blotch above pectoral fin base; vertebrae 53–56;
    Lakes Sorell and Crescent, Tasmania .... ***Galaxias auratus*** p. 62; Fig. 10.8

14 Pelvic fin rays 5 .... ***Paragalaxias julianus*** p. 76; Fig. 10.20
    Pelvic fin rays usually 6 15

15 Open laterosensory pores on head very large; two pores present on each side
    below lower jaw .... ***Paragalaxias dissimilis*** p. 72; Fig. 10.17
    Open laterosensory pores on head small; very rarely pores present beneath lower jaw 16

16 Usually 10–12 dorsal fin rays; 15 principal caudal rays; fin forked; 40–41 vertebrae
    .... ***Paragalaxias mesotes*** p. 75; Fig. 10.19
    Usually 13–14 dorsal fin rays and 14 principal caudal rays, caudal fin truncate;
    37–39 (rarely 40) vertebrae .... ***Paragalaxias eleotroides*** p. 74; Fig. 10.18

CLIMBING GALAXIAS *Galaxias brevipinnis* Günther

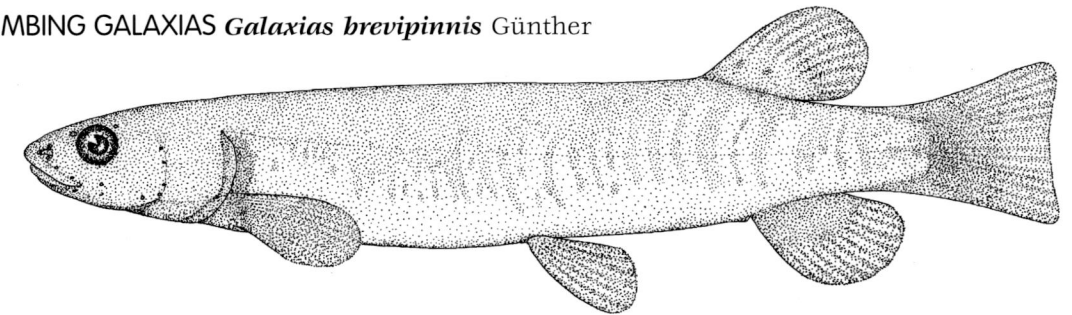

10.1 Climbing galaxias *Galaxias brevipinnis*. (R.M. McDowall)

FAMILY GALAXIIDAE

Description A relatively large and elongate species, trunk almost tubular, head large and flattened dorsally, especially in small specimens; mouth large, reaching well below eyes, lower jaw distinctly shorter than upper, jaws with well developed canine teeth laterally. Fins thick and fleshy, especially at bases, dorsal fin (9–11 rays) high and rounded, but short-based, middle rays the longest; anal fin (9–13 rays) similar in shape, anal origin distinctly behind dorsal origin; pectoral (13–16 rays) and pelvic fins large and round, low on body, facing downwards. Tail on a long and rather slender caudal peduncle, fin truncate. Gill rakers (11–15) of moderate length, stout; two long pyloric caeca. Vertebrae 54–62.

Colour Greyish brown to dark olive, darker on back, paling somewhat on sides, often with a distinct blue-black blotch above pectoral fin base, pattern on trunk varying from bold chevron-shaped bands to irregular and variable blotches, or spots either arranged in rows along the chevrons or irregularly dispersed. Belly a dull silvery olive. Golden iridescence evident on back and sides in bright sunlight.

Size Is known to reach 278 mm, but commonly to 150–170 mm.

Distribution Found in Australia in coastal drainages from about Sydney, south and west to Adelaide, where suitable habitats remain, but distribution probably fragmented by habitat deterio-

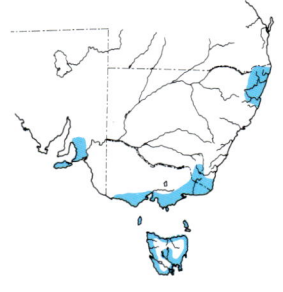

ration over much of range; widespread in Tasmania, both in coastal drainages as well as numerous landlocked populations in inland lakes; also on King and Flinders Islands in Bass Strait. Penetrates long distances inland.

Recent records from the inland Murray-Darling system presumably reflect emigration of fish from lakes of the Snowy River power scheme, which transfers water from the Snowy into the Murray. Climbing galaxias is known also from New Zealand, the Chatham, Auckland and Campbell Islands (where it is called koaro).

Conservation status Distribution has probably contracted with extensive deforestation and land development; there is evidence in Australia (as in New Zealand) of high vulnerability to predation by rainbow trout in lakes and displacement by brown trout in streams, as in Lakes Eucumbene and Tali Karng.

Natural history Not well known. Is a secretive and solitary species that inhabits clear, tumbling, bouldery streams, most often those in the headwaters that flow through forest, though it seems to avoid the dark, low pH (acid) waters found in some forest streams; is common down to sea level

Climbing galaxias *Galaxias brevipinnis*. (N. Armstrong)

in southern Tasmanian streams that do not contain introduced species, is also present in lakes and their tributaries and occurs among rocks and cover on lake beds. Breeds during autumn and winter, the eggs having been found among litter of the forest floor beyond the normal limits of stream flow, presumably spawned there when the streams were in flood. The fertilised eggs are of unknown size, numbering several to many thousand (up to 23,000 reported). The newly hatched larvae are thought to be swept downstream and into the sea (in diadromous populations), where they live for 5–6 months and return as slender, transparent 'whitebait' juveniles 40–50 mm long during spring. These may move into river estuaries in large, mixed-species shoals but soon become bottom-dwelling and probably then slowly make their way upstream to find suitable rearing and feeding habitats. Commonly becomes landlocked in inland lakes, where the life cycle is similarly structured, though spawning may occur in the spring and the larvae are pelagic in lakes rather than in the sea. Shoals of juveniles can be found around lake margins, where they are a forage fish for introduced trout.

Is an aggressive upstream migrant and is well known for ability to climb vertical waterfalls and rock faces tens of metres high, as long as these are moist (hence the common name 'climbing galaxias', though it is by no means the only galaxiid species capable of doing this). It adheres to the damp rock surface using its large, downward-facing pectoral and pelvic fins and wriggles upwards with lizard-like movements. This fish is a generalised invertebrate carnivore, feeding on the wide variety of insects and other life forms in its habitat, especially mayfly and caddis larvae and amphipods; the diet includes some surface drift insects such as millipedes and beetles, in spite of the undershot lower jaw that appears best adapted for feeding from the substrate.

Utility The climbing galaxias is a part of the Tasmanian whitebait fishery, originally based primarily on *Lovettia sealii* (see Chapter 11); this fishery was closed after the 1973 season and remained closed until 1990 owing to collapse and a failure to recover. When the fishery was reopened for recreational fishing in a limited number of rivers in 1990, the intention was to focus harvest on species of *Galaxias* such as jollytail and avoid exploitation of Tasmanian whitebait and other galaxiid species such as climbing galaxias; the season was timed to restrict their exploitation.

Is easy to keep in captivity and can be an active and engaging species, though its propensity for climbing means that aquaria have to be carefully covered to prevent escape.

Similar species Resembles Pedder galaxias, Clarence galaxias, mountain galaxias (with which there was much early confusion) and Swan galaxias.

Other names Common: Cox's mountain galaxias, Pieman galaxias; known as koaro in New Zealand. Scientific: *Galaxias coxii*, *G. affinis* (mainland Australia), *G. weedoni*, *G. parkeri* (Tasmania); *G. niger*, recently described from a Tasmanian lake, is considered a landlocked stock of this species by some Tasmanian biologists; there are numerous scientific names for both diadromous and landlocked stocks of this species in New Zealand.

Literature Fulton, 1990; Green, 1979; Koehn & O'Connor, 1990; Koehn & O'Connor, 1992; McDowall & Frankenberg, 1981; Merrick & Schmida, 1984; Morison & Anderson, 1990; O'Connor & Koehn, 1991; Williams, 1975.

MOUNTAIN GALAXIAS
Galaxias olidus Günther

Description A small, variable, but usually rather stocky galaxiid; head small, snout blunt, mouth large, reaching well below eyes, jaws about equal, with lateral canine teeth weak or lacking. Fins are small, thick and fleshy at bases; dorsal fin short (8–12 rays); anal fin (9–13 rays) with origin well back, usually behind middle of dorsal, both somewhat high and rounded, with middle rays longest. Pectoral fins (15–18, mostly 14–16 rays) and pelvic fins small and rounded, fins positioned very low on sides; tail truncate to weakly forked. Gill rakers (10–15) short and stout; usually only 1 short pyloric caecum. Vertebrae 49–59.

Colour Widely variable, yellowish green to brown on back, belly olive to silvery white, the back and upper sides variously and profusely marked with darker grey-olive speckling, blotching or banding. No dark blotch above bases of pectoral fins.

Size Reaches a length of 135 mm, but often only 60–70 mm.

Distribution Very widespread in eastern Australia at moderate and high elevations (to about 1800 m altitude) in rivers draining the Great Dividing Range and associated mountains, in both coastal

FAMILY GALAXIIDAE

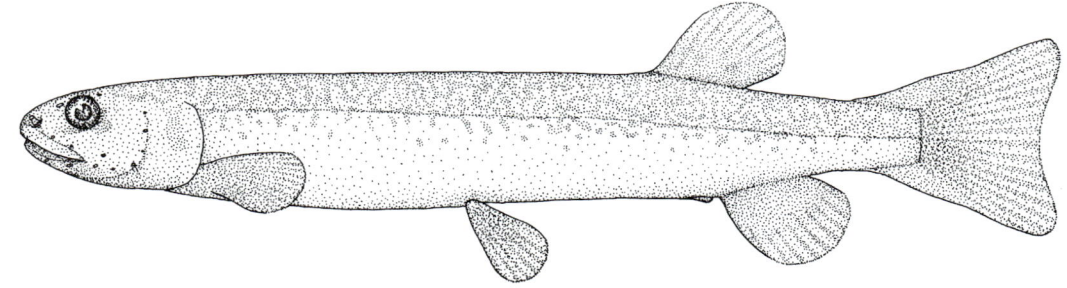

10.2 Mountain galaxias *Galaxias olidus*. (R.M. McDowall)

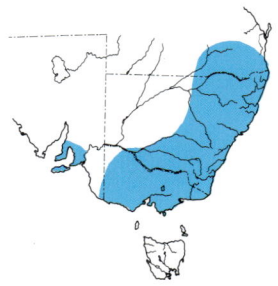

drainages and the inland Murray-Darling drainages. Occurs from southern Queensland (in western drainages, tributaries of the upper Darling River), to eastern South Australia.

Conservation status Remains very widespread and locally abundant through its extensive range; however, has proved sensitive to introduction of trout to its habitat and may disappear rapidly in the presence of trout; in some areas is found only where trout have not become established, such as upstream of falls. The Australian Society for Fish Biology lists *Galaxias fuscus* as valid and distinct from *G. olidus* and designates it as 'endangered'.

Natural history Occurs primarily in small streams, and sometimes small tarns and ponds, at higher elevations where water temperatures remain cool in summer. There it is found in small, loose shoals mostly in pools and runs, but may be solitary among substrate rocks and around stream margins. Spawning occurs during spring, sometimes extending through summer and even into autumn. Some males and occasional females mature at the end of their first year (0+), and all fish are mature at the end of their second (1+). Longevity may reach four years, but differential sexual mortality means that all 3+ fish are females. The ripe fish move into riffles near their normal adult habitats to spawn, and the eggs are laid in masses adhering

Mountain galaxias *Galaxias olidus*. (R.H. Kuiter)

to the undersides of boulders and cobbles in riffles. Eggs are few in number, 44–384 reported in females 55–88 mm long. The fertilised eggs are about 2.3 mm in diameter, take about 3 weeks to hatch (at 13–15°C), and the newly hatched larvae are 9–10 mm long. There is no seagoing migration and the young may be found with or near the adults, forming loose shoals in pools.

Feeds on a wide variety of aquatic insects, crustaceans, molluscs and worms; also feeds avidly on terrestrial insects and spiders, especially where there is vegetation overhead.

Utility None, other than being robust, easily kept and attractive in captivity.

Similar species Is very similar to other galaxiids, such as climbing galaxias, Clarence galaxias, Pedder galaxias and Swan galaxias, the last three occurring only in Tasmania.

Other names Common: many, including ornate mountain galaxias, and inland galaxias. Scientific: has been redescribed many times under many names, such as *Galaxias schomburgkii*, *G. ornatus*, *G. bongbong*, *G. findlayi*, *G. kayi*, *G. oconnori* and *G. fuscus*. Some observers consider the last a valid species, and though fish comparable to types of *G. fuscus* are distinctively coloured, the separate species status of these populations remains contentious and unresolved. Apart from colour differences, nothing that distinguishes *G. fuscus* has been described.

Literature Armstrong, 1993; Cadwallader, 1979; Cadwallader *et al.*, 1980b; Fletcher, 1979; Koehn & O'Connor, 1990; McDowall & Frankenberg, 1981; Marshall, 1989; Merrick & Schmida, 1984; O'Connor & Koehn, 1991.

CLARENCE GALAXIAS *Galaxias johnstoni* Scott

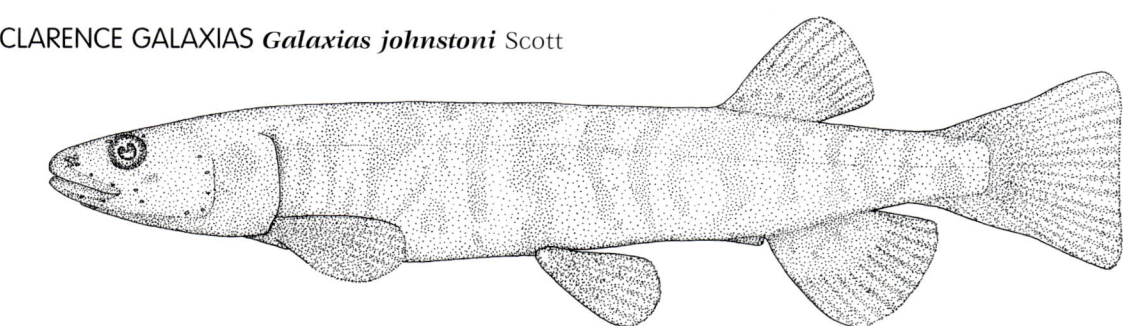

10.3 Clarence galaxias *Galaxias johnstoni*. (R.M. McDowall)

Description Closely resembles the mountain galaxias; a moderately stout species with bluntish head, a large mouth reaching well below eyes, jaws equal, lateral canine teeth weak or lacking; fins small, with 8–10 dorsal rays, 9–11 anal rays, fins rounded, middle rays longest; anal fin origin well back, below middle of dorsal fin. Pectorals with 10–13 rays; tail truncate to emarginate. Gill rakers (9–15) short and stout; two pyloric caeca of moderate length. Vertebrae 49–52.

Colour Dark greyish brown on back and sides, silvery olive below, the sides marked with bold, irregular brownish bands and blotches.

Size Grows to 140 mm and is commonly more than 70 mm.

Distribution Known only from Clarence Lagoon and a small lagoon in the Wentworth Hills nearby. Was formerly present also in rivers and streams of the upper Clarence River (Derwent River system), but range has contracted in recent decades.

Conservation status Range in parts of the Clarence River/Clarence Lagoon drainage has contracted in recent decades to the extent that it is now described as 'endangered' by the Australian Society for Fish Biology.

Introduced brook char also occur in Clarence Lagoon but are thought not to have adversely affected the galaxiid population. Management objectives are to maintain a healthy population of brook char to discourage introduction of potentially more aggressive trout species. The Wentworth Hills population is secure at this time.

Natural history Has been found in both flowing and still waters. Spawns during spring following migrations into inflowing creeks. The eggs are reported adhering to rocks in the streams and take about 2 months to hatch, followed by a planktonic larval stage lasting about the same time. A marine larval-juvenile whitebait stage can safely be discounted for this species. Lives for at least 4 years. Juvenile diet is dominated by planktonic crustaceans and terrestrial insects, while the adults take benthic crustaceans.

Utility None, though would probably do well in captivity, like many other galaxiids.

Similar species Resembles climbing galaxias, Pedder galaxias, Swan galaxias and mountain galaxias, the first three all occurring in high-elevation Tasmanian waters and are likely to cause confusion.

Other names Common: none. Scientific: none.

Literature Fulton, 1990; McDowall & Frankenberg, 1981; Merrick & Schmida, 1984; Sangar & Fulton, 1991.

PEDDER GALAXIAS *Galaxias pedderensis* Frankenberg

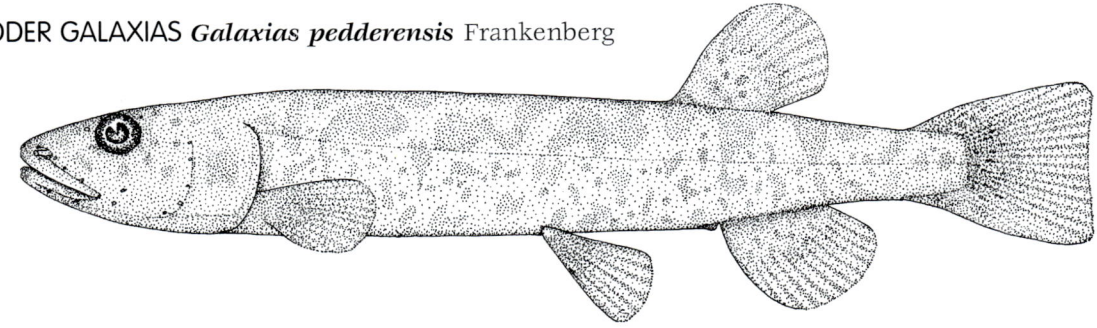

10.4 Pedder galaxias *Galaxias pedderensis*. (R.M. McDowall)

Description Resembles the preceding two species, but is more slender and elongate, head strongly depressed, with upper eye margins close to head profile. Fins small, 8–11 dorsal rays, 9–13 anal rays and 11–14 pectoral rays; dorsal and anal fins high and rounded, with middle rays longest. Tail truncate to emarginate. Gill rakers (10–17) short and stout; 0–2 pyloric caeca of moderate length. Vertebrae 51–54.

Colour Striking. Greenish brown, the back and sides covered with profuse brownish and off-white highly contrasting bands and blotches, fragmenting to finer vermiculations and spotting on lower sides; colour pattern may extend onto fleshy fin bases in larger fish; belly off-white to silvery. Golden iridescence evident in some lights.

Size Grows to 160 mm, commonly exceeding 70 mm.

Distribution Originally present only in the Lake Pedder area in south-west Tasmania. Lake Pedder was engulfed in a huge hydro impoundment in the 1970s (the new Lake Pedder), where the species now occurs (but see Conservation Status).

Pedder galaxias *Galaxias pedderensis*. (R.M. McDowall)

Conservation status Was very abundant in 'old' Lake Pedder and initially adapted well to habitats available in 'new' Lake Pedder; populations crashed during the 1980s, possibly as a result of predation by brown trout in the lake, which quickly grew to enormous size; however, coincidentally, there developed lacustrine populations of climbing galaxias in the lake, so that causes for the decline in Pedder galaxias cannot presently be isolated. Is now rare in Lake Pedder and as a result is regarded as Tasmania's most endangered fish. Efforts are under way to induce captive breeding and eventually to establish populations in other waters. Is classified as 'endangered' by the Australian Society for Fish Biology.

Natural history Occurs among rocks and boulders around the lake margins and also among cover in streams entering the lake. Becomes sexually mature at two years and may live for five to six years; spawning occurs in early spring, when water temperature begins to rise, the females producing 150–1200 relatively large eggs (1.9 mm in diameter before spawning). Wild spawning is not described, but in captivity eggs were deposited under flat rocks. Artificially fertilised eggs hatched in less than 30 days. Larvae are about 10 mm on hatching, and larvae and juveniles have been reported from the lake during summer. Feeds on small aquatic and terrestrial insects and aquatic crustaceans.

Utility None.

Similar species Resembles climbing galaxias, Clarence galaxias, Swan galaxias and mountain galaxias.

Other names Common: none. Scientific: none.

Literature Frankenberg, 1968; Fulton, 1990; Gaffney *et al.*, 1992; Hamr, 1982, 1991, 1992; McDowall & Frankenberg, 1981; Merrick & Schmida, 1984; Sanger & Fulton, 1991.

SWAN GALAXIAS *Galaxias fontanus* Fulton

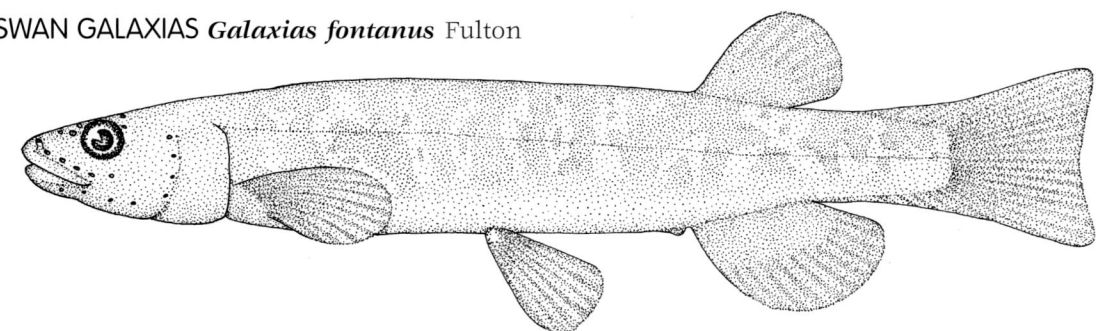

10.5 Swan galaxias *Galaxias fontanus*. (R.M. McDowall)

Description A moderate-sized species similar to the Clarence and mountain galaxias; head flattened, eyes at upper head profile, interorbital broad and flat. Fins small, dorsal 7–9 rays, anal 9–11 rays, these fins rounded, middle rays longest; anal origin below dorsal origin. Pectorals with only 10–11 rays; all fins somewhat fleshy-based. Tail slightly forked to emarginate. Jaws about equal, with strong lateral canine teeth. Gill rakers (11–14) very short and stout; no pyloric caeca. Usually 5–6 pores in series from below eyes to anterior nostril (only 4 in other galaxiids). Vertebrae 50–53.

Colour Brownish olive, back and sides covered with dense irregular brownish blotching or band-

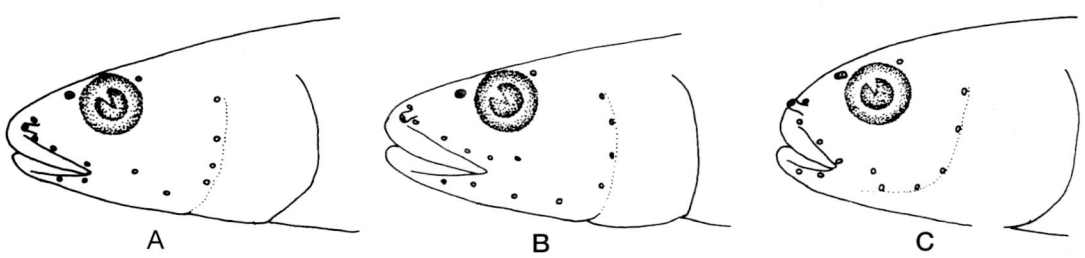

10.6 Head pores on galaxiids: **A** a typical *Galaxias*; **B** *Galaxias fontanus*; **C** *Galaxias parvus*. (R.M. McDowall)

FAMILY GALAXIIDAE

Swan galaxias *Galaxias fontanus*. (G.R. Allen)

ing; fins generally without markings; belly creamish to silvery white.
Size Grows to about 135 mm, often 65 mm or more.
Distribution Known from a few localities  in the upper reaches of the Swan and Macquarie Rivers in eastern Tasmania and is the only galaxiid distinctive to the east.
Conservation status Being known from very few localities makes this species vulnerable, this being accentuated by the shoaling habits of the juveniles, which makes them easy prey to introduced brown trout. As a result, there is little overlap in the distributions of the Swan galaxias and brown trout. Is now regarded as 'endangered' by the Australian Society for Fish Biology.
Natural history Is little known, but essentially a riverine species that has no seagoing juvenile stage. Lives for at least 3 years, first spawning at 2, during spring; eggs take about 8 weeks to hatch. Shoals of juveniles, 15–25 mm long, occur in the rivers over the spring and recruit into the subadult population in early summer. Is an opportunistic carnivore (like its relatives), including a large proportion of terrestrial food in its diet at times.
Utility None.
Similar species Climbing galaxias, Clarence galaxias, Pedder galaxias, mountain galaxias.
Other names Common: none. Scientific: none.
Literature Fulton, 1978, 1990; McDowall & Frankenberg, 1981; Merrick & Schmida, 1984; Sanger, 1993; Sanger & Fulton, 1991.

SPOTTED GALAXIAS
Galaxias truttaceus (Valenciennes)

Description A large and stout-bodied species, rather deeper-bellied than most galaxiids; head long, broad and deep, mouth large, reaching to front of eyes, jaws equal, no enlarged canine teeth laterally in jaws. Fins quite large, dorsal (9–11 rays) and anal (12–15 rays) high and rounded, with middle rays longest, origin of anal directly below origin of dorsal; pectoral fins expansive (13–16 rays); tail slightly forked. Gill rakers (13–17) slender and of moderate length; pyloric caeca reduced to vestiges. Vertebrae 56–62.
Colour Brownish to deep olive, paling to brownish grey on sides, silvery olive on belly; trunk

GALAXIIDS

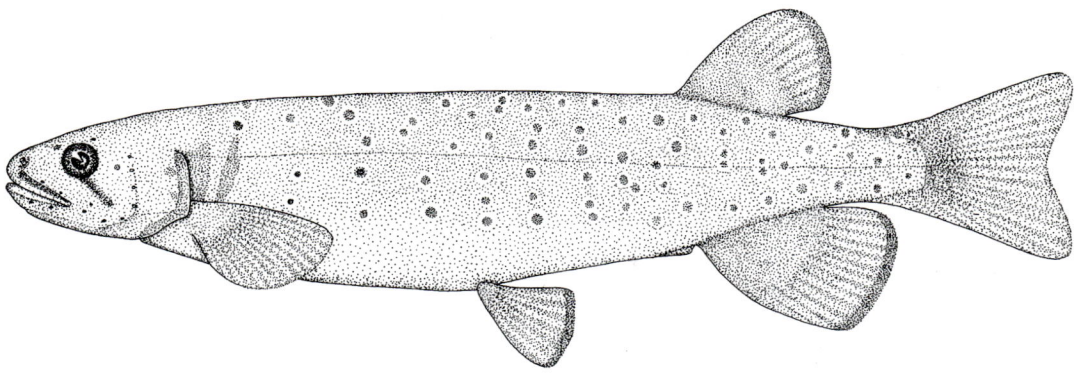

10.7 Spotted galaxias *Galaxias truttaceus*. (R.M. McDowall)

covered with many round spots, sometimes in vertical rows, each spot surrounded by a paler halo; two distinct blue-black blotches above pectoral fin bases. Fins brownish to olive, dorsal, anal and caudal fins golden to bright orange on outer halves, with dark fringes on dorsal, anal and pelvic fins. A distinct, diagonal stripe passing back and down through eyes. On entering streams from the sea, the whitebait juveniles are transparent but have black spots along the back; as they settle into fresh water they become dusky and develop dark vertical bars across the flanks; these soon fragment and are replaced with the spots characteristic of subadults and adults.

Size A large galaxiid, reaching to over 200 mm, commonly 120–140 mm.

Distribution Found in coastal drainages of Southern Victoria, on the islands of Bass Strait and widely in Tasmania on all coasts; also occurs in Western Australia. Landlocked populations also occur in some inland Tasmanian lakes.

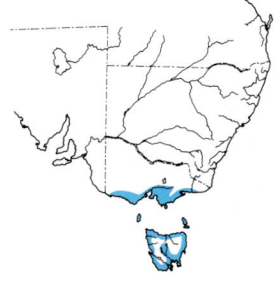

Conservation status Remains locally abundant, but range has probably been fragmented by deforestation in some areas.

Natural history Is a species of low-elevation

Spotted galaxias *Galaxias truttaceus*. (N. Armstrong)

streams, where it is found in the pools, most often from within marginal cover, under logs or overhung banks, or behind boulders; also found in lakes and lagoons, often at higher elevations, again occurring within cover. Riverine populations spawn in autumn–winter, the spawning site as yet undescribed; a downstream spawning migration has been suggested. Landlocked stocks delay spawning until spring, their eggs being found among the stems of aquatic plants. The eggs are about 1.3 mm in diameter and numerous, with 1000–16,000 in females, 72–142 mm long. The eggs take about 4 weeks to hatch. The newly hatched larvae are 6.5–9.0 mm long, and are swept to sea when they hatch (or into lakes in landlocked populations). The whitebait juveniles return to fresh water during spring (as a part of the 'whitebait' migrations), when 45–65 mm long.

Spotted galaxias are carnivores, living on a wide range of aquatic and terrestrial invertebrates found in their habitat. Much food is taken from the drift, both in mid-water (particularly caddis larvae and mayflies) and at the surface (beetles, spiders, bugs ants, etc).

Utility Formerly contributed to the Tasmanian whitebait fishery (along with Tasmanian whitebait, see Chapter 11, and climbing galaxias, see p. 53) and is now, with other Tasmanian species of *Galaxias*, expected to form the basis for the fishery since it was reopened in 1990. Is an attractive fish that is easily kept in aquaria, but requires cool water.

Similar species Could be confused with golden galaxias and perhaps saddled galaxias, though haloed spots are distinctive.

Other names Common: sometimes called spotted mountain trout, mountain trout or trout minnow. Scientific: *Galaxias ocellatus, G. scopus*; specific name sometimes misspelled *truttaceous*.

Literature Caughley, 1992; Fulton, 1990; Humphries, 1989, 1990; Koehn & O'Connor, 1990; McDowall & Frankenberg, 1981; Merrick & Schmida, 1984; Scott, 1941; Sloane, 1984; Williams, 1975;

GOLDEN GALAXIAS *Galaxias auratus* Johnston

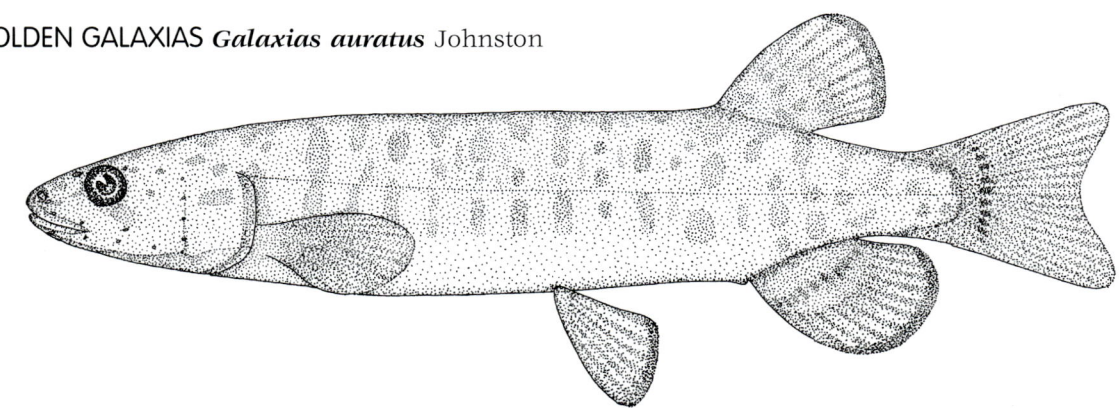

10.8 Golden galaxias *Galaxias auratus*. (R.M. McDowall)

Description A large, rather streamlined but stout-bodied species, head long, tapering to a slender snout; mouth large, reaching back to about eyes, jaws about equal, with enlarged canine teeth laterally. Fins well developed, thick and fleshy at bases, 7–10 dorsal rays, 11–12 anal rays, middle rays the longest in both; anal fin origin a little back from dorsal origin; 14–18, usually 15–16, pectoral rays. Tail slightly forked. Gill rakers (16–20) stout, of moderate length; pyloric caeca reduced to weak shoulders or lacking. Vertebrae 53–56.

Colour Gold to olive green on back paling to bronze-gold on sides, silvery grey on belly; back and sides covered profusely with vertical grey-olive bands in small fish, these fragmenting to a series of round spots along back and distinctly oval spots on sides. Fins amber to pinkish orange; fringes of dorsal, anal and pelvic fins black.

Size Reaches 237 mm and commonly grows to 140 mm.

Distribution Found only in Tasmania in Lakes Sorell and Crescent and associated rivers.

Conservation status Remains in high abundance in the few known habitats and is under no identified threat,

Golden galaxias *Galaxias auratus*. (G.R. Allen)

despite predation by introduced trout.
Natural history Is a species of still or gently flowing waters. Life cycle is completed in fresh water, with spawning during spring. The eggs are deposited among substrates of rocky lake shallows. Like other galaxiids, is a predatory carnivore, feeding on aquatic insect larvae, molluscs and crustaceans that share its habitat.
Forms an important food for trout introduced into the lakes; unlike many galaxiids, has proved capable of thriving in spite of this predation over more than a century.
Utility As just noted, is important as a forage species for trout; makes an attractive aquarium species.
Similar species Resembles spotted galaxias and saddled galaxias.
Other names Common: none. Scientific: none.
Literature Fulton, 1990; McDowall & Frankenberg, 1981; Merrick & Schmida, 1984.

SADDLED GALAXIAS
Galaxias tanycephalus Fulton

Description Another rather stout-bodied species reaching a moderate size; head long and tapering to a long, slender snout, mouth large and reaching back to about eyes; jaws equal, with slight enlargement of lateral canine teeth. Fins well developed, rather thick and fleshy at bases, 8–10 dorsal rays and 11–12 anal rays, middle rays in both longest; anal origin a little behind dorsal origin. Pectorals with 14–17 rays. Tail distinctly forked. Gill rakers (16–21) strong, of moderate length; pyloric caeca reduced to weak shoulders or absent. Vertebrae 53–56.

Colour Green-olive on back with saddle-like greyish bars across back and sides, belly silvery olive; sometimes spotted rather than barred, and sides sometimes with a distinctly purplish sheen. Fins olive to amber, without black fringes.
Size Reaches 147 mm, but few more than about 70 mm.
Distribution Occurs only in Arthurs and Woods Lakes, in high-elevation central Tasmania.

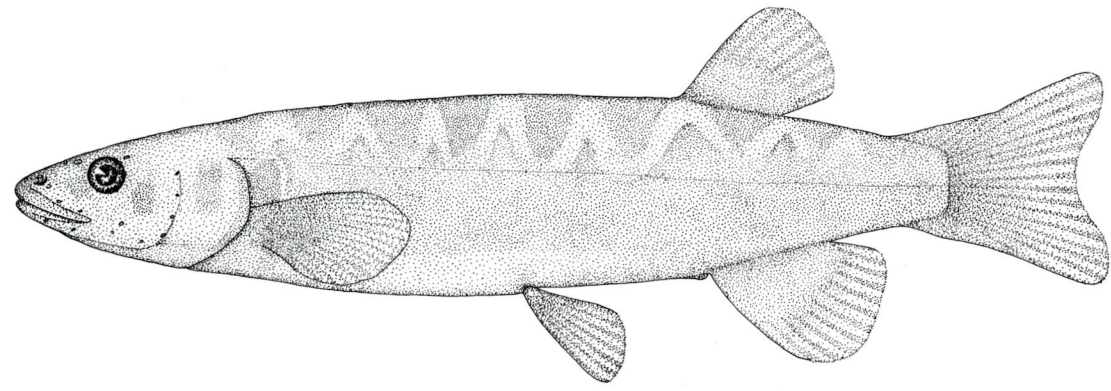

10.9 Saddled galaxias *Galaxias tanycephalus*. (R.M. McDowall)

Conservation status Has never been found in large numbers and is regarded as 'vulnerable' by the Australian Society for Fish Biology; the small number of known habitats contributes to its vulnerability.
Natural history Not well known, but is found among bouldery debris around the margins of lakes. Is reported to have an extended breeding season, with young found throughout the year, though a breeding peak may occur in summer. Life history is certainly confined to fresh water. Diet consists largely of crustaceans, planktonic species in the larvae, and benthic ones, plus some aquatic insects, in adults.
Utility None recognised, though may do well in aquaria, like many of its congeners.
Similar species Resembles spotted galaxias and golden galaxias, but confusion is not likely.
Other names Common: none. Scientific: none.

Literature Fulton, 1978, 1990; McDowall & Frankenberg, 1981; Merrick & Schmida, 1984; Sanger & Fulton, 1991.

Saddled galaxias *Galaxias tanycephalus*. (W. Fulton)

TASMANIAN MUDFISH
Galaxias cleaveri Scott

Description An elongate, tubular species with a short, bluntly-rounded head; mouth of moderate size, the jaws about equal, with moderate development of canine teeth laterally; eyes very small. Anterior nostrils long, projecting forwards over upper lip. Fins thick and fleshy at bases, small dorsal (9–11 rays) and anal (10–13 rays), rather low, rays rather even in length; anal origin a little behind dorsal. Pectoral fins small and paddle-shaped (13–14 rays); pelvic fins very small. Tail rounded, with flanges running forward along caudal peduncle almost to bases of dorsal and anal fins. Gill rakers (9–13) short and stout; two long pyloric caeca. Vertebrae 56–60.

Colour Greenish brown on back and sides, belly greyish, sides and bases of fins liberally marked with irregular, darker brown stripes and blotches.
Size Grows to 140 mm, commonly 80 mm or more.
Distribution Present at low elevations all around the coast of Tasmania, on Flinders Island in Bass Strait, and also in southern Victoria (recent records from the Otways and Wilsons Promontory).

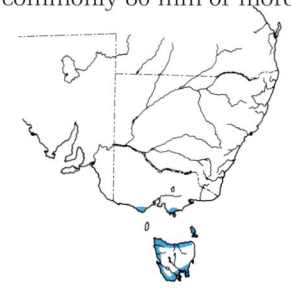

Conservation status Is quite widespread and common, although, as in most lands, swamp drainage and reclamation threaten populations.

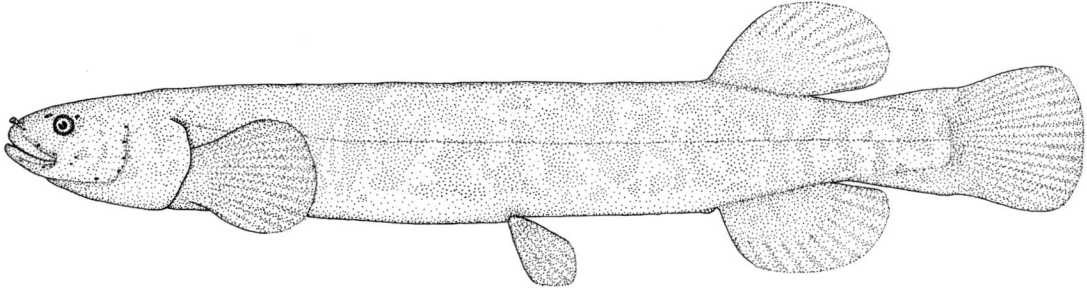

10.10 Tasmanian mudfish *Galaxias cleaveri*. (R.M. McDowall)

The isolated localities in Victoria perhaps represent fragments of a more continuous distribution. Is a protected species in Victoria but is not listed in the Australian Society for Fish Biology list of threatened species.

Natural history Found mostly in still waters, in heavily vegetated mud-bottomed swamps and drains, where it is mostly nocturnal; is known to be able to aestivate if it can find moist refuges when free water in habitat dries up, the fish then being found beneath logs, stones, etc. It was discovered when a stump was exploded out of a dried-out swamp. Spawning occurs in winter, with the newly hatched larvae moving to sea for a brief period of 2–3 months; juveniles make up part of the spring migrations of the multispecies Tasmanian whitebait fishery; the whitebait juveniles are smaller than other species, around 30–40 mm.

Food of the Tasmanian mudfish is unreported, but it presumably lives on aquatic and perhaps terrestrial insects and other small animals.

Utility Though it is a part of the Tasmanian whitebait fishery, numbers of mudfish involved are likely to be very small. Is easily kept in captivity, but is secretive and nocturnal.

Similar species Distinctive, though sometimes possibly confused with common jollytail or swamp galaxias.

Other names Common: mud trout, mud galaxias. Scientific: variously called *Saxilaga cleaveri*, *S. anguilliformis*, and *Galaxias upcheri*.

Literature Fulton, 1990; Jackson, 1982; Koehn & O'Connor, 1990; Koehn & Raadik, 1991; McDowall & Frankenberg, 1981.

Tasmanian mudfish *Galaxias cleaveri*. (R.H. Kuiter)

FAMILY GALAXIIDAE

SWAMP GALAXIAS *Galaxias parvus* Frankenberg

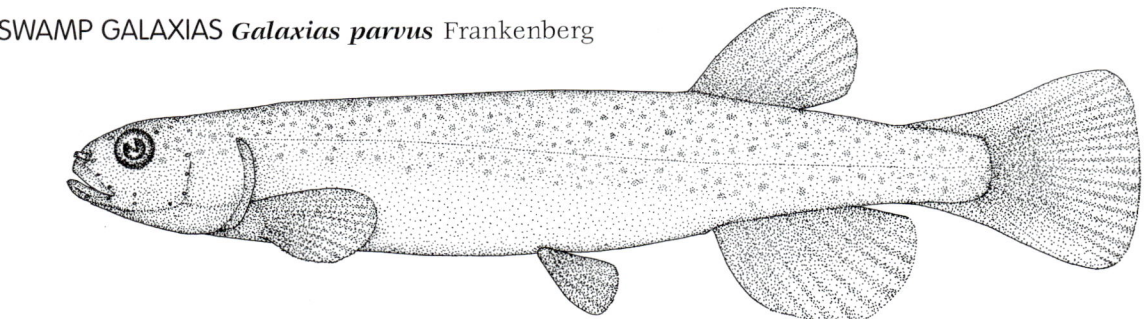

10.11 Swamp galaxias *Galaxias parvus*. (R.M. McDowall)

Description A small, chunky, blunt–headed fish, trunk tubular, tapering to a long caudal peduncle. Head of moderate size, mouth small, reaching to about eyes; jaws about equal, lacking lateral canine teeth. Eyes of moderate size. Anterior tubular nostrils long, projecting forward over upper lip. Only three pores present in pre-opercular row (Fig. 10.6C—usually four). Fins small, fleshy-based, dorsal (9–10 rays) and anal (11–12 rays) rounded, middle rays longest, anal origin a little further back than dorsal. Pectoral fins (11–14 rays) small and fan-shaped; pelvic fins with only 5–6 rays (7 in most *Galaxias*). Tail rounded. Gill rakers (9–13) short and stout; no pyloric caeca. Vertebrae 44–49.

Colour Back and sides greyish to yellowish brown, becoming grey-green on lower sides and whitish on belly; back and sides with numerous small and irregular dark grey-brown spots and blotches; colour patterning unusually finely detailed for *Galaxias*. Operculum has greenish iridescence.

Size Known to reach about 100 mm, but seldom more than 70 mm.

Distribution Found only in south-western Tasmania in swamps and streams of the headwaters of the Gordon River (Lake

Swamp galaxias *Galaxias parvus*. (R.M. McDowall)

Pedder) and the Huon River.
Conservation status Became more abundant after the flooding of Lake Pedder and remains abundant in the region (unlike the Pedder galaxias, see p. 58); however, the Australian Society for Fish Biology lists it as 'potentially threatened'. Breeding in captivity has occurred.
Natural history Occurs in swamps, pools, backwaters and lake margins among rocks and vegetation; probably spawns in spring, the life cycle being completed in fresh water. Males may mature at age 1 but females probably not until two years old and about 55 mm long. There are 75–500 eggs, about 2 mm in diameter before spawning, and they take 26–30 days to hatch. The larvae are about 7 mm long and metamorphose at about 20 mm length when 30–45 days old. Diet consists of terrestrial insects and aquatic insects and crustaceans.
Utility None.
Similar species Distinctive, though somewhat resembles Tasmanian mudfish.
Other names Common: initially rather prosaically called dwarf galaxias. Scientific: none.
Literature Frankenberg, 1968; Fulton, 1990; McDowall & Frankenberg, 1981; Merrick & Schmida, 1984.

COMMON JOLLYTAIL *Galaxias maculatus* (Jenyns)

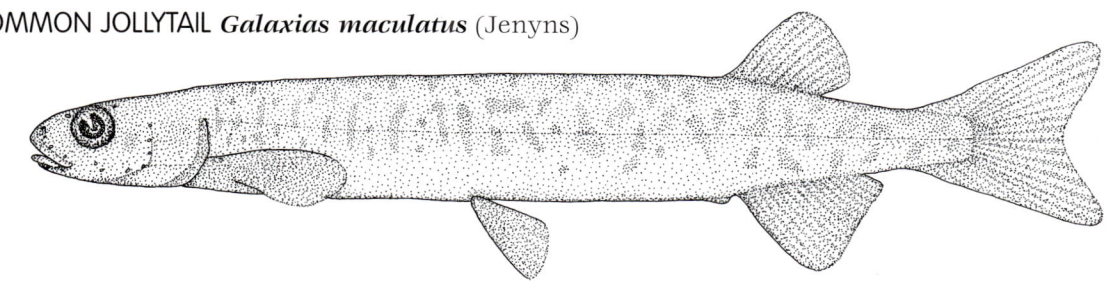

10.12 Common jollytail *Galaxias maculatus*. (R.M. McDowall)

Description A smallish, slender, streamlined species. Head small, bluntly pointed, eyes large; mouth small, reaching back only to front of eyes; jaws equal, lacking lateral canine teeth. Fins thin and membranous, dorsal fin small (7–11 rays) rounded, anal (12–17) longer-based and lower, rather angular, with anterior rays longest; anal origin directly below dorsal origin. Tail distinctly forked, with a very slender caudal peduncle. Pectoral fins small (11–15, usually 12–14 rays). Gill rakers (13–20) long and slender; no pyloric caeca. Vertebrae 50–62.
Colour Translucent grey-olive to amber, back and sides with irregular and variable greenish grey blotches or spots; sometimes little pattern, at other times profuse spotting; mid-lower sides may be vivid green. Belly, gill covers and eyes bright silvery; fins largely unpigmented.
Size May reach 190 mm, but is not often more than about 100 mm.
Distribution Found at low elevations in streams draining to the coast in southern Queensland, New South Wales and Victoria, south and west to about Adelaide; also widespread at low elevations on all coasts of Tasmania, including King and Flinders Islands. Known also from southern parts of Western Australia, as well as in Lord Howe

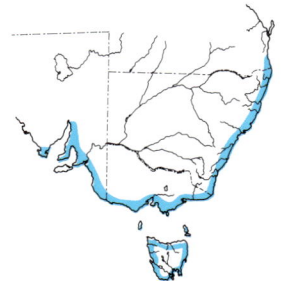

Island, New Zealand, Chatham Islands, Patagonian South America (Chile and Argentina) and the Falkland Islands. Possibly the most widely dispersed freshwater fish known.
Conservation status Remains widespread and abundant throughout range; seems to withstand both predation and habitat degradation better than most galaxiids.
Natural history Is the best known galaxias, largely on account of its importance to fisheries in New Zealand. Commonly found in still or gently-flowing streams and rivers, around lake and lagoon margins, usually in small to moderate shoals. Tolerates a wide range of habitat conditions, including salinities well in excess of full sea water. Adults migrate downstream on new or full moons, mostly during autumn; the eggs are small (about 1 mm in diameter) and number several thousand. Spawns among terrestrial vegetation on the margins of river estuaries when inundated at high spring tides, the eggs being left stranded out of

FAMILY GALAXIIDAE

Common jollytail *Galaxias maculatus*. (G. Schmida)

the water (for 2 or more weeks, until a later set of spring tides). The adults mostly die after spawning. The eggs hatch when the water returns and reinundates the vegetation; the larvae (about 7 mm long) go to sea, spend the winter there and migrate back as slender, transparent juveniles 45–50 mm long after about 5–6 months (Fig. 10.12). They re-enter rivers in huge shoals on the rising tide and move upstream into adult habitats to feed and grow. They typically reach maturity and spawn the following autumn at 1 year of age, though females may occasionally delay maturation until the second or even third year. Landlocked stocks occur in many lakes in Victoria and some in Tasmania. In Lake Modewarre (Victoria), the spawning migration has changed from downstream to one in which the mature fish leave the lake and move upstream into a tributary on a flood.

Feeds on a wide range of small aquatic insects, crustaceans and molluscs and terrestrial insects, especially midge (chironomid) larvae, but any small animals are taken from the surface, mid-water or the substrate.

Utility Was once a significant component in the commercial and recreational Tasmanian whitebait fishery; the fishery was reopened in 1990 as a purely recreational one and it has again become a major part of the fishery, particularly in the light of management intentions of avoiding Tasmanian whitebait, *Lovettia sealii* (see p. 78); is the principal species in the New Zealand whitebait fishery.

Possibly forms a significant food for brown

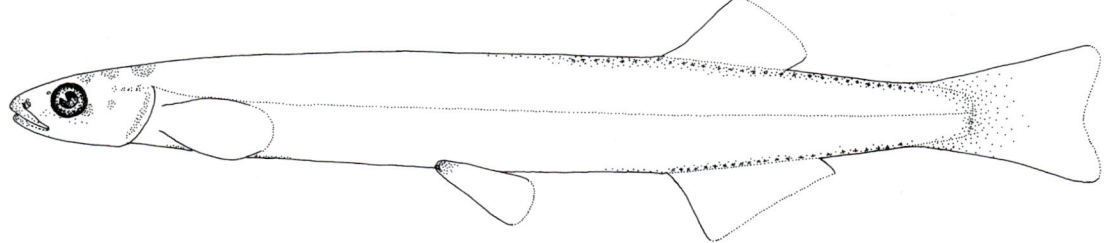

10.13 Whitebait juvenile of common jollytail *Galaxias maculatus*. (R.M. McDowall)

GALAXIIDS

trout and other predatory fishes inhabiting lowland rivers and estuaries during the spring and summer.

Is easily kept, active, and attractive in captivity, though females may become eggbound during autumn and often then die.

Similar species Closely resembles Murray jollytail, the two species occurring together at the mouth of the Murray River; can also be confused with mountain galaxias.

Other names Common: common galaxias, spotted minnow; inanga in New Zealand, puyen in Chile and Argentina. Scientific: many, including *Galaxias attenuatus, Galaxias attenuatus scriba, Austrocobitis attenuatus,* and *Galaxias parrishi.*

Literature Chessman & Williams, 1975; Fulton & Pavuk, 1988; Koehn & O'Connor, 1990; McDowall, 1968; McDowall & Eldon, 1980; McDowall & Frankenberg, 1981; McDowall *et al.* 1975, 1994; Merrick & Schmida, 1984; Pollard, 1971a, b, 1972a, b, 1973.

MURRAY JOLLYTAIL *Galaxias rostratus* Klunzinger

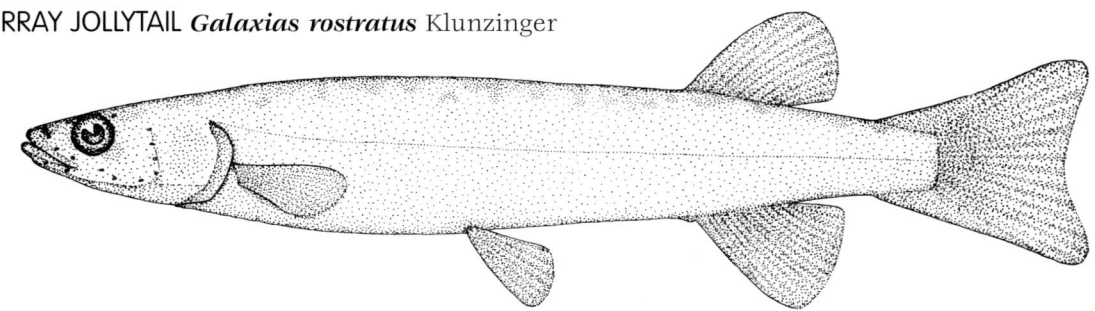

10.14 Murray jollytail *Galaxias rostratus.* (R.M. McDowall)

Description Closely resembles common jollytail but has top of head flattened, mouth larger, reaching well below the large eyes, jaws equal or lower, protruding slightly; dorsal fin with 8–10 rays, middle rays longest; anal with 11–15 rays, anterior rays longest. Anal origin directly below dorsal origin. Pectorals small, with 11–13 rays. Gill rakers (14–19) of moderate length; no pyloric caeca. Vertebrae 52–57.

Colour Greenish olive on back and sides with

Murray jollytail *Galaxias rostratus.* (R.H. Kuiter)

irregular darker grey-green blotches, but often has little marking of any sort; bright silvery below lateral line; eyes also silvery. Fins colourless.
Size Grows to about 120 mm, but seldom exceeds 100 mm.

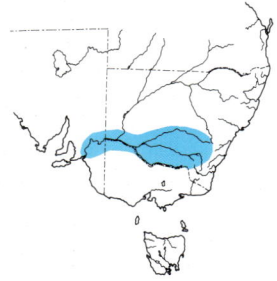

Distribution Known only in the Murray-Darling system, where it is widespread but intermittent and locally abundant at low elevations (Murray and Murrumbidgee, with an isolated record from the upper Darling).
Conservation status Distribution appears rather local and intermittent, though whether this is caused by habitat suitability or habitat degradation is not known. Concern for this species resulted in its listing as 'restricted' by the Australian Society for Fish Biology.

Natural history Found mostly in still and gently-flowing waters, in lakes, lagoons, billabongs and backwaters. Occurs in shoals in mid-water. Spawns at cool temperatures (9–14°C), the eggs small (about 1.5 mm in diameter) and numbering several thousand. They are spawned randomly and settle on the bottom, hatching in about 9 days, the larvae 6–7 mm long. Probably feeds on a wide range of aquatic insects and crustaceans.
Utility May be a food fish for other larger predatory fishes in the Murray; otherwise is of little importance.
Similar species Closely resembles common jollytail (which is coastal) and also mountain galaxias (which occurs at higher elevations).
Other names Common: flathead galaxias, flathead jollytail. Scientific: long known as *Galaxias planiceps*.
Literature Llewellyn, 1971; Koehn & O'Connor, 1990; McDowall & Frankenberg, 1981; Merrick & Schmida, 1984.

EASTERN LITTLE GALAXIAS *Galaxiella pusilla* (Mack)

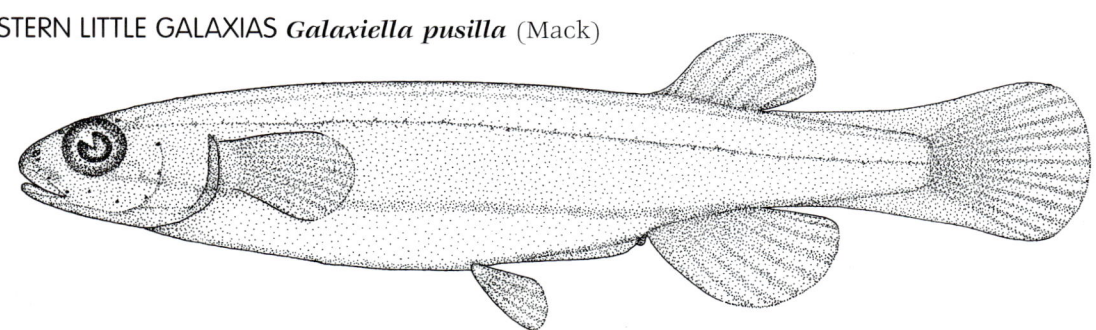

10.15 Eastern little galaxias *Galaxiella pusilla*. (R.M. McDowall)

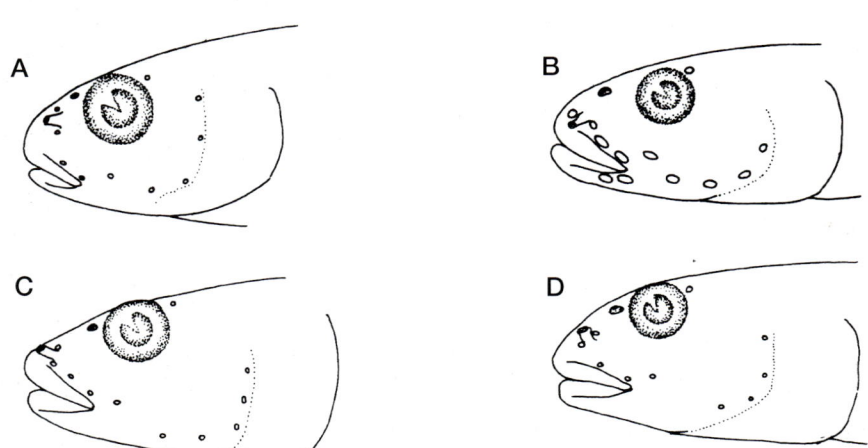

10.16 Head pores in galaxiids: **A** *Galaxiella pusilla*; **B** *Paragalaxias dissimilis*; **C** *Paragalaxias mesotes*; **D** *Paragalaxias eleotroides*. (R.M. McDowall)

GALAXIIDS

Eastern little galaxias *Galaxiella pusilla*, male. (R.M. McDowall)

Description A very tiny, somewhat stocky fish, trunk deepened at belly, compressed behind vent; head small, snout short and blunt, jaws equal, mouth small, reaching eyes; no lateral canine teeth in jaws. Fins small and membranous, dorsal with 5–8 rays, anal with 7–10 rays, middle rays longest in both; anal origin distinctly in front of dorsal origin. Tail rounded, with only 12–14 rays; very strong flanges along the long caudal peduncle, almost joining tail to dorsal and anal fins. Pelvic fins very small (4–6 rays); pectorals paddle-shaped (10–14 rays). Gill rakers (14–17) long and slender; no pyloric caeca; no open pores beneath lower jaw (Fig. 10.16A). Vertebrae 36–40.

Sexual dimorphism Colouration differs between sexes (see Colour, below). Female grows much larger than male. Pectoral–pelvic and pelvic–anal intervals are greater in females than in males; also ventral abdominal keel of males is better developed than in females.

Colour More or less transparent olive-amber on back, with three longitudinal black stripes along trunk; male has a brilliant orange stripe between the middle and lower dark stripes. Belly silvery white. Female lacks orange stripe.

Size Females may reach 40 mm, but males only 34 mm.

Distribution Found in southern Victoria from Gippsland west to the eastern corner of South

Eastern little galaxias *Galaxiella pusilla*, female. (R.M. McDowall)

Australia (about Mount Gambier); also present on Flinders Island and in the east of the north coast of Tasmania.

Conservation status Distribution seems patchy and is likely to have suffered from pervasive wetland drainage, which has variously destroyed, reduced or fragmented habitats, but seems a prolific little fish and remains locally abundant. Is totally protected in Victorian waters and is listed as 'potentially threatened' by the Australian Society for Fish Biology.

Natural history Occurs mostly in still or gently flowing waters in the shallows around the margins of creeks, drains and swamps, usually heavily overgrown with aquatic macrophytes or emergent plants. An association with yabbies (*Geocharex* sp.) has been suggested, with the fish taking refuge in yabby burrows when frightened. Its habitat may dry up seasonally (autumn–early winter) and it seems likely to be able to aestivate for several months when this happens, possibly again in association with yabby burrows. Spawns in spring in pairs, the females laying the tiny, adhesive eggs one by one on aquatic plants; a male follows to fertilise each. A female carries between about 65 and 250 eggs, each less than 1.0 mm in diameter, expanding to 1.1–1.3 mm when fertilised and water-hardened. Females appear to spend several days, perhaps as much as 2 weeks, laying small batches of eggs, attended by various males from the population. No elaborate courtship has been observed; rather the male just nudges the abdomen of the female. Larvae hatch in 2–3 weeks and are about 4.5 mm long. There is only 1 year class in the populations, so it is evidently an annual species, with the adults dying after spawning. Life cycle wholly in fresh water.

Feeds mostly in the water column on a wide variety of largely aquatic invertebrates—chironomid larvae, copepods, cladocerans, ostracods, as well as collembola and range of terrestrial arthropods.

Utility A handsome and active, if small, species that is easily kept in captivity, and seems to breed there freely.

Similar species Quite unlike other fishes in area.

Other names Common: dwarf galaxias. Scientific: sometimes included in *Galaxias* or *Brachygalaxias* (a Chilean genus); subspecies have been recognised by some workers but appear to have little significance.

Literature Backhouse, 1983; Backhouse & Vanner, 1978; Beck, 1985; Humphries, 1986; Koehn & O'Connor, 1990; McDowall, 1978a, b; McDowall & Frankenberg, 1981; Merrick & Schmida, 1984.

SHANNON PARAGALAXIAS *Paragalaxias dissimilis* (Regan)

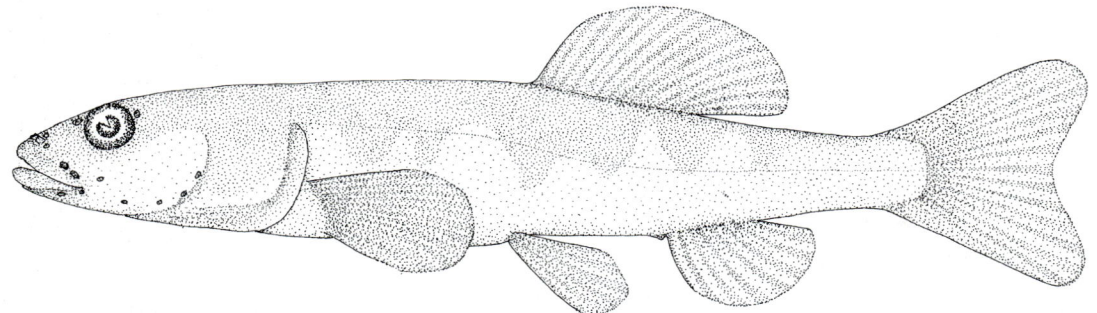

10.17 Shannon paragalaxias *Paragalaxias dissimilis*. (R.M. McDowall)

Description A small, bullet-shaped fish, head long and somewhat flattened, snout long, mouth large, reaching a little beyond anterior eye margins. Jaws about equal, no lateral canine teeth in jaws. Dorsal fin rounded, quite long (13–14 rays), middle rays longest; origin about above pelvic fins; anal fin small (7–9 rays), rounded. Tail forked (14–15 rays). Pectoral fins large (12–14 rays); 6 pelvic fin rays. Gill rakers (15–18) long and slender; no pyloric caeca; open pores on head very large, especially those above upper jaw; pores usually present below lower jaws (Fig. 10.16B. Vertebrae 41–44.

Colour Variable; olive-brown to grey-green or almost black on back, the sides with indistinct, bold, dark botches merging with dark colouration

on back. Olive-gold iridescence on gill covers and belly. Fins olive and unmarked.
Size Known up to 75 mm, but commonly to only 50 mm.

Distribution Known only from Great Lake, Shannon Lagoon, the Shannon River, which connects these two waters, and Penstock Lagoon, all on the Central Plateau of Tasmania. The two lagoons are artificial impoundments and their populations may be derived from that in Great Lake.
Conservation status Remains abundant in habitats despite trout predation.
Natural history Found during the day among cover around the lake margins, under rocks, debris or vegetation; possibly emerges at night to feed in mid-water. Shares lakes with Great Lake paragalaxias but is more common around the lake margins. Swims freely in mid-water in captivity and seems well-adapted to life in mid-water. Spawning occurs in summer (December–January), the eggs being relatively large for such a small fish (1.8–2.0 mm in diameter in ripe females); fecundity is accordingly quite low, with about 40–180 eggs in females, 42–60 mm long. Both sexes attain maturity at the end of the first year and may live for up to 3 years. The eggs are deposited in clusters under rocks in the shallows of lake margins and take several weeks to develop. On hatching, the larvae are about 8.5 mm long and probably live in the open waters of lakes for about 6 months, feeding and growing there, before joining the adults in the substrates of lake shallows. The life cycle is completed in fresh water. The diet consists of a wide range of aquatic animals, especially chironomid and caddis larvae and various crustaceans such as copepods, ostracods and cladocerans.
Utility Possibly a forage fish for trout; used rarely as bait by anglers.
Similar species Resembles other species of *Paragalaxias*.
Other names Common: Shannon galaxias. Scientific: *Paragalaxias shannonensis*.
Literature Fulton, 1982, 1990; McDowall & Fulton, 1978a; McDowall & Frankenberg, 1981; Merrick & Schmida, 1984.

Shannon paragalaxias *Paragalaxias dissimilis*. (R.M. McDowall)

FAMILY GALAXIIDAE

GREAT LAKE PARAGALAXIAS *Paragalaxias eleotroides* McDowall and Fulton

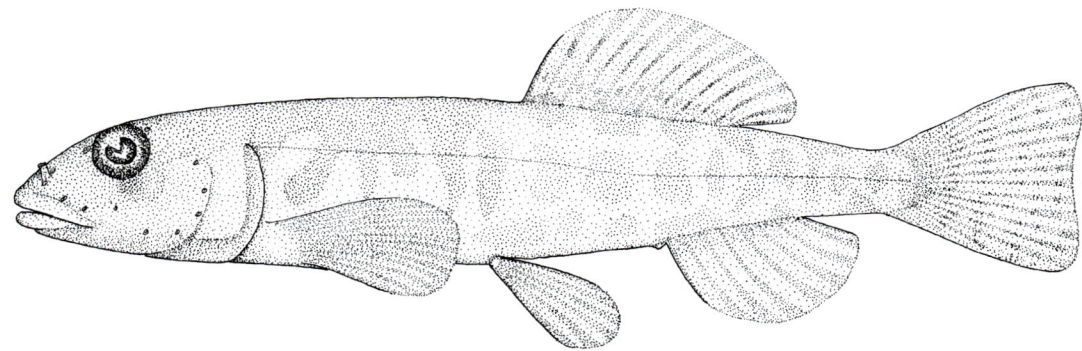

10.18 Great Lake paragalaxias *Paragalaxias eleotroides.* (R.M. McDowall)

Description A small, stout species, belly rather flattened and back more arched; head blunt, snout with steep profile, mouth of moderate size, jaws about equal, without lateral canine teeth and reaching back to about eyes; mouth low on head. Eyes small, high on head and close together. Dorsal fin of moderate size, rounded, middle rays longest (12–14 rays), origin above pelvic fin bases. Anal fin small (6–8 rays); pectoral fins low on sides (13–14 rays); 6 pelvic fin rays. Tail truncate to emarginate (14 rays). Open pores on head small, none below lower jaw (Fig. 10.16D); gill rakers (9–12) very short and stout; no pyloric caeca. Vertebrae 37–40.

Colour Light brownish gold with no obvious bands, but irregular, diffuse blotching or speckling on sides and back, belly silvery green; eyes golden. Fins amber to olive, with quite distinct black specks along rays.

Size Known to reach only 59 mm, commonly up to 40 mm.

Distribution Known only from Great Lake and Shannon Lagoon on the Central Plateau of Tasmania.

Conservation status In spite of sharing its principal habitat with brown and rainbow

Great Lake paragalaxias *Paragalaxias eleotroides.* (R.M. McDowall)

trout, is not reported to be under threat.
Natural history Found on the lake floor, among cover—under rocks, debris and vegetation; largely benthic in habit, resembling eleotrids in behaviour, propping itself up with its pectoral fins, head elevated, and darting from place to place. Shares lakes with Great Lake paragalaxias but is more benthic and widespread across the lake floor. Life cycle spent entirely in fresh water. Reaches maturity at end of first year and some probably live for 2 years. Spawning is probably during spring (October–November). The eggs are about 1.8–2.0 mm in diameter and number only about 45–150 in females, 40–53 mm long. Little else is known. The diet consists of a wide range of aquatic animals, especially chironomid and caddis larvae, and various crustaceans such as copepods, ostracods and cladocerans.
Utility None; perhaps preyed upon by introduced trout.
Similar species Resembles other species of *Paragalaxias*.
Other names Common: Great Lake paragalaxias. Scientific: has been included in descriptions with *P. dissimilis* and *P. shannonensis*.
Literature Fulton, 1982, 1990; McDowall & Frankenberg, 1981; McDowall & Fulton, 1978a; Merrick & Schmida, 1984.

ARTHURS PARAGALAXIAS *Paragalaxias mesotes* McDowall and Fulton

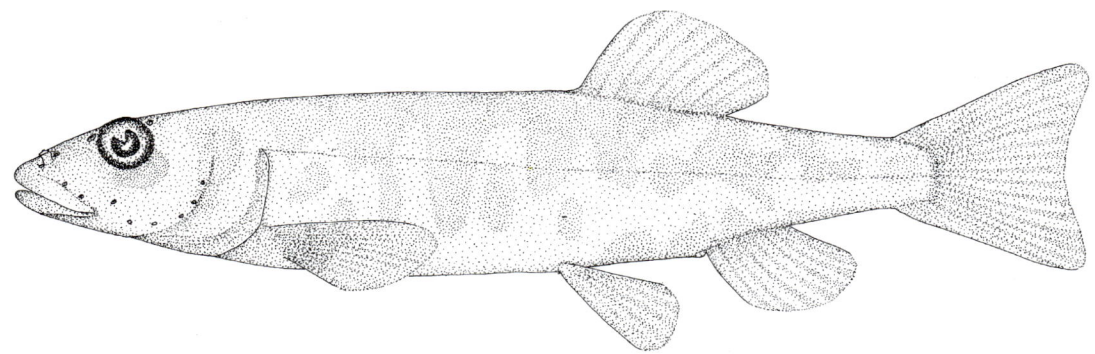

10.19 Arthurs paragalaxias *Paragalaxias mesotes*. (R.M. McDowall)

Description A small, stout, rather bullet-shaped species, somewhat flattened below, back more arched. Head long, sloping to a blunt, long snout, mouth large, reaching a little beyond front of eyes; jaws equal, lacking canine teeth. Eyes of moderate size, high on head, somewhat projecting above head profile. Dorsal fin smallish (9–12 rays), rays about equal in length or anterior rays longer; origin a little behind pelvic fin bases; anal also small (7–9 rays), rounded, middle rays longest. Pectoral fins low on sides (13–15 rays); 6 pelvic fin rays. Tail slightly forked to emarginate, on a long, slender caudal peduncle. Open pores on head small, usually none below lower jaw (Fig. 10.16C); gill rakers (12–15) moderately long, slender; no pyloric caeca. Vertebrae 40–41.

Colour A handsome, boldly coloured fish, dark greenish grey on back, this colour spreading down on sides as irregular bands that become fragmented into isolated patches, paler colouration of trunk grey to olive-gold, belly silvery grey. Fins amber, with some darker colouration along rays.

Size Known to reach 80 mm and not uncommonly more than 60 mm.

Distribution Known only from Arthurs Lake, Woods Lake and the Lake River below Woods lake in the eastern part of the Central Plateau of Tasmania.

Conservation status Has never been found commonly (since discovery in the mid-1970s), and status is a matter for concern; it is listed as 'restricted' by the Australian Society for Fish Biology.

Natural history Occurs among rocks and boulders, less often among marginal vegetation around the lake shores; seems to occur only in still waters. Very little is known, but the whole life cycle is obviously in fresh water. The diet consists of a wide range of aquatic animals, especially chironomid and caddis larvae, and various crustaceans

FAMILY GALAXIIDAE

Arthurs paragalaxias *Paragalaxias mesotes*. (R.M. McDowall)

such as copepods, ostracods and cladocerans.
Utility None, apart from being a handsome fish in captivity.
Similar species Resembles other species of *Paragalaxias*.

Other names Common: none. Scientific: none.
Literature Fulton, 1982, 1990; McDowall & Frankenberg, 1981; McDowall & Fulton, 1978a; Merrick & Schmida, 1984; Sanger & Fulton, 1991.

WESTERN PARAGALAXIAS *Paragalaxias julianus* McDowall and Fulton

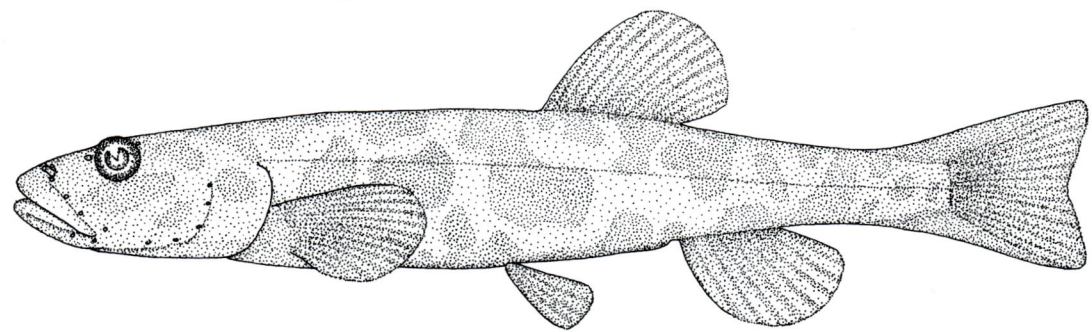

10.20 Western paragalaxias *Paragalaxias julianus*. (R.M. McDowall)

Description A small, stout species, bullet-shaped, tapering to a slender tail. Head long and somewhat flattened, tapering to a long, blunt snout, mouth reaching to about front of eyes, jaws equal and lacking canine teeth. Eyes small to moderate, high on head, and raised somewhat above head profile. Dorsal fin small (9–11 rays), origin a little behind pelvic fin bases; anal fin also small (7–8 rays), both fins rounded, middle rays longest. Tail slightly forked (15 rays). Pectoral fins small, fan-shaped (11–14 rays); 5 pelvic fin rays. Gill rakers (12–15) reduced to very short stubs; no pyloric caeca; open pores on head small, present below lower jaw. Vertebrae 40–43.

GALAXIIDS

Western paragalaxias *Paragalaxias julianus*. (W. Fulton)

Colour Brownish to black-olive markings on back and sides varying from broken irregular patches to bold and distinct bars, background colour yellowish olive, belly silvery olive; colour varies with habitat.

Size The largest species of *Paragalaxias*, reaching about 100 mm in length and commonly more than 60 mm.

Distribution Known from lakes in the upper reaches of the Ouse, James and Little Pine Rivers, all on the high-altitude central plateau of Tasmania.

Conservation status Little known but seems abundant in known habitats, which are highly remote, this providing some security from human-induced decline.

Natural history Found abundantly among and beneath rocks on lake beds, particularly where rocks are set apart. Little is known of reproduction, though the life cycle is certainly completed in fresh water. The diet consists of a wide range of aquatic animals, especially chironomid and caddis larvae, and various crustaceans such as copepods, ostracods and cladocerans.

Utility None.

Similar species Resembles other species of *Paragalaxias*, particularly Arthurs paragalaxias.

Other names Common: none. Scientific: none.

Literature Fulton, 1982, 1990; McDowall & Fulton, 1978b; McDowall & Frankenberg, 1981; Merrick & Schmida, 1984.

11
Family Aplochitonidae
Tasmanian whitebait

R. M. McDOWALL

The family Aplochitonidae is a small one, containing only the Tasmanian whitebait (genus *Lovettia*, one species) and the South American peladillos (genus *Aplochiton*, two species). Though the two genera are included in the one family, they may not be closest relatives, and the family may not be a natural group. These fish are thought most closely related to the galaxiid fishes (family Galaxiidae, see Chapter 10), and these two southern families to the Northern Hemisphere smelts (Osmeridae) and trouts and salmons (Salmonidae, Chapter 12).

The single species of *Lovettia* is found only in Tasmania; it is characterised by having the dorsal fin about mid-length, just behind the level of the abdominal pelvic fins, a small adipose fin, a lateral line, no scales, and 16 principal rays in the caudal fin.

TASMANIAN WHITEBAIT *Lovettia sealii* (Johnston)

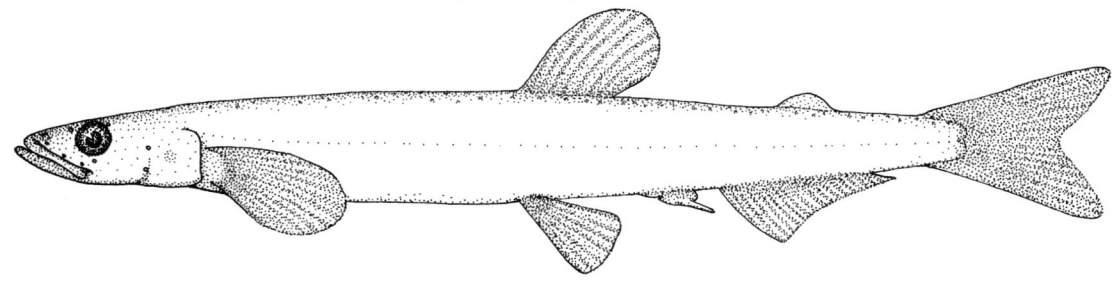

11.1 Tasmanian whitebait *Lovettia sealii*, female. (R.M. McDowall)

Description A small, elongate and very slender fish, with dorsal fin (7–9 rays) just behind level of pelvic fins; there is a small, low adipose fin attached to back along entire base, a moderately forked tail fin, a low, long-based anal fin with 16–19 rays, and paddle-shaped pectoral fins with 10–11 rays. Snout quite sharply pointed, long and flattened, and mouth large. The upper jaw has a pair of large fangs towards front. There are no scales, but a lateral line is present, though it is indistinct.

Colour When leaving the sea is mostly transparent, with a silvery band along sides. Pigmentation develops as the pre-spawning adults move upstream, becoming increasingly dusky grey and eventually almost black. There are no distinctive markings.

Sexual dimorphism There are pronounced differences between the sexes in mature adults. Oviducts of females open on a large, rather spoon-shaped genital papilla that lies between vent and front of anal fin. In the male, however, openings for testes migrate forwards during maturation, so that they open to the exterior just behind head, below and between pectoral fins, via a large, flat papilla. The opercular membranes of the male also expand ventrally, becoming large, rounded flaps covered with small tuberosities or papillae that may spread upwards onto head. These give lower surfaces of head a roughness. Pectoral and pelvic fins of males are much larger than those

of females (Fig. 11.2).
Size Is a small fish, known to reach 77 mm, but is mostly less than 60 mm.
Distribution Found only in coastal seas and rivers of Tasmania, mainly along the north coast and but also in rivers of the west and far south-east.
Conservation status Being once the basis for a significant fishery, was formerly extraordinarily abundant; though it is now much less common, it is not endangered.
Natural history Spends most of its life at sea. Mature adults migrate into low-elevation river estuaries in huge shoals during spring. These are mixed-species shoals that may include whitebait juveniles of several amphidromous species of *Galaxias* (see Chapter 10) and mature, anadromous adults of Tasmanian smelt (see Chapter 13). Spawning occurs during spring when the adults are about a year old. These migrate into lowland, estuarine reaches of Tasmanian rivers in huge shoals; they spawn in estuaries, the eggs being deposited on logs, submerged branches, stones, etc., after which the spent adults die. The eggs, only a few hundred per female, are about 1 mm in diameter. They hatch after 2–3 weeks, and the young are carried downstream in the river's flow, and to sea. The larvae live and grow there for the following year before returning to rivers to spawn, completing the cycle. A few whitebait have been found in the stomachs of marine fish, but very little is known about their life at sea.
Utility A commercial fishery for Tasmanian whitebait has been known since the 1930s and developed most vigorously during the early 1940s. It is not entirely based on *Lovettia*, but also includes young of several species of *Galaxias* and occasional Tasmanian smelt (*Retropinna tasmanica*) that migrate upstream with them (see Chapters 10 and 13). The fishery, which was always much more productive in Tasmania's northern rivers, reached a peak production of more than 480,000 kg in 1947. However, it was already in decline as in that year catch for effort was less than in the previous year. Catch slumped rapidly and alarmingly, prompting a one-year closure in 1949; it was reopened in 1950 but continued to decline, dropping to about 1570 kg in 1955 and 1010 kg in 1972. With the decline in catch, the proportion of *Lovettia* declined and various *Galaxias* species became more

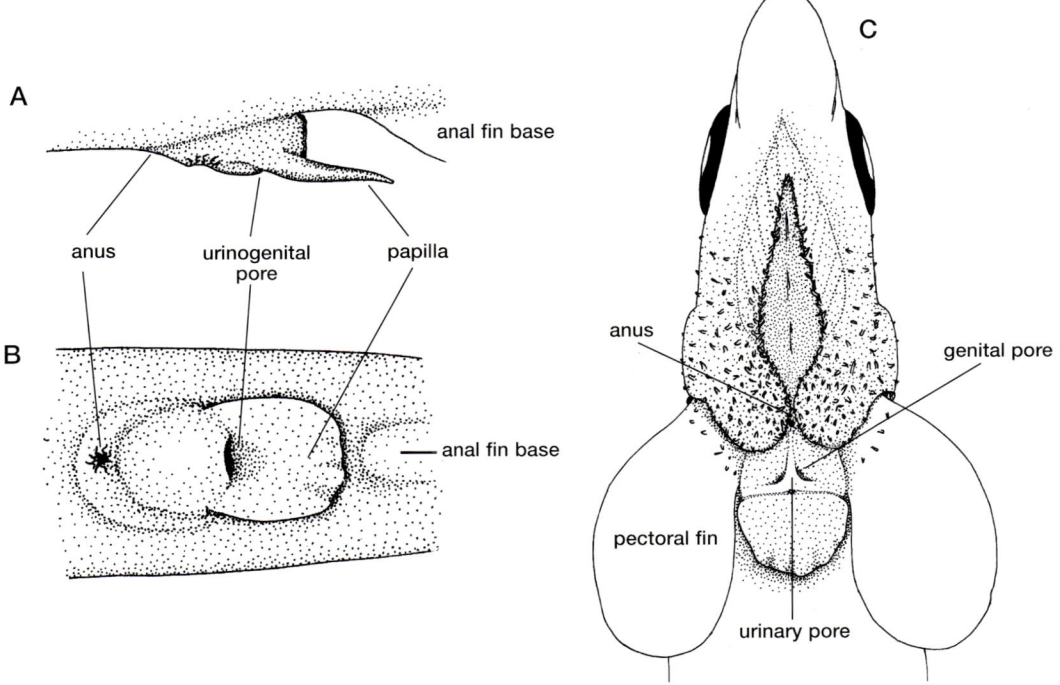

11.2 Genital papilla of Tasmanian whitebait *Lovettia sealii*: **A** lateral view of female; **B** ventral view of female; **C** ventral view of male. (R.M. McDowall)

common. Catch was either canned or frozen. The fishery was closed after the 1973 season in an attempt to allow stocks to recover, and it remained closed until reopened for a trial recreational season for a few weeks in 1990. The new fishery was strictly regulated and was intended to focus exploitation on *Galaxias* rather than *Lovettia*. The season was described as 'a trial fishery with no chance of overexploitation'. Fishing was allowed for only a few weeks on a limited number of rivers and at a time when the peak runs would not be occurring. There were daily and seasonal catch limits. Modest catches were taken, enough to justify further trial seasons on a strictly recreational basis in later years.

With the loss of the fishery, some investigation was undertaken into farming of Tasmanian whitebait—though a successful and economic industry seems unlikely, given the very small growth achieved by the fish.

In addition to being taken by humans for food, whitebait provide a valuable food for predatory estuarine fishes, including sea-migratory brown trout that have been introduced into Tasmanian rivers; decline in the fishery generated concern among trout anglers about loss of a valued food for trout.

Is not known as an aquarium fish and is unlikely to have any potential.

Similar species Resembles sea-migratory young of various Tasmanian species, including several species of *Galaxias* (see Chapter 10), grayling (*Prototroctes maraena*, see Chapter 14) and Tasmanian smelt (see Chapter 13).

Other names None. Though the species name is sometimes incorrectly spelt *seali*; *sealii* is the original spelling and is correct.

Literature Blackburn, 1950; Fulton, 1984; Fulton & Pavuk, 1988; Lynch, 1965a, b; McDowall, 1971; Scott, 1971; Whitley, 1957a.

12
Family Salmonidae
Salmons, trouts and chars

P.E. DAVIES AND R.M. McDOWALL

Though relatively small, the family Salmonidae contains some of the world's most important freshwater angling and aquaculture fishes. Salmonids are native to cool and cold waters around the Northern Hemisphere, some species being restricted to fresh water, others spending various proportions of their lives in the sea. Several of the better-known species have been introduced into many parts of the Southern Hemisphere—South America, South Africa, New Zealand and Australia—where there are suitable cool waters. Five species occur in Australia.

The family is characterised by a dorsal fin high on the back and further forward than the pelvic fins, an adipose fin, small scales, a lateral line, and axillary processes on pelvic fins.

Key to salmonids

1. Flesh along base of teeth in lower jaw a dusky grey-black; anal fin long-based, fin base longer than longest ray; usually 13–19 rays . . . . . . **Oncorhynchus tshawytscha** p. 89; Fig. 12.7
   Flesh along base of teeth in lower jaw not dusky; anal fin shorter based, fin base shorter than longest ray; usually 8–12 rays 2

2. Leading edge of pectoral, pelvic and anal fins white, with a contrasting black line behind; trunk dark, with paler spots or markings. . . . . **Salvelinus fontinalis** p. 86; Fig. 12.4
   Leading edge of pectoral, pelvic and anal fins without distinctive contrasting colouration; if whitish, no contrasting black line; trunk paler, with darker spots 3

3. Spots numerous and prominent on tail; often a pink to red flush along head and sides. . . . .
   . . . . **Oncorhynchus mykiss** p. 87; Fig. 12.5
   No spots on tail, or few, small and inconspicuous; no pink or red flush along sides. 4

4. Anal fin when pressed against caudal peduncle may reach tail base; dark spots on body usually surrounded by paler halos, often pale or reddish-orange spots along and below lateral line; teeth on head of vomer, those on shaft strong and adherent; mouth large, reaching behind eye . . . . **Salmo trutta** p. 81; Fig. 12.1
   Anal fin when pressed against caudal peduncle does not reach base of tail; dark spots on body not surrounded by paler halos, no pale or reddish spots on sides; no teeth on head of vomer, those on shaft weak and deciduous; mouth smaller, reaching to below eye . . . .
   . . . . **Salmo salar** p. 85; Fig. 12.3

BROWN TROUT
Salmo trutta Linnaeus

Description A quite thick-bodied and shallow species with a big head and mouth, eyes moderate to large; mouth extends back below eyes, becoming increasingly large with growth, and eyes relatively smaller; dorsal fin (12–14 rays) high on back, in front of pelvic fins; 13–14 pectoral rays; adipose fin well developed. Caudal peduncle relatively deep and tail only slightly forked, if at all. Scales of moderate size, with a distinct lateral line (110–120 scales). A very large number of pyloric caeca on stomach. Vertebrae 56–61.

FAMILY SALMONIDAE

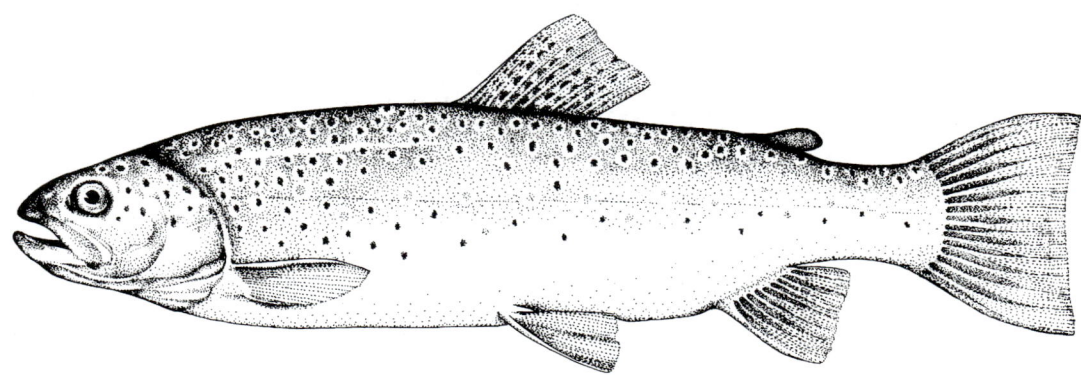

12.1 Brown trout *Salmo trutta*. (C. Kroger)

Sexual dimorphism Mature and spawning males develop elongated jaws, the lower with a markedly upturned tip (kype) that fits into a groove in the upper.

Colour Highly variable, depending on age, habitat and life history. Sea-run fish are generally silvery, darker and somewhat olive on back, with indistinct dark spots, occasionally bright silver, with no or few spots. Lake fish also tend to be silvery, with bold, dark, brownish spots, back brownish olive. River fish are often much darker and more distinctly coloured, back deep brown to olive, with darker brown spots on dorsal and adipose fins and back, sides and gill covers. Below lateral line some pale spots occur, as well as sometimes bright red spots surrounded by pale halos. Edge of adipose fin, leading edge of pelvic fins and lower edge of tail may have an orange or reddish flush.

Size Reaches a large size, at least 1400 mm fork length and weights of more than 20 kg; is known to reach at least 900 mm and 14 kg in Australia.

Introduction and distribution Is native to Europe, from Iceland and Scandinavia south to Spain and northern Africa, and eastward to the Black and Caspian Seas.

Was introduced into Australia in the 1860s and spread by a combination of stocking and migration. Is restricted to cooler waters in Australia, mostly in the highlands above about 600 m from northern New South Wales to the south coast of Victoria, descending to lower altitudes where there are large lakes and reservoirs or where swiftly flowing waters remain cool (generally < 25°C summer maxima). Populations also maintained by stocking in the Adelaide region, South Australia, with several self-sustaining populations

Brown trout *Salmo trutta*, large lake fish. (R.M. McDowall)

SALMONS, TROUTS AND CHARS

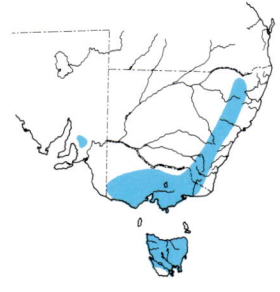

in small streams. In Tasmania it is widespread and abundant down to sea level in all major drainages except the Davey River and Bathurst Harbour streams. Frequently not found in sections of drainages blocked by significant barriers to migration (e.g. waterfalls). Low genetic diversity of Tasmanian stocks suggests that only the initial introduction from southern England was successful.

Natural history Occurs commonly in cool, well-oxygenated waters, usually in streams with moderate to swift flow, but also in cool, clear lakes and in estuaries and adjacent coastal areas. Spawns in autumn and winter (April–August), often after a flood. Spawning fish frequently migrate upstream into small tributaries or feeder streams with gravel beds but may spawn locally in a river or on a lake shore if gravel is present in bars or pockets with sufficient water movement. Larger-scale migrations of spawning fish into estuaries are also observed. Spawning takes place only when sufficient depth and velocity of water occur over a bar or patch of gravel. When spawning, pairs form, the females excavating a series of depressions in stream bed by beating against the gravel with her tail while the male drives away intruders. The male joins the female above the depression and eggs and milt are released, the eggs settling in the depressions.

After spawning, the female covers the eggs by dislodging gravel upstream from the pocket. The resulting small gravel mound, which contains a number of egg pockets, is known as a redd. Redds may contain between 500 and 3000 eggs, depending on the size of the female. Eggs are large, 4–5 mm in diameter, bright orange in colour. Annual spawning migrations into spawning streams are a significant feature of several lake populations, with notable runs at Great Lake and Lake Sorell, Tasmania, of around 15,000 and 80,000 fish respectively.

The eggs take several weeks to develop (typically 6–20 weeks, depending on water temperature) and the young hatch and stay in the gravel for some time (as alevins), absorbing their yolk. It is at this stage (September to early November) that they are highly susceptible to mortality from declining river flows. After emerging to begin feeding, they may form small shoals in slow, shallow water along stream edges or in backwaters. The young (parr, Fig. 12.2) have a series of bold, dark blotches along their sides. They are initially gregarious but soon become increasingly solitary and territorial through their first two years, moving to deeper water. They feed on a wide variety of animals—aquatic crustaceans, molluscs, insects and small fishes, as well as terrestrial insects that fall onto the water from overhanging vegetation. Is believed to have had significant impacts on several species of native fish, particularly galaxiids.

Utility Is one of the angler's great fishes, populations being maintained solely for angling in some streams and lakes only by liberation of hatchery-

Brown trout *Salmo trutta*, large river fish. (R.M.McDowall)

FAMILY SALMONIDAE

Brown trout *Salmo trutta*, small stream fish. (R.H. Kuiter)

reared stock. Anglers targeting this species number in the hundreds of thousands in south-eastern Australia alone. In Tasmania, the fishery is worth some $30–35 million in expenditure per year. It has proved to be the most successful introduced salmonid species, quickly establishing self-supporting populations in many areas. Many popular angling streams and lakes have populations entirely maintained by natural recruitment, though they are susceptible to the effects of dry years.

If water is cool and well aerated, brown trout are not difficult to keep in captivity; however, they rarely settle and remain nervous. Problems with poor hatching and rearing success in part preclude them from being a commercially viable aquaculture species.

Similar species Other trouts and salmons, particularly Atlantic salmon.

Other names Common: sea trout, Englishman. Scientific: *S. fario*.

References Allen, 1951; Davies, 1989; Davies *et al.*, 1989; Elliott, 1994; Frost & Brown, 1967; Jackson, 1975, 1978a, 1980, 1981; Jackson & Williams, 1980; Kailola *et al.*, 1993; Lake, 1957; McKeown, 1934, 1937, 1955; Nicholls, 1958a, b, c; Ovenden *et al.*, 1993; Parrott, 1932; Scott & Crossman, 1973; Sloane, 1983; Staley, 1966; Weatherley, 1958; Weatherley & Lake, 1967.

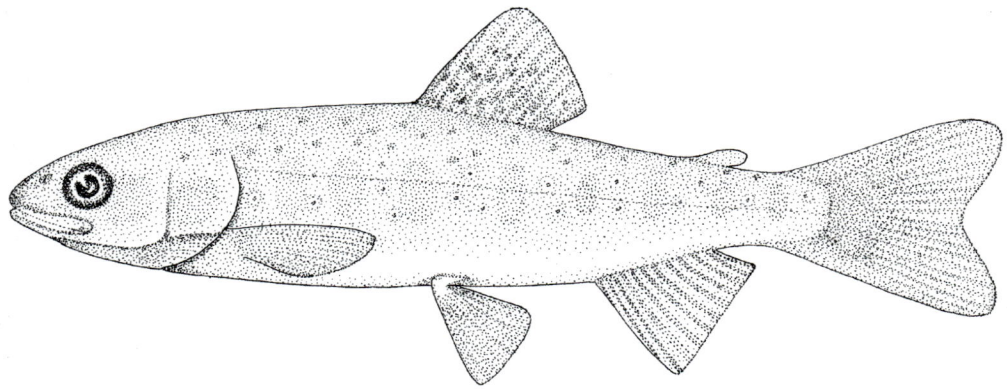

12.2 Parr of brown trout *Salmo trutta*. (R.M. McDowall)

ATLANTIC SALMON *Salmo salar* Linnaeus

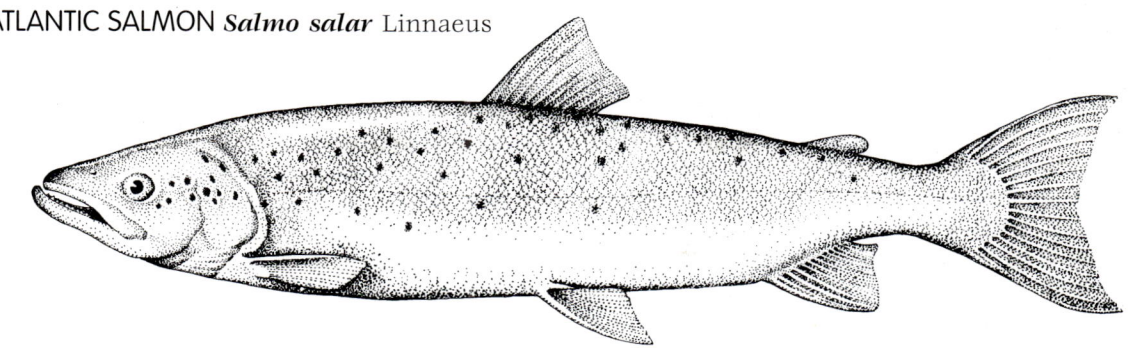

12.3 Atlantic salmon *Salmo salar*. (C. Kroger)

Description Closely resembles brown trout, and certain identification can be a problem. Caudal peduncle longer and more slender; anal fin when pressed against caudal peduncle does not reach base of tail. Tail usually more deeply forked. As in other salmonids, dorsal fin (11–12 rays) further forward than pelvic fins; anal fin with 9–10 rays. Teeth absent on head of vomer; those on shaft may be deciduous. Mouth smaller, not reaching back past eyes. Scales small (110–125 along lateral line); vertebrae 58–60; gill rakers 17–21.
Sexual dimorphism See brown trout.
Colour Back blue to silvery blue, varying to brownish olive, sides silvery, belly silvery white. Small to largish black spots on back and sides, cheeks, gill covers and dorsal fin base. Spotting often not very bold, spots X-shaped. No red colour evident. In fresh water is often slim and silvery, with deeply forked tail and slender tail fin base.
Size Reaches at least 1500 mm and 40 kg in Europe; common sizes in Australia (mainly sea cage farm escapees) 1–5 kg, but landlocked fish are generally smaller, no more than 1–3 kg. Broodstock for Tasmanian aquaculture has been reared to 12 kg.
Introduction and distribution Occurs naturally in cool and cold waters flowing into the North Atlantic Ocean, from northern Spain through Eastern Europe to Iceland, Greenland and south along the coast of North America as far as the state of Connecticut. No self-sustaining sea-run populations are recorded in Australia. Stocks were liberated intensively in Tasmania and Victoria between 1864 and 1870 with little success. Established in Burrinjuck Dam, New South Wales (1963–64), and in Lake Jindabyne but maintained in those storages by annual stocking. Well established in Tasmania in the aquaculture industry since 1984, with frequent escapes from sea cages to local estuaries in southern and western Tasmania. Atlantic salmon are reared in government and private freshwater hatcheries in Victoria, and populations established from escapees have been recorded in the Rubicon and Latrobe Rivers.

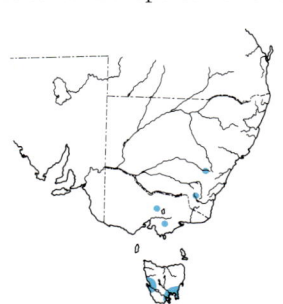

Natural history Like other trouts and salmon, requires cool or cold waters. In Europe most populations are seagoing. Young live in fresh waters for up to 3 years before migrating to sea as silvery smolts. Australian landlocked and seagoing populations are dependent on liberation or escape of fish reared in hatcheries or sea cages. If and when spawning occurs, resembles that of brown trout in almost all details. Virtually nothing is recorded of natural history of Atlantic salmon in Australian waters.
Utility Regarded as a great sporting fish in Europe and was introduced into Australia for this reason. Forms the basis of an intensive aquaculture industry, based mainly in Tasmania. Current (1994) production is estimated at around 5000 tonnes at the farm gate, with an estimated total value of $80–100 million (including value-adding). Young are reared in freshwater hatcheries and transferred as smolts to floating sea cages on coastal and/or estuarine sites for intensive rearing.
Similar species Brown trout particularly, also other trouts and salmons.
Other names Common: none. Scientific: none.
References François, 1965; Jones, 1959; Kailola *et al.*, 1993; Ovenden *et al.*, 1993; Scott & Crossman, 1973.

FAMILY SALMONIDAE

BROOK CHAR
Salvelinus fontinalis (Mitchill)

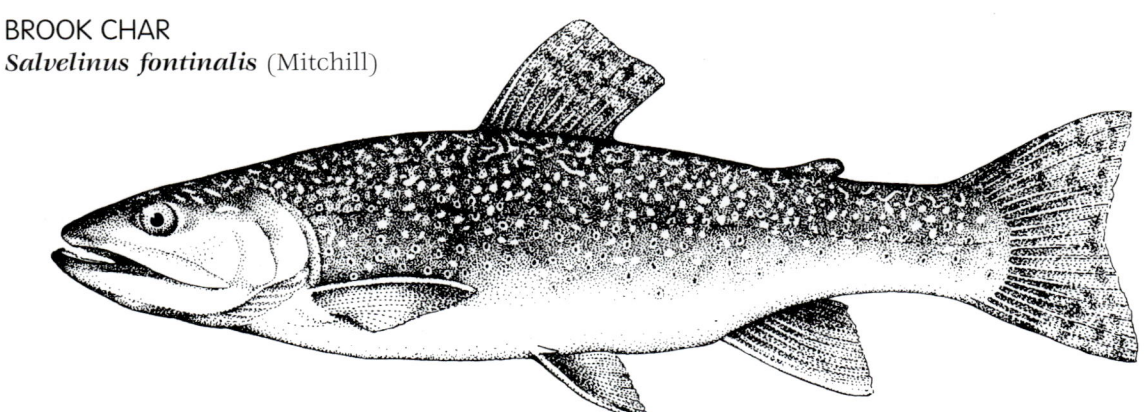

12.4 Brook char *Salvelinus fontinalis*. (C. Kroger)

Description A rather slender and elongated salmonid that resembles brown trout in general form; 10–14 dorsal rays, 9–12 anal rays, and 11–14 pectoral rays; mouth very large, reaching back beyond eyes; teeth present on head of vomer but not on shaft (Fig. 12.6 C, D). Scales small (110–132 along lateral line); vertebrae 58–62; gill rakers 14–22.
Sexual dimorphism As for brown trout.
Colour A very handsome species with bright colouration in well-grown adults; back dark olive green with irregular vermiculations and dense paler markings, sides somewhat paler with red spots surrounded with blue halos; belly silvery white. Dorsal fin and tail olive green with irregular markings, pectoral, pelvic, and anal fins dusky grey with a bold white stripe along leading edge followed by a contrasting black stripe. Fish in breeding condition have a flush of bright orange red along sides and belly, and the lower fins become reddish.
Size Known to reach a length of 850 mm and a weight of 6.5 kg; is mostly a small fish and in stream populations may reach only 200–300 mm.

Brook char *Salvelinus fontinalis*. (R.H. Kuiter)

SALMONS, TROUTS AND CHARS

No published figures for Australia, although reaches nearly 4 kg in Clarence Lagoon, Tasmania, and up to 2 kg in waters in the Snowy Mountains.
Introduction and distribution Occurs naturally in waters along the east coast of North America from northern Canada, south into the United States, mostly only in the mountains in the southern parts of range.

Was introduced from North America in the 1870s but generated little early interest. Stocks were first liberated in Tasmania in the early 1900s and in the 1970s into streams on the mountain tablelands of New South Wales. There are no self-sustaining populations in NSW other than in one stream in New England. Stocked annually in Lake Jindabyne with fish reared in the Gaden hatchery. Reintroduced into Tasmania and persists there as viable populations in Clarence Lagoon and lakes of the Tyndall Ranges, with a high genetic diversity. Populations also maintained by stocking in South Australia.

Natural history A cool- and cold-water fish occurring both in streams and clear, pure lakes. Stream populations have a strong tendency to move up into very small tributaries, where they mature at small size—and so are often neither accessible nor of much interest to anglers; does not usually coexist well with other salmonids, which may explain why success is limited in Australia. Breeding resembles that of brown trout. Food includes the usual diverse array of aquatic insects, crustaceans and molluscs, some small fishes, as well as terrestrial insects.
Utility Sporting qualities less well known than those of other salmonids; tends to be easily caught, like rainbow trout, but does not fight as well. Its beautiful colouration adds to its attraction as a sporting fish.

Probably the most rewarding salmonid to keep in captivity; settles down quickly in an aquarium, becoming quite tame. However, the water must be kept cool.
Similar species Other trouts and salmons.
Other names Common: brook trout, fontinalis. Scientific: none.
References Anon., 1970; McAfee, 1966a; Ovenden *et al.*, 1993; Scott & Crossman, 1973.

RAINBOW TROUT
Oncorhynchus mykiss Walbaum

Description Resembles brown trout in most features, but rather deeper bodied and more compressed. Dorsal fin (10–12 rays) high on back, in front of pelvic fins; anal with 8–12 rays; pectorals with 14–16; a well developed adipose fin; tail slightly forked to almost truncate. Head of moderate size, the mouth large, eyes of moderate size. Scales small (115–150 along lateral line); vertebrae 61–66; gill rakers 17–22.
Sexual dimorphism Male has enlarged jaws with the tip of the lower jaw forming a kype, as in brown trout.
Colour Variable, lake fish predominantly silvery; back dark, either greenish olive or a deep steely blue, with many small, dark spots. Often a rosy pink stripe along sides and on gill covers. River fish and those on spawning migrations become

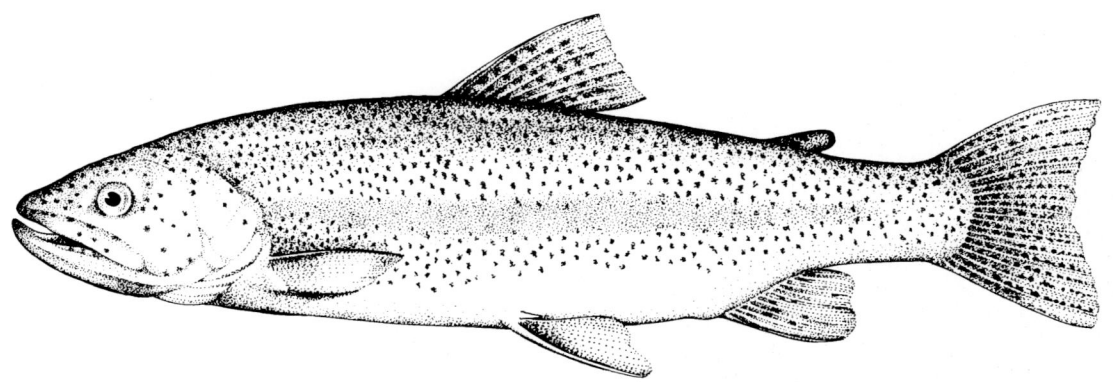

12.5 Rainbow trout *Oncorhynchus mykiss*. (C. Kroger)

FAMILY SALMONIDAE

Rainbow trout *Oncorhynchus mykiss*, large lake fish. (R.M. McDowall)

more intensely coloured; the pink stripe intensifies to deep crimson, lower fins becomes reddish and lower sides and belly a deep smoky grey. Spots on sides become bolder and more conspicuous.
Size Is known to reach 1120 mm and more than 20 kg, although there is an unconfirmed record of one weighing more than 25 kg; reported to reach about 775 mm and nearly 8 kg in Australian waters. However, fish larger than about 600 mm and 4 kg are exceptional in most waters.

Rainbow trout *Oncorhynchus mykiss*, small river fish. (R.H. Kuiter)

SALMONS, TROUTS AND CHARS

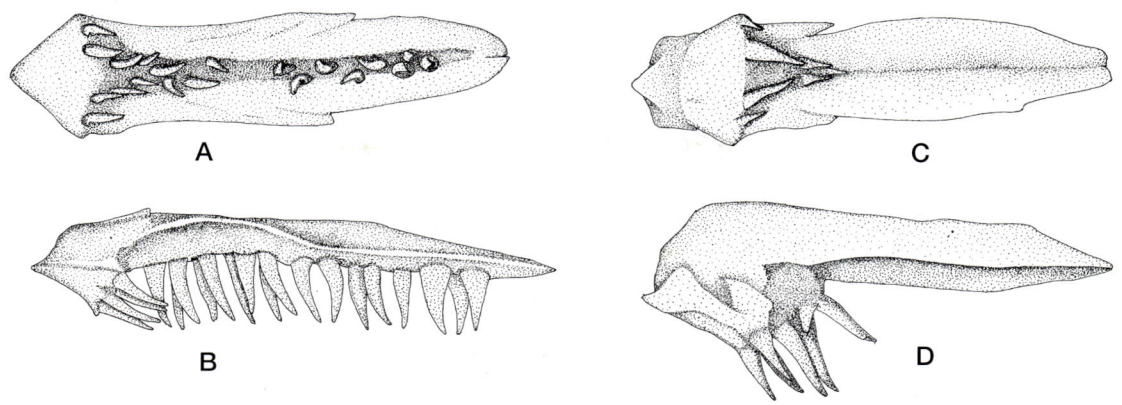

12.6 Vomerine teeth in species of *Salmo*: **A** lateral view; **B** ventral view; and in *Salvelinus*: **C** lateral view; **D** ventral view. (after Hubbs and Lagler, 1958)

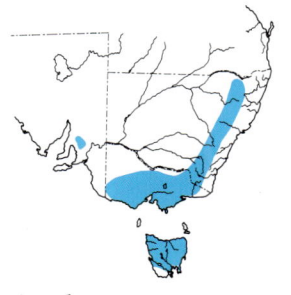

Introduction and distribution Is native to the Pacific coast of North America from Alaska to Mexico, primarily in rivers draining into the Pacific Ocean; also along the Pacific Coast of north-eastern Siberia, particularly the Kamchatka Peninsula.

Australian rainbow trout are derived from stocks obtained from New Zealand in the 1890s (where the species had been introduced from California). The species now occurs in high country from northern New South Wales to Victoria, in hills near Adelaide, South Australia, and in Tasmania—range generally similar to that of brown trout but with a more restricted local distribution and abundance; is frequently maintained by stocking.

Natural history Life history is essentially similar to that of brown trout; temperature tolerances of rainbow trout are a little higher. The rainbow tends to be more successful in lakes than rivers and streams. Spawning is somewhat later, in winter and early spring (August–October). Other details are similar to those of brown trout, although eggs are smaller and develop more quickly. Feeds on a wide range of aquatic insects, crustaceans, molluscs and fishes, as well as terrestrial insects.

Where rainbow and brown trout occur together the brown usually dominates.

Utility Perhaps for its size the most exciting trout to catch; more easily hooked than the brown and a better fighter once hooked. Is easily hatched and reared, given appropriate conditions of cool, clean and well-oxygenated water. Is of significance to the well-established Australian salmonid aquaculture industry (mainly based in Victoria and Tasmania), both in freshwater hatchery and in sea cage farms, although not as popular commercially as Atlantic salmon.

Is easily kept in captivity but takes some time to settle down; needs cool, well-oxygenated water.

Similar species Other trouts and salmons.

Other names Common: steelhead. Scientific: *Salmo gairdneri* and *S. irideus*.

References Davies & Sloane, 1986; McAfee, 1966b; Scott & Crossman, 1973; Sloane, 1983; Tilzey, 1972, 1977; Weatherley & Lake, 1967.

QUINNAT SALMON
Oncorhynchus tshawytscha (Walbaum)

Description Generally trout-like; however, snout tends to be rather more pointed than in other salmonids and the mouth larger, reaching back well beyond eyes. Dorsal fin with 11–15 rays; anal fin long-based (15–20 rays) and low, the longest ray being much shorter than the length of the fin base. Pectorals with 14–17 rays. Scales very small (130–165 along lateral line); vertebrae 62–65; gill rakers 23–27.

Sexual dimorphism Male develops an elongated and hooked lower jaw, rather different from the upturned tip (kype) in the trouts.

Colour Predominantly silvery, back a greenish olive with many small, black spots, also on dorsal and adipose fins and tail. Inside mouth, alongside

89

FAMILY SALMONIDAE

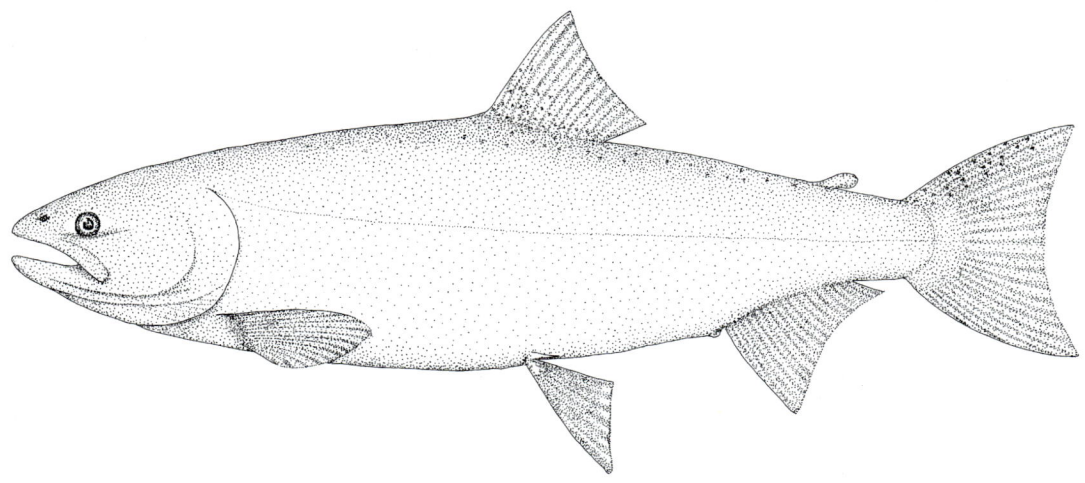

12.7 Quinnat salmon *Oncorhynchus tshawytscha.* (R.M. McDowall)

the teeth in lower jaw, is a dusky grey-black; eyes olive gold.

Size Sea-run quinnat reach a large size, at least 1500 mm in length and 57 kg in weight. Landlocked quinnat, as occur in Australia, may reach 850 mm and 8.5 kg.

Introduction and distribution Native to the west coast of North America from northern California to Alaska, also north-eastern Asia from northern Japan to Kamchatka.

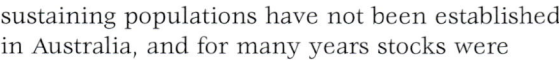

Australian stocks originated in New Zealand, where the species was introduced and anadromous stocks are established. Self-sustaining populations have not been established in Australia, and for many years stocks were maintained with New Zealand ova. However, populations are now maintained in lakes by releases of hatchery-reared brood fish. About 150,000 were released into 16 Victorian lakes in the 1930–40s. They grew well only in Lakes Bullen Merri and Purrumbete, and further liberations were made to Lake Purrumbete in the 1970s. Despite further trial stockings in the 1980s in other Victorian waters, stocking is now essentially restricted to Bullen Merri and Purrumbete, which do not support natural recruitment. These lakes sustain a significant recreational fishery for quinnat salmon. Hatchery brood stocks are maintained in Snobs Creek hatchery, Victoria, and fish are now also reared in several private Victorian hatcheries.

Natural history Normally spends most of its life at sea, but this is not true of Australian populations that are landlocked, occurring in surface, open waters of lakes into which they are liberated.

Quinnat salmon *Oncorhynchus tshawytscha.* (C.A. Barnham)

A spawning migration of mature to ripe fish was reported from a tributary of Lake Purrumbete, but the stream bottom was unsuitable for spawning. Feeds on a variety of available aquatic animals, including small fishes.

Utility In North America, in particular, is an important fish for both anglers and commercial fishermen. A fine angler's fish, introduced into Australia for that reason. Can be maintained in captivity with little difficulty if water is cool, but is seldom kept, apart from breeding stocks in hatcheries.

Similar species Other trouts and salmons.

Other names Common: king salmon, spring salmon, chinook salmon. Scientific: none, although name commonly misspelled—not surprisingly.

References Barnham, 1977; Butcher, 1947; Finlay, 1972; Healey, 1991; Parrott, 1971; Scott & Crossman, 1973; Vronisky, 1973.

13
Family Retropinnidae
Southern smelts

R.M. McDOWALL

The family Retropinnidae is small, with just four species, two each in Australia and New Zealand. Retropinnids are slender, silvery fishes closely related to the southern graylings (family Prototroctidae, see Chapter 14), these two southern families having close affinities with the northern osmerid smelts (family Osmeridae). All these families share the peculiar feature of emitting a distinct cucumber-like odour when freshly caught—though the name smelt has nothing to do with this smell, but rather is an old Anglo-Saxon word for silvery. There is one mainland Australian species of *Retropinna* and another in Tasmania. The family's fishes are characterised by being small, shiny and silvery and having a dorsal fin well back, over the anal fin, a small adipose fin, thin scales and no lateral line, and 18 principal caudal fin rays (16 branched).

Key to smelts

Ratio of lateral scale rows to vertebrae high, more than 1.4; vertebrae 52–55; slender-bodied; known only in Tasmania . . . . **Retropinna tasmanica** p. 95; Fig. 13.2
Ratio of lateral scale rows to vertebrae lower, less than 1.2; vertebrae 45–53 but mostly less than 52; deeper-bodied; mainland Australia . . . . **Retropinna semoni** p. 92; Fig. 13.1

AUSTRALIAN SMELT *Retropinna semoni* (Weber)

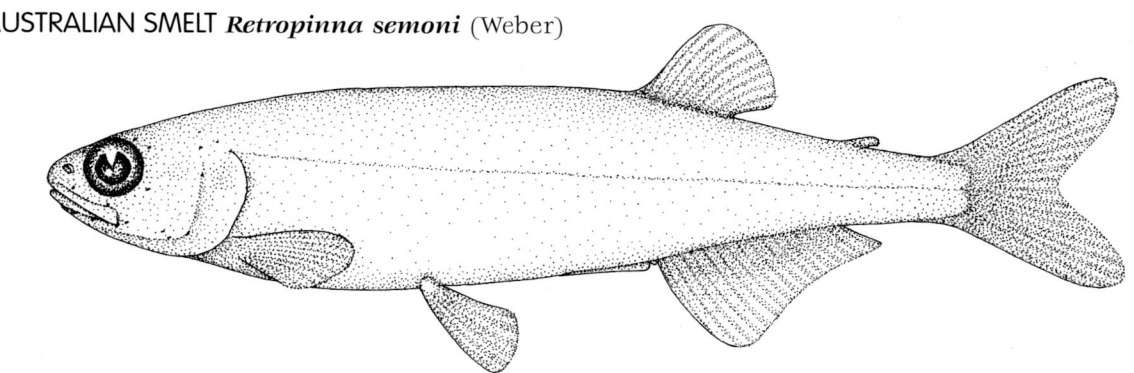

13.1 Australian smelt *Retropinna semoni*. (R.M. McDowall)

Description A small, compressed, moderately deep-bodied fish with large eyes and a somewhat rounded snout. Dorsal fin small and high (7–11 rays), well back on trunk, originating just a little in front of level of anal fin; anal fin longer and lower, with rays towards rear much shorter (13–19 rays). Adipose fin very small. Pectoral fins low, behind head (8–12 rays). Caudal peduncle very slender, supporting a moderately forked tail. Scales present on trunk but absent from head, small, very thin and cycloid and easily dislodged; about 50–70 along side (but very hard to count

SOUTHERN SMELTS

Australian smelt *Retropinna semoni*, male. Note large fins. (R.H. Kuiter)

precisely). There is a tendency for scales to fail to develop in some inland drainage populations. No lateral line, though there is often a mid-lateral pigment row that looks like a lateral line. Gill rakers (16–25) long and slender; vertebrae 45–53. Freshly caught smelt may have a strong cucumber-like odour.

Sexual dimorphism Differences between sexes develop as fish approach sexual maturity. The fins of males become greatly enlarged and quite spectacular, especially in populations where maturity is reached at a small size, pectoral and pelvic fins being particularly enlarged. Nuptial tubercles (small horny nodules, Fig. 2.16) develop on scales and fin rays in both sexes, though they are more abundant and widespread in males than in females.

Colour Bright silvery, back and upper sides somewhat olive, a little darker than lower sides and belly; belly silvery white; eyes silvery; top of head and snout often dusky and fins largely unpigmented, though there may be small melanophores along fin rays. A purplish sheen occurs along mid-sides.

Size A tiny species that may reach 100 mm, but rarely exceeds 75 mm, and in most populations reaches only 50–60 mm. Size reached less in northern than in southern populations, while those in inland drainages particularly may reach maturity at only 40 mm.

Distribution One of the most widespread species in Australia's south-east, the Australian smelt occurs in coastal drainages from about the

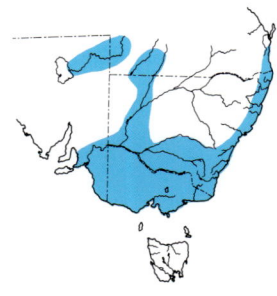

Fitzroy River in southern Queensland, south and west as far as eastern South Australia; also in tributaries of the Murray-Darling system that join the main river from the south-east, though not those joining from the north-west; however, occurs in Cooper Creek drainages that flow towards or into Lake Eyre.

Australian smelt were introduced into Tasmania during the 1960s, but success was elusive; the species persisted only in a farm dam and was never released into wild habitats.

Conservation status Widespread and abundant throughout its range and is of no present conservation concern.

Natural history Australian smelt occur in still and gently flowing waters of south-eastern Australia in great abundance, often as large roving shoals of small fish. Though other species of retropinnid spend part of their lives at sea (being anadromous), this is certainly not an obligatory aspect of the life cycle and may not occur at all.

Upstream migrations of juvenile and adult smelt (15–40 mm long) have been documented during the day at the Torrumbarry fishway on the Murray River. In both sexes only the left gonad is present. They reach maturity towards the end of their first year and may live for two or more years. Mature fish develop nuptial tubercles the purpose of which is unknown, though it presumably relates to contact between sexes during spawning. Spawning takes place when water temperature reaches about 15°C and occurs from mid-winter to autumn in Queensland but mostly during spring and summer further south. The eggs are tiny, about 1 mm in diameter; they number between about 100 and 1000 in females between about 35 and 55 mm length. They are demersal and adhesive and sink to the substrate, where they become attached to vegetation, debris or sediment. They hatch in about 10 days, though timing depends on water temperatures. The larvae are tiny, less than 5 mm long. Diet consists of a wide variety of small aquatic insects and crustaceans, varying with types of food most freely available.

Utility Because of its great abundance in some parts of its range, is an important forage species for larger, predatory fishes; it was for this reason that it was introduced into Tasmania, having the advantage over Tasmanian smelt of not needing to spend part of its life at sea. Use of the species in this way may have been prompted by successful use of New Zealand's common smelt (*R. retropinna*) in stimulating recovery of famous rainbow trout fisheries in lakes there. Unsuccessful attempts were also made to introduce this species to Papua New Guinea.

Is extremely fragile and requires considerable care when being transferred into captivity. In essence, it is necessary to transfer fish to the captive container without touching them with anything rougher than a soft, fine-meshed net, and even then mortality may be high. Once established, however, it can be kept without much difficulty and is an attractive, active, bright silvery fish.

Similar species Is very similar to the Tasmanian smelt and to small grayling, as well as small trouts and salmons, various clupeids, atherinids and mullets—in general to a diversity of small, silvery fish.

Other names Common: none. Scientific: a distinct species, *R. victoriae*, was formerly recognised in Victoria but is now regarded as a synonym of *R. semoni*. The name *R. richardsoni* appears in some early Australian literature but was applied to New Zealand populations of *R. retropinna*.

Literature Jolly, 1967; Koehn & O'Connor, 1990; Lake, 1971, 1978; McDowall, 1979; Mallen-Cooper, 1994; Merrick & Schmida, 1984; Milward, 1969; Milton & Arthington, 1985.

Australian smelt, *Retropinna semoni* female. Note small fins. (R.M. McDowall)

TASMANIAN SMELT *Retropinna tasmanica* McCulloch

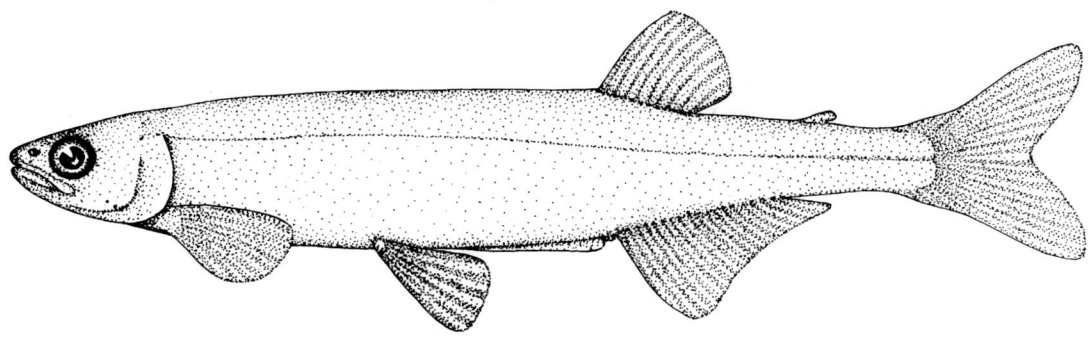

13.2 Tasmanian smelt *Retropinna tasmanica.* (R.M. McDowall)

Description A small, compressed and rather slender fish with large eyes and a pointed snout. Dorsal fin high, middle rays longest (10–12 rays), and well back on trunk, a little in front of level of vent and anal fin; is followed by a small adipose fin. Anal fin longer (16–21 rays), rays at front much the longest, margin concave. Pectorals small, with 9–12 rays. Caudal peduncle very slender; tail moderately forked. Scales small and thin, easily dislodged, covering entire trunk but absent from head. Gill rakers (18–23) long and slender; vertebrae 52–55. Freshly caught fish have a distinct cucumber-like odour.
Sexual dimorphism Like other smelts, males at maturity have much enlarged fins, especially pectoral and pelvic fins, and have more abundant and more widely distributed nuptial tubercles (Fig. 2.16).
Colour Bright silver, especially on sides, with back somewhat darkened to an amber/olive colour, belly silvery white; top of head somewhat dusky and fins more or less colourless, perhaps with a few small melanophores along the fin rays.
Size Known to grow to 67 mm, but mostly smaller, 55–60 mm.
Conservation status Little has been reported, and status is really not known; may have suffered with other species in the 1940s and later decline in the Tasmanian whitebait fishery, though there is no documentation.
Distribution Widespread in Tasmania at low elevations and close to coast, especially in the north and south-west, but also in eastern and western drainages.
Natural history Essentially unstudied and little

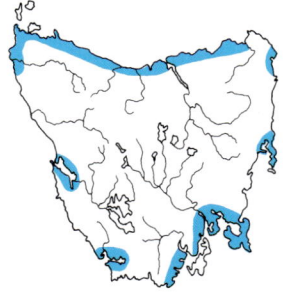

known. Is probably anadromous and enters coastal rivers during spring among shoals of Tasmanian whitebait (*Lovettia sealii* and various *Galaxias* species). These may be of two year classes, the older ones mature and approaching spawning. Spawning site undescribed but is probably in the lower reaches of rivers, with eggs demersal and developing on the sandy substrates of these rivers. Eggs tiny, less than 1 mm in diameter. The larvae probably hatch and go to sea, where most growth takes place.

Smelt are predators, feeding on diverse, small, aquatic insects and crustaceans—mosquito, beetle and caddis larvae, amphipods and the like.
Utility Tasmanian whitebait was primarily *Lovettia sealii* (see p. 78), but a few smelt occurred among the shoals of fish sought by fishers. Is probably a significant food for large, predatory fishes in lowland rivers, especially sea-migratory and estuarine brown trout introduced into Tasmanian rivers.

Like all retropinnid smelts, is very fragile and needs very careful handling for successful transfer to captivity; it is not known as a significant aquarium species.
Other names Common: none. Scientific: none.
Literature Blackburn, 1950; Fulton, 1990; Lynch, 1969; McDowall, 1979.

14
Family Prototroctidae
Southern graylings

R.M. McDOWALL

The family Prototroctidae is small, with just two species, one in south-eastern Australia and the other in New Zealand, though the latter is considered extinct. They are closely related to the southern smelts (family Retropinnidae, see Chapter 13), the two southern families having affinities with the northern temperate smelts (family Osmeridae). The common name 'grayling' is rather a misnomer, since the true graylings (genus *Thymallus*) are salmonids, though still fairly closely related to these southern families. The family is characterised by having a dorsal fin high on the back, a little behind the level of the pelvic fins, a small adipose fin, thin scales, no lateral line, a distinctive horny sheath forming the margins of the lower jaw, and two complete loops in the intestine. The southern graylings have a strong cucumber-like odour when first taken from the water.

AUSTRALIAN GRAYLING *Prototroctes maraena* Günther

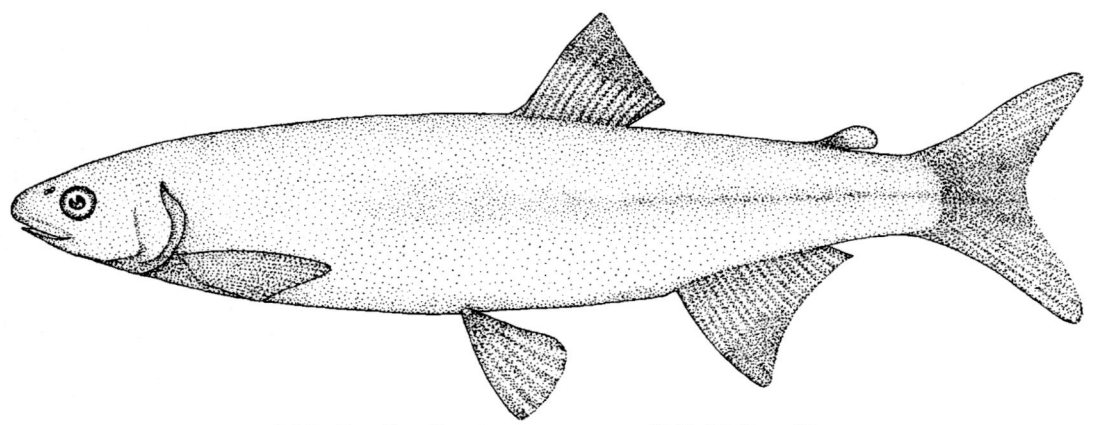

14.1 Grayling *Prototroctes maraena*. (R.M. McDowall)

Description A smallish species that is slender and compressed, with a small head, large eyes and a somewhat bluntly pointed snout. A small, short-based dorsal fin (9–13 rays) is high on back and originates just behind level of pelvic fins. It is followed by a small adipose fin. The modest tail is forked. Anal fin longer and lower than dorsal (16–20 rays, those towards the rear much shorter than those anteriorly). Along belly in front of vent is a weak, horny, abdominal keel. Teeth in upper jaw form a very uniform, comb-like row and are bluntly pointed. They bite onto an unusual horny shelf that surrounds lower jaw. Lower jaw quite sharply pointed, shorter than upper, and its teeth are few, well spaced, and sit well within upper jaw when mouth is closed. Cycloid scales thin, of moderate size (68–84 along side) and easily dislodged. Vertebrae 62–65; gill rakers 20–26, of moderate length.

Colour Probably variable with degree of maturity, from silvery, with back olive grey and belly whitish, through olive green on back to brownish on back, with a darker midlateral streak and belly a silvery yellow, fins greyish.

Size Is reported to have grown to over 300 mm, but mostly smaller, commonly 170–180 mm; the largest recent record is of 253 mm. Small grayling are less often reported, though confusion with smelt (*Retropinna* spp.) is a possible reason.

Distribution Known in streams and rivers draining to the sea from eastern and southern flanks of the Great Dividing Range, from as far north as the Grose River in central New South Wales, south and west to the Hopkins River in western Victoria; probably penetrates inland to elevations up to 1000 m, having been recorded in tributaries of the Snowy River draining the slopes of Mount Kosciusko. Grayling are absent from the inland Murray-Darling system. Occurs widely in Tasmania, particularly in northern and eastern rivers, and occasional western rivers, again at lower elevations. There is also a record from King Island, lying between Victoria and north-western Tasmania.

Conservation status For several decades was described as one of Australia's rarest and most vulnerable fishes. This was probably mostly a result of the lack of study and concern that the Australian species might rapidly follow its New Zealand counterpart into extinction—an extinction that has never been satisfactorily understood or explained. Has been totally protected in Tasmania for many years, although this protection is not afforded in mainland Australia. In the last 10–20 years has proved to be rather more common and widespread than hitherto believed, with more than 300 taken below a newly constructed dam in the Shoalhaven River central NSW in November 1976 and more than 1000 taken from the Tambo River in eastern Victoria during a 10-month study there. However, though the species appears considerably more common than was earlier believed, it is still listed as 'potentially threatened' by the Australian Society for Fish Biology. Its need to migrate to and from the sea to complete its life cycle makes it vulnerable to depletion as a result of barriers to upstream and downstream migration.

Natural history Knowledge has increased significantly in the past decade or two as a result of some intensive studies.

Occurs most commonly in clear, gravelly streams with a moderate flow, and from estuarine reaches substantial distances inland, as long as

Grayling *Prototroctes maraena*. (R.H. Kuiter)

FAMILY PROTOTROCTIDAE

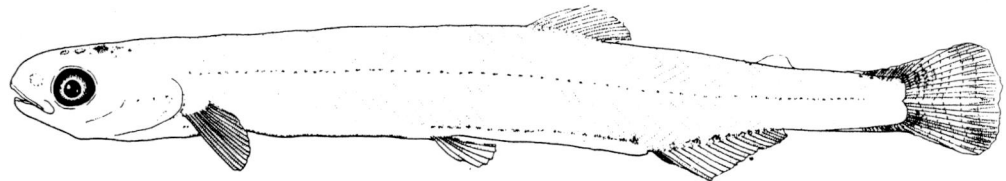

14.2 Migratory juvenile of grayling *Prototroctes maraena*. (B. Washington)

impenetrable barriers to upstream movement do not interfere with migration. It occurs in fast-moving shoals and seems a shy fish that flees when disturbed; lives for up to 3 years, with most reaching 2. Spawning takes place during autumn, probably during April and May. Males may spawn at age 1, ripe males developing nuptial tubercles on the scales and along the fin rays; these are small horny nodules that make surface of body rough and probably have functions relating to spawning behaviour; they disappear after spawning. Females do not mature until age 2. Both sexes have only the left gonad. Eggs are small, less than 1 mm in diameter in ripe unspawned eggs and 1.1 mm after fertilisation. They are very numerous, with 25,000–67,000 eggs in females 170–200 mm long. The eggs are amber-coloured, and demersal, probably shed and settle among gravel of stream bed. The spawning site and behaviour are not known. The eggs take 12 days to hatch at about 16°C, but probably longer at lower temperatures. The larvae are about 4.5 mm long and are positively phototropic; this means that they will actively swim towards surface of water, which will result in them being swept downstream and to sea. Larval life is marine, and juveniles return to rivers from the sea during spring, almost certainly at about 6 months of age. The rest of their life is spent in rivers, the fish reaching about 80 mm at age 1, 150 mm at age 2, and 190 mm at age 3; a few may live longer than 3, fish of maximum size perhaps being 5.

Is an opportunistic omnivore, feeding on a mixed diet of aquatic algae and insects, including insects found among the algae as well as some other aquatic and terrestrial insects. It has a double loop in the intestine, an unusual feature in salmoniform fishes, and this adaptation is certainly related to the need for a longer intestine to facilitate digestion of algae.

Utility Was once a popular angling fish, especially before its numbers declined, its distribution became more restricted and trout became widespread in southern Australia and Tasmania. Has sweet and fine white flesh but is quite small and has numerous bones. Is no longer sufficiently abundant to attract attention from anglers.
Is sensitive to handling, but if introduced into captivity with care, can be kept in larger aquaria.

Similar species Whitebait juveniles and young of smelt resemble juvenile grayling entering rivers from the sea; adult smelt and various atherinids, and small trouts and mullet, may be confused with small grayling, while larger trout and mullet may be confused with larger adult grayling.

Other names Common: cucumber mullet, cucumber herring or Yarra herring. Scientific: none.

Literature Allport, 1870; Bacher & O'Brien, 1989; Bell *et al.*, 1980; Berra, 1982; Berra & Cadwallader, 1983; Bishop & Bell, 1978a; Jackson, 1976; Koehn & O'Connor, 1990; McDowall, 1974, 1976; Saville-Kent, 1886; Stead, 1903.

15
Family Cyprinidae

Carps, minnows, etc.

A.R. BRUMLEY

The family Cyprinidae is one of the largest of all fish families, with more than 1700 species found in temperate and tropical waters of Europe, Africa, Asia and North America. Carps of various sorts are the best known, but the family also includes many other well-known fish types, including minnows, daces, bitterlings, danios and rasboras, many of which are important to the aquarium trade. Generally, they are small fishes, most characterised by the absence of teeth in the jaws or palate (other fish may also lack jaw teeth), a protrusible upper jaw, and sometimes barbels. Australia has no indigenous cyprinids, but four species are established as self-maintaining populations in inland rivers and associated water bodies, mostly in the south-east. Carp and goldfish are the most widely recognised, though tench and roach are also present. Carp may be the most common fish in the waters of south-eastern Australia and often has the greatest abundance.

Key to cyprinids

1. Barbels present at corners of mouth 2
   No barbels at corners of mouth 3

2. Two barbels at each corner of mouth; scales large, less than 40 along lateral line .... ***Cyprinus carpio*** p. 101; Fig. 15.3
   One barbel at each corner of mouth; scales very small, about 100 along lateral line .... ***Tinca tinca*** p. 104; Fig. 15.4

3. About 26–34 scales along lateral line; dorsal fin III–IV, 14–20; gill rakers long and slender .... ***Carassius auratus*** p. 99; Fig. 15.1
   About 40–42 scales along lateral line; dorsal fin III, 9–11; gill rakers short and stout .... ***Rutilus rutilus*** p. 105; Fig. 15.5

GOLDFISH
Carassius auratus Linnaeus

Description A small, plump, deep-bodied and moderately compressed fish, with a large, blunt head, moderately large eyes and a small, toothless, protrusible mouth; pharyngeal teeth not molar-like, arrangement 4:4. Dorsal fin (III–IV, 14–20), long-based and high at front, the last of the several spines at front of fin strongly serrate on hind edge (Fig. 15.2). Anal fin small (II–III, 5–7), below rear of dorsal fin. Tail moderately forked. Pelvic fins abdominal, 7 rays. Pectorals low behind head, with 16–18 rays. Scales large and cycloid (26–34 along lateral line), covering trunk but absent from head; many (40–46) long gill rakers. Vertebrae 27–28.

Colour In the wild, usually olive bronze to deep gold, darker and brownish on back, paling on sides, and silvery white on belly. Fins dark olive bronze. The well-known orange-red or pearly 'fancy' varieties of the aquarium trade are occasionally seen in wild populations.

Size Commonly grows to 100–200 mm but may reach 400 mm and 1 kg; goldfish this big typically look very obese.

FAMILY CYPRINIDAE

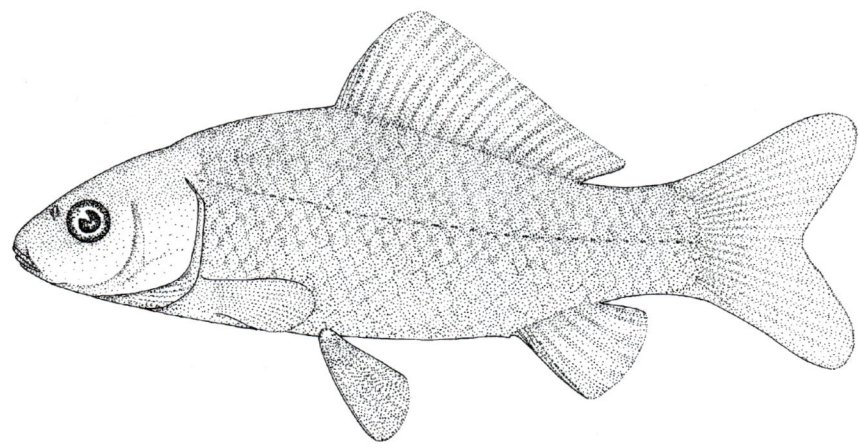

15.1 Goldfish *Carassius auratus*. (R.M. McDowall)

Introduction and distribution Is native to eastern Asia, but now has almost worldwide range because of its use as a captive ornamental fish, throughout eastern Asia and the western world, for many centuries.

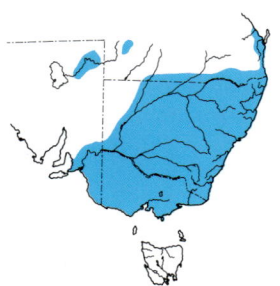

Was imported into Australia from the 1860s as an ornamental fish and is now widespread and locally present throughout New South Wales, Victoria and South Australia, in coastal streams as well as the inland Murray-Darling system and Cooper Creek; also present in Western Australia, making it perhaps the most widely distributed fish in Australian fresh waters.

Natural history Is a species of still and sluggish waters, where it is well able to survive at relatively high temperatures and low oxygen concentrations. Spawns during summer. Goldfish in Victoria mature early, with ripe/spent fish as small as 100–150 mm. Males develop tubercles on head and body (sometimes called 'pearl organs'); ripe females becoming very fat, their abdomens distended with up to several hundred thousand eggs in very large goldfish. Overseas studies suggest that females release chemicals that attract males. The eggs are small (1 mm in diameter) and are laid among aquatic plants; they hatch in about a week. The young attach themselves to aquatic plants for a few days until the yolk sac is

Goldfish *Carassius auratus*. (R.M. McDowall)

CARPS, MINNOWS, ETC

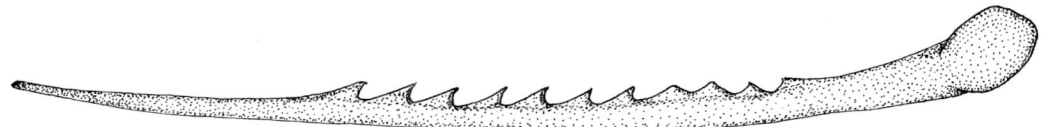

15.2 Serrate spine from dorsal fin of goldfish *Carassius auratus*. (R.M. McDowall)

completely absorbed. Goldfish feed on a wide variety of plant materials, organic detritus and small aquatic insects and other animals. Studies in South Australian and Victorian waters showed that small crustaceans were the dominant food, with plant material consumed by smaller goldfish, and by larger ones during the winter.

Utility Has high value as an aquarium and pond fish, being very widely known because of the ease with which it can be kept in captivity, also because of bizarre fin and body shapes, protruding eyes and highly variable and bright colours that can occur in carefully bred aquarium stocks. Is highly compatible with other fish species in captivity. Is widely used as a test species in physiological and toxicity testing.

Is used as a bait for catching Murray cod. However, use as bait should not be encouraged as it may lead to wider distribution of this exotic fish, which may have environmental impacts, though this question is unstudied.

Similar species Resembles common carp and possibly young of some inland percichthyids such as Murray cod, Macquarie perch and silver perch, etc.

Other names Common: carp, crucian carp, Prussian carp. Scientific: *Carassius carassius*, a polyploid form of *Carassius*, is sometimes listed from Australian waters, though its presence there has never been authenticated.

Literature Baker, 1933; Banarescu & Coad, 1991; Brumley, 1991; Howes, 1991; Hume *et al.*, 1983b; Lake, 1967a; McKay, 1984; Tilzey, 1980.

CARP *Cyprinus carpio* Linnaeus

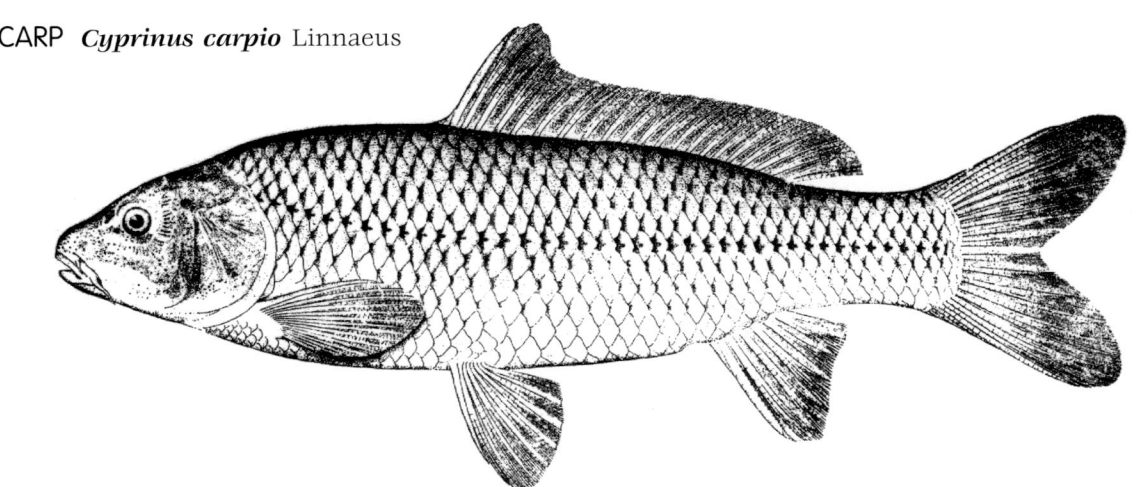

15.3 Carp *Cyprinus carpio*.

Description An elongated fish with an arched back but not deep belly. Head of moderate size, triangular, with a long, blunt snout; eyes small. Mouth small to moderate, lips thick, two barbels at each corner of mouth, posterior pair largest; pharyngeal teeth molar-like, arrangement 1,1,3: 3,1,1. Dorsal fin origin in front of mid body length, above abdominal pelvic fins, long-based and rather low (III–IV, 15–24); anal fin small (III, 5); spines in both fins strongly serrate; pectoral fin I, 15–16. Tail forked. Scales large and thick, cycloid (33–40 along lateral line); some variants, known as mirror carp, are only partly scaled, with very large scales in scattered patches or lines.

Colour Variable, usually olive green to silvery grey on back, paling to silvery yellow on belly; juveniles silvery grey. Fins all dark, brownish olive, often with red colouration on extremities of anal, lower tail, and sometimes other fins. Koi carp, with orange, yellow, white and black markings

FAMILY CYPRINIDAE

Carp *Cyprinus carpio*. (R.H. Kuiter)

and blotches, are an ornamental strain of carp.
Size Can attain a very large size, with reports of carp up to 1200 mm long and weighing 60 kg. Reaches 10 kg in south-eastern Australia, but 4–5 kg is more usual.

Introduction and distribution Is a native of Asia, from where it was spread by humans through Europe, and is now established on all continents except Antarctica; it can be considered the world's most widely distributed freshwater fish.

Three strains of carp have been introduced into Australia; an ornamental strain was released near Sydney in 1850–60 and a Singaporean strain of koi was accidentally released in the Murrumbidgee area in 1876. A third hybrid 'Boolara' strain was imported for aquaculture in Victoria in 1961, and though this was banned, some fish escaped into Lake Hawthorn, near Mildura, in 1964. These bred, spread up the Murray and Darling Rivers and interbred with koi in the Murrumbidgee to produce a stock of carp with broad genetic make-up. They dispersed rapidly, their spread being hastened by floods, especially in the mid-1970s. As a result, the carp now occurs throughout the large inland Murray-Darling system, including all its impoundments. Spread into Queensland has resulted in large populations there. It also occurs in the Shoalhaven River in coastal New South Wales as well as in the Gippsland Lakes, where it grows well in brackish conditions. Fears have been expressed that carp might spread along the coast, but it cannot tolerate sea waters. Carp have recently been discovered in Lake Crescent and several other waterways in the vicinity, in Tasmania. How they got there is not known, but they are a matter of considerable concern, and intensive studies are underway to determine whether extermination or containment is possible.

Natural history Is usually found in still or gently flowing waters, especially where aquatic vegetation is prolific; has great tolerance of low oxygen levels, which enables it to live in stagnant waters from which other fish are excluded.

Carp in Victorian waters mature early, males at 1 year, and females at around 2 years and only

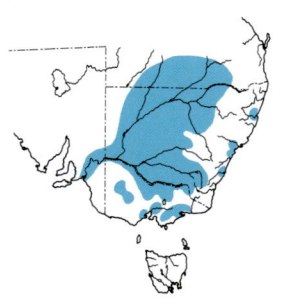

125–150 mm length. Fecundity is very high, a female over 6 kg having up to 1.5 million tiny eggs. Spawning season depends on water temperature, occurring in spring (September–December) at about 17–25°C. Eggs are deposited during several days on any fibrous plant matter in shallow water. They hatch after only a few days and the young grow rapidly in warm waters with prolific plankton; they can reach 180 mm in one year and 500–700 mm in two years.

Diet is highly varied and includes diverse aquatic animals such as molluscs, crustaceans, insect larvae and seeds. Food is taken in by suction, filtered out by the long, fine gill rakers and crushed and shredded by the molar-like pharyngeal teeth. When food in the water column is unavailable, especially during winter, soft plant matter and detritus are sucked from the substrate. This method of feeding leads the carp to be accused of generating high turbidity in its habitat and affecting native animals and plants. Studies in Victoria suggest that this is not always the case. In high densities, carp may damage soft-leaved plants; this is likely to occur in shallow wetlands. Changes in such habitats in New South Wales have occurred and it is possible that carp are responsible.

Utility Has been used in aquaculture almost throughout human history, being cultured in China at least from 475 BC. In Japan, koi are symbolic of courage and energy. Is important economically in many parts of the world and continues to be important in captive fish production because of its potentially rapid growth in fertile waters, its diverse and cheap diet and its ability to tolerate adverse water conditions. This is especially true in Europe and Asia. Is prized by anglers in many countries. However, is regarded as a pest in North America, Australia and New Zealand. It is a noxious fish in Victoria, and keeping carp alive is illegal. However, in spite of this, interest in angling for carp is growing in Australia, where they can be taken with bait and lures at all times of the year.

Commercial harvesting of carp occurs in Victoria, with over 1100 tonnes taken during 1991–92. Larger carp are eaten by people, though flesh quality depends on preparation and cooking. A limited market exists for small carp that are frozen into blocks and distributed for crayfish bait. Despite some removal of carp, their numbers and distribution continue to increase, and there is concern among some anglers, commercial fishers, irrigators, conservationists and ecologists about ways of controlling the species. As a benthic feeder, carp may be responsible for resuspension of sediments and nutrients, and this may be contributing to blue-green (cyanobacterial) algal blooms in south-eastern Australia. New methods for the control of carp are needed.

Carp *Cyprinus carpio*, head showing barbels. (R.H.Kuiter)

Similar species Goldfish. Hybrids of carp and goldfish are common in Victorian waters, though are not always recognised. They can be distinguished by the number of scales in lateral line (29–34), number of dorsal fin rays (17–20) and number of gill rakers on lower gill arch (17–22); barbels are often reduced in size. Pharyngeal teeth show an arrangement that is intermediate (either 1,4:4,1 or 2,4:4,2). Hybrids of goldfish and koi have been described from New South Wales. Though fish hybrids are often sterile, ripe and spent hybrids of both sexes have been observed in a Victorian lake, indicating that they are fertile and have spawned. Stomach analyses of carp, goldfish and hybrids from the same lake showed that the diet of hybrids was intermediate between that of carp and goldfish.

Other names Common: European carp, common carp, koi. Scientific: none.

Literature Banarescu & Coad, 1991; Brumley, 1991; Brumley *et al.*, 1987; Gehrke & Harris, 1994; Howes, 1991; Hume *et al.*, 1983a, b; Lake, 1967a, b; Mulley & Shearer, 1980; Persson, 1991; Sibbing, 1991; Weatherley & Lake, 1967.

FAMILY CYPRINIDAE

TENCH *Tinca tinca* Linnaeus

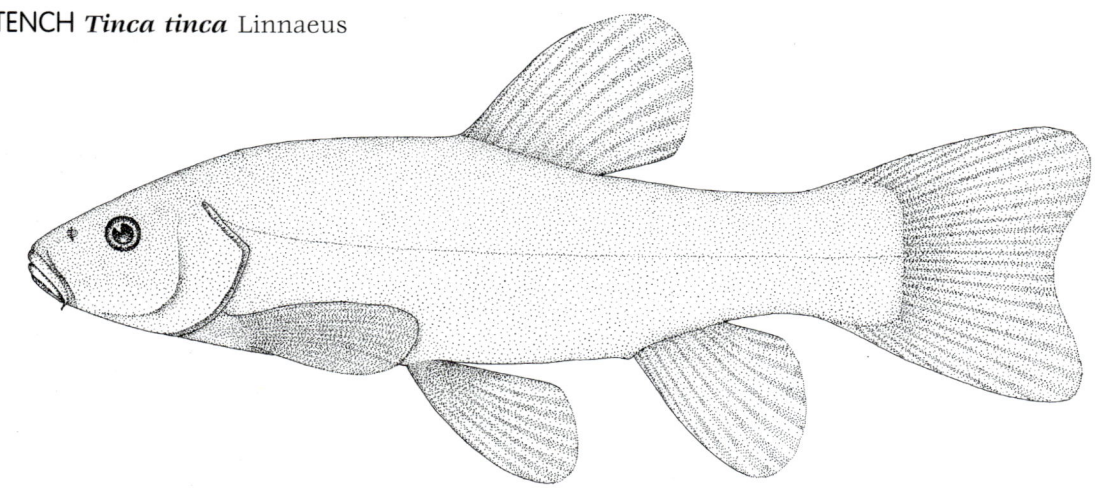

15.4 Tench *Tinca tinca*. (R.M. McDowall)

Description A moderately thick-set fish with a long, blunt snout and small eyes; a small barbel occurs at each corner of mouth; a single dorsal fin (III, 8), high on back, short-based, high and rounded; anal fin (III, 6–8) similar in shape, but smaller; 17–19 pectoral rays, leading ray spinous, broad and flattened. Scales very small, 91–105 along lateral line; 37–39 vertebrae; 12–15 long, stout gill rakers.

Colour Varies from dark olive to pale golden tan or silvery; mostly greenish olive, with greenish gold iridescence on cheeks, gill covers and sides of body. Fins grey or sometimes pinkish; eyes a distinctive brick-orange.

Size Mostly 100–300 mm but recorded up to 700 mm; a weight of nearly 9 kg reported in New South Wales.

Introduction and distribution A native of Europe. Was introduced to Australia in 1876 and occurs widely through the Murray-Darling system

Tench *Tinca tinca*. (R.H. Kuiter)

CARPS, MINNOWS, ETC

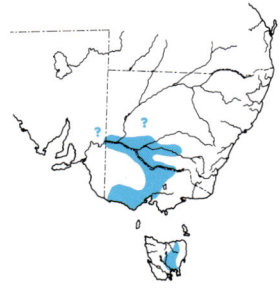

as well as in coastal rivers, in Victoria and South Australia; is occasional in southern New South Wales. Also present in Tasmania, particularly though not exclusively in the Derwent River basin.

Natural history A rather secretive fish that typically inhabits sluggish or still waters, often occurring among weed or in deep sheltered holes. Is tolerant of low oxygen levels. In Europe, spawns in spring and summer at temperatures of 15–22°C and can produce 3–4 batches of tiny, yellowish eggs at intervals of about 14 days at warm temperatures. These are laid over immersed vegetation and are apparently poisonous to predators. Diet is mainly aquatic insects, crustaceans and to a lesser extent molluscs. The small eyes and sensory barbels suggest that food is found by feel and taste rather than sight.

Utility Was introduced to Australia by acclimatisation societies as an angling species and is highly favoured by so-called 'coarse fish' anglers, especially in Tasmania. Is said to have good-eating white flesh, though coarse fishers tend to hold their catches in 'keep nets' and liberate them at the end of a day's fishing. The tench was once commercially fished in the Murray but declined in abundance when carp became dominant in the 1970s. Although used in aquaculture in Europe and Africa and reared for stocking (in Belgium), this is not true in Australia.

Similar species Goldfish and common carp.
Other names Common: none. Scientific: none.
Literature Banarescu & Coad, 1991; Brumley, 1991; Howes, 1991; Lake, 1967a; Mills, 1991; Weatherley, 1959, 1962; Weatherley & Lake, 1967.

ROACH *Rutilus rutilus* (Linnaeus)

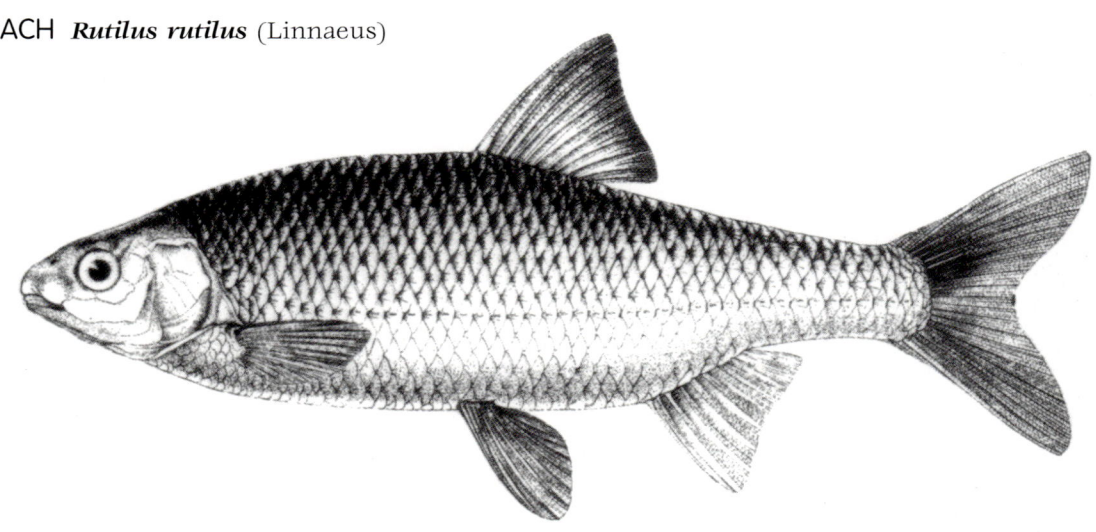

15.5 Roach *Rutilus rutilus*. (J.A. Langhorne)

Description Relatively deep-bodied, with a high arched back and small head and mouth. Dorsal fin short-based (III, 9–11), high on back above pelvic fins; anal fin (III, 9–11) similar in shape. Trunk tapers to a relatively slender caudal peduncle with a distinctly forked tail. Small axillary processes at bases of pelvic fins. Scales of moderate size (40–45 along lateral line) and firmly attached. Lateral line curves distinctly downwards, following belly profile. No barbels. Gill rakers short and stout.

Colour A handsome fish, back olive green, sides silver and belly silvery white. Dorsal fin and tail greyish brown, other fins orange to orange red; eyes bright red.
Size Is known to reach 450 mm, but more usually 150–200 mm; little is known of size reached in Australia.
Introduction and distribution Was introduced from Europe between 1860 and 1880 for angling and liberated into selected streams in southern Victoria, including the Yarra, where it is still

FAMILY CYPRINIDAE

Roach *Rutilus rutilus*. (R.H. Kuiter)

found. Range does not appear to be expanding. Is a native of Europe, where it is very widespread.
Natural history Is virtually unstudied in Australia. Is a day-active fish that swims in shoals in ponds, lakes and slow-flowing rivers, especially where aquatic plants are prolific. Spawning occurs among vegetation during increasing day length in spring when water temperature reaches about 15°C. Males develop nuptial tubercles on the head, fins and scales; eggs are small (to about 1.5 mm) and adhere to plants among which they are laid. Larvae hatch in 2–3 weeks and attach themselves to plants for a few days before becoming free-swimming. Feeds on a wide variety of foods, mainly aquatic invertebrates, supplemented by higher plants, which are crushed and shredded with the pharyngeal teeth.
Utility Is a popular 'coarse fish' for anglers in Europe, but has little importance in Australia, where it is little-known and seldom fished for. Is a handsome fish in captivity and easily kept.
Similar species None.
Other names Common: none. Scientific: none.
Literature Banarescu & Coad, 1991; Brumley, 1991; Howes, 1991; McKay, 1984; Wheeler, 1969.

ANOTHER CYPRINID SPECIES

A further cyprinid species, the rosy barb, *Puntius conchonius*, was reported in the literature as having self-maintaining populations in the wild. It was apparently accidentally released from aquarium use into Norman Creek, a small urban stream in Brisbane. Populations became established between 1970 and 1972 and a reproducing population persisted until 1984. The fish remained in large open pools and did not spread along the stream. Successive dry years diminished the pools, littoral aquatic plants were replaced with grasses, and the rosy barb population apparently died out. These events are important in recognising the potential of aquarium fish to be released and become established in natural waters. Exotic fish continue to be imported, most requiring relatively warm waters. As a result the threats from aquarium escapes or releases are most pertinent in northern Australian waters. However, the principle remains that aquarium species may pose a threat to native ecosystems.
Literature Brumley, 1991; McKay, 1984.

16
Family Ariidae
Salmon or fork-tailed catfishes

D.A. POLLARD AND M.A. RIMMER

Members of this large family of catfishes are distributed worldwide in tropical and sub-tropical waters. Though regarded primarily as a marine and estuarine group, many of the species in the Australo-Papuan region are found in freshwater habitats. Until recently the taxonomy of these Australo-Papuan ariids has been confused. Most species in this family are very similar in appearance, and identification is difficult, but only one species is found in the fresh waters of south-eastern Australia. The family is characterised by three pairs of barbels around the mouth, dorsal and pectoral fins with serrated pungent spines, an adipose dorsal fin and a forked tail. All known members of the family are mouth-brooders.

BLUE CATFISH *Arius graeffei* Kner & Steindachner

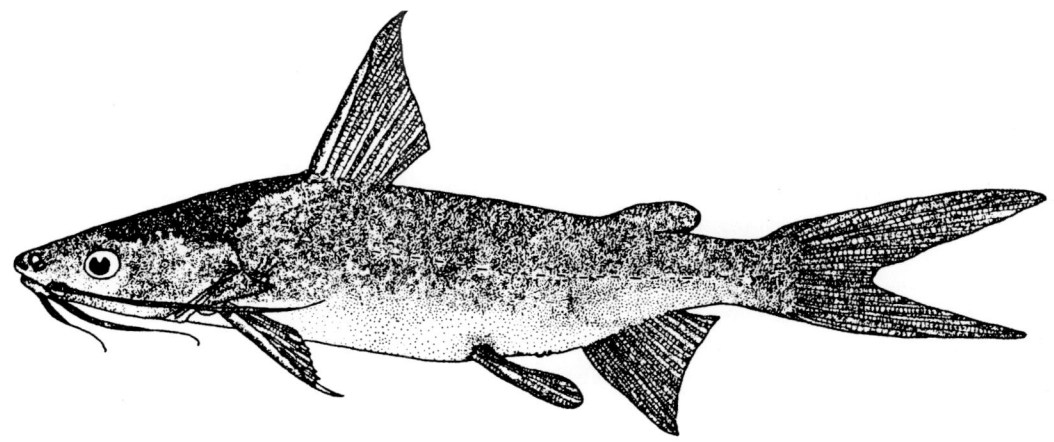

16.1 Blue catfish *Arius graeffei*

Description Body robust, with a broad, flattened head, deepest just behind head and tapering towards tail. Skin tough and lacking scales, but head armoured with bony shields dorsally. Mouth wide, with 6 long slender barbels, 1 at each upper corner and 4 on lower lip. Spiny dorsal fin (I, 7) placed high on back and a well-developed adipose fin above anal fin (15–19). Stout spines occur at front of pectoral fins (I, 10–11); tail deeply forked.
Sexual dimorphism Pelvic fins of adult females are longer and more rounded than those of adult males and have a hook-like thickening (termed a 'clasper') on the dorsal surface.
Colour Dark grey to dusky blue on back, paling to silvery white below; lacks distinctive markings. Piebald specimens have been recorded from northern Australia.
Size Grows to over 500 mm in length and 3 kg in weight.
Distribution Found in coastal drainages of northern Australia, from northern New South Wales through Queensland and the Northern

FAMILY ARIIDAE

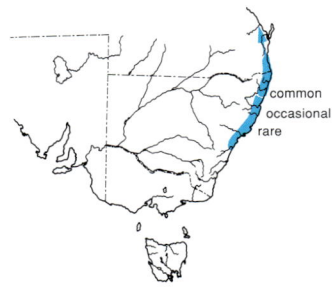

Territory to the far west of Western Australia; also in southern Papua New Guinea. In New South Wales this species is abundant only in the Richmond and Clarence Rivers, although it has been recorded as far south as Port Jackson.

Conservation status Remains abundant throughout its broad range.

Natural history Breeds from early November to early December in northern New South Wales. Both sexes reach sexual maturity at 250–350 mm in length and 400–1000 g in weight. Can complete its life cycle in fresh water, but it is not known whether it can also do so in estuarine or marine habitats. Populations living in estuarine to marine waters are reported to undertake extensive anadromous migrations associated with breeding. Female blue catfish spawn a small number (40–120) of large (11–14 mm diameter) yellow eggs and thousands of small (< 3 mm diameter) paler eggs. The large eggs are fertilised at spawning, the small eggs are infertile. The functional significance of these small eggs, which are spawned in an adhesive mass together with the large eggs, is unknown. Claspers on the pelvic fins of females increase in size as the breeding season approaches, suggesting that they may be used during spawning to channel or hold the egg mass.

The developing eggs and larvae are incubated orally by the male. Brooding males may carry up to 123 eggs and are readily identified by their distended branchial region. The oral epithelium of brooding males thickens to cover the palatine tooth patches, reducing the chance of palatine teeth abrading and damaging eggs or larvae in the mouth. Brooding lasts 6–8 weeks, and brooding males do not feed during this time.

Larvae hatch 4–5 weeks after fertilisation, and hatching occurs over 5–7 days. Yolk-sac larvae begin feeding soon after hatching, presumably ingesting plankton entering the male's mouth. A wide range of organisms is eaten, including aquatic microcrustaceans, insect larvae and filamentous algae. Yolk absorption is completed before release of juveniles from the male's mouth. Juveniles grow to 50–60 mm in total length and about 1.6 g in weight before being released.

Adults eat a wide range of prey, most of the diet being made up of macrocrustaceans, usually prawns, and fish; like other ariids, it will also scavenge dead fish and other animals.

Utility Large specimens provide reasonable sport on rod and line. The flesh is edible, despite anecdotal claims that catfish in general are 'poisonous'. Ariids are important food fishes in South-East Asia, and some of the larger Australian species (*Arius midgleyi*, *A. leptaspis* and *A. thalassinus*) are marketed in limited quantities in Australia as 'silver cobbler' and 'golden cobbler'. Some *A. graeffei* are also marketed from rivers in northern New South Wales.

Ariids should be handled with care, because the serrated dorsal and pectoral spines can inflict painful wounds. Juveniles make interesting and hardy aquarium fishes.

Similar species Unlikely to be confused with other fishes except the much larger tropical marine and estuarine giant salmon catfish *Arius thalassinus*, which is occasionally found in estuaries in northern New South Wales.

Other names Common: freshwater fork-tailed catfish. Scientific: *Neoarius australis*, *Arius australis*. This species has previously been misidentified as *Arius* (*Hexanematichthys*) *leptaspis* (e.g. in the previous edition of this book), but the latter is now recognised as a distinct species.

Literature Allen, 1989; Cribb, 1988; Grant, 1991; Kailola, 1983; Lake, 1978; Lake & Midgley, 1970; Merrick & Schmida, 1984; Rimmer, 1985a, b, c; Rimmer & Merrick, 1983; Rimmer & Midgley, 1985; Stead, 1934; Sumpton & Greenwood, 1990; Whitley, 1957.

17
Family Plotosidae
Eel-tailed catfishes

D.A. POLLARD, T.L.O. DAVIS AND L.C. LLEWELLYN

Members of this family live in both salt and fresh waters in the tropics and subtropics of the Indo-West Pacific region. Many, particularly in Australia and Papua New Guinea, are adapted to living permanently in freshwater habitats. These freshwater species tend to be very similar to one another and their taxonomy remains confused. However, they all have a tapering and rounded tail, unlike the forked tail of the related salmon catfishes (family Ariidae).

Key to eel-tailed catfishes

1. Second dorsal fin long, extending well forwards on back, almost as far as anal fin; 1st dorsal fin with 1 spine and 6 rays . . . . ***Tandanus tandanus*** p. 109; Fig. 17.1
   Second dorsal fin short, extending forwards on back nowhere near as far as anal fin 2

2. First dorsal fin with one spine and four rays . . . . ***Neosilurus argenteus*** p. 112; Fig. 17.2
   First dorsal fin with one spine and five or six rays . . . . ***Neosilurus hyrtlii*** p. 113; Fig. 17.3

FRESHWATER CATFISH *Tandanus tandanus* Mitchell

17.1 Freshwater catfish *Tandanus tandanus*. (F. Olsen)

Description A stout, robust fish with posterior of body compressed; head large and flattened below, with eyes of moderate size. The down-turned mouth has thick, fleshy lips, with tubular nostrils in front border of upper lip, pointing forwards. There are four pairs of barbels around mouth. First dorsal fin (I, 6) short-based, high on back, just behind head and preceded by a strong, serrated spine. Second dorsal fin originates just behind origin of anal fin and is continuous with tail and anal fin (total rays 140–150). Pectoral fins (I, 10), located low on body, also have strong, serrated spines. Skin smooth and tough, with no scales.
Sexual dimorphism At around 1 year of age,

FAMILY PLOTOSIDAE

Freshwater catfish *Tandanus tandanus*, large adult. (R.H. Kuiter)

the sexes can be distinguished by shape of urino-genital papilla, which is triangular in females and cylindrical in males.

Colour When less than 150 mm long, freshwater catfish are grey or brown dorsally and laterally, usually mottled with dark brown to black blotches, fading to whitish below. This mottling usually disappears in larger fish, which vary from olive green to brown, reddish brown or even purplish in colour above, fading to whitish below.

Size Has been recorded up to 900 mm and 7 kg; however, fish over 2 kg are not common.

Freshwater catfish *Tandanus tandanus*, small juvenile. (R.M. McDowall)

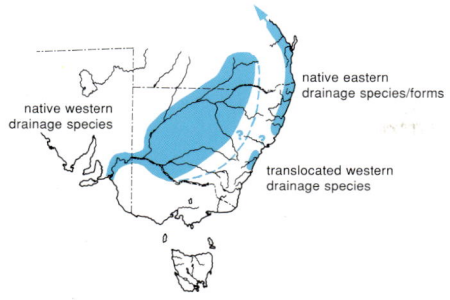

Distribution Widespread throughout the Murray-Darling system in inland Queensland, New South Wales, Victoria and South Australia, but not upstream of Wagga Wagga in the Murrumbidgee River or Mulwala in the Murray River, probably owing to cool water discharges from large dams; also present in many coastal rivers of eastern Australia from around Sydney (see below) to north of Cairns.

Conservation status Remains widely abundant, though becoming scarcer in some of its previous inland, riverine habitats.

Natural history Usual habitat is near the bottoms of lakes and slow-flowing rivers. May reach maturity at 3 years, and all fish mature by 5 years. Spawns when temperatures rise during spring and summer. Spawning is not stimulated by flooding. The number of eggs increases with size, ranging from about 2800 to 20,600 eggs in females between 390 and 530 mm long. They build a circular to oval nest, generally around 0.6–2.0 m in diameter, from pebbles and gravel. During courtship, both male and female circle and weave above the nest. The female then arches her body, agitates her pelvic fins and releases eggs about 30 cm above nest. They are spherical, non-adhesive and a light greenish yellow in colour. Male fertilises eggs, which sink to the bottom, settling in gravel of nest. One of the adults, usually the male, remains at nest until the eggs hatch in about 7 days. The larvae are about 7 mm long when they hatch, and barbels appear on the young after about 3 days. They grow quickly, reaching 90 mm by their first winter and 500 mm in their sixth.

Young catfish feed on zooplankton and small insects (especially chironomid midge larvae); catfish longer than about 100 mm also eat small fish such as western carp gudgeon; adults eat shrimps and crayfish in summer and midge larvae in winter. They also eat molluscs, and dragonfly, caddis and mayfly larvae, as well as terrestrial invertebrates such as beetles, earthworms and termites exposed during flooding. They are thus mainly carnivorous bottom-feeders.

Movements of catfish in rivers are generally limited, and most remain in the one locality. They do not migrate for spawning, unlike some other inland species, such as golden perch and Murray cod.

Utility Provide good sport when hooked on light gear and are a popular inland angling species. The flesh is of excellent quality, contrary to widespread prejudice against 'catfishes' in general. They breed and grow well in ponds and farm dams and can be stocked with silver perch. They show potential for artificial propagation and fish farming, reaching an acceptable table size in 2-3 years; do not survive well below 4°C.

Similar species Populations in the Hawkesbury and Hunter Rivers north of Sydney are, as far as is known, derived from stocks translocated from westward-flowing tributaries of the Murray-Darling system. Those in coastal rivers from the Manning River northwards, however, are thought to be native to these eastward-flowing drainages. Recent electrophoretic studies of proteins of some more northerly populations indicate that catfish inhabiting coastal rivers from the Manning to the Bellinger Rivers may be a different species and those from the Nymboida River (Clarence River system) northwards to at least the Tweed River a subspecies, different from the widespread inland form (D. Jerry, K. Bishop and R. Watts, personal communication). In south-eastern Australia, species of the related genus *Neosilurus* occur only in inland waters of north-western New South Wales and south-western Queensland and northwards from the Mary River in the coastal drainages of south-eastern Queensland.

Other names Common: tandan, freshwater jewfish, dewfish, eeltail catfish, kenaru. Scientific: none.

Literature Allen, 1989; Davis, 1977a, b, c, d; Gerry, 1994; Grant, 1991; Keenan *et al.*, 1994; Lake, 1966, 1967a, b, c; Merrick & Midgley, 1981; Merrick & Schmida, 1984; Reynolds, 1983; Scott *et al.*, 1974; Whitley, 1957.

FAMILY PLOTOSIDAE

CENTRAL AUSTRALIAN CATFISH
Neosilurus argenteus Zietz

Description An elongate species with a tapering tail, a small and rather flattened head, bluntly pointed snout and a small somewhat downturned mouth surrounded by four pairs of long, slender barbels. Eyes small. Anterior dorsal fin (I, 4) located just behind head, short-based, high and flag-like, with a stout, serrated, venomous spine anteriorly. Most of the back between anterior dorsal fin and tail finless, the 2nd dorsal fin very small and continuous with the rounded tail and anal fin (total rays about 120). Anal fin long, extending well forwards on lower surface, and quite deep. Pectoral fins, like the dorsal, have venomous spines, and generally 7 rays. Skin smooth, with no scales.

Colour Body silvery white, fins yellowish to orange.

Size Reputedly reaches about 200 mm in length.

Distribution Was originally described from Central Australia and is variously listed as occurring in inland rivers of South Australia, large areas of the Northern Territory and rivers entering the Gulf of Carpentaria.

Conservation status Remains abundant and widespread.

Natural history Virtually nothing is known of its natural history, although related species have large, benthic eggs.

Utility Is probably too small to be of much sporting or culinary value, though larger, plotosid catfishes are well regarded as food fish. Small specimens make hardy and attractive aquarium fishes. The dorsal and pectoral fin spines are venomous and can inflict painful wounds.

Similar species Several other undescribed species of *Neosilurus* occur together with this species and *N. hyrtlii* in various parts of the Lake Eyre internal drainage system in north-eastern South Australia, including Cooper Creek. These and further species of *Neosilurus* resemble each other rather closely, but this silvery species is generally paler in colour than those others in the internal drainage systems of north-eastern South Australia.

Other names Common: none. Scientific: *Porochilus argenteus, Plotosus argenteus.*

Literature Allen, 1989; Brown, 1992; Glover, 1989, 1990; Grant, 1991; Lake, 1966; Merrick & Schmida, 1984; Orr & Milward, 1984; Pollard, 1974; Scott *et al.*, 1974; Taylor, 1964; Whitley, 1957.

Central Australian catfish *Neosilurus argenteus*. (P. Parker)

HYRTL'S TANDAN
Neosilurus hyrtlii Steindachner

Description Generally similar to *Neosilurus argenteus*, but has a stouter body, 5–6 soft dorsal rays and 10–11 soft pectoral rays.

Colour Body colouration varies from dark brown to a pale yellowish brown, olive green or light grey, with the ventral surface and fins generally paler and/or yellowish in colour.

Size Reaches at least 400 mm in length.

Distribution Is widespread throughout coastal drainages of northern Australia, extending northwards from the Pilbara and Kimberley regions in Western Australia and eastwards and southwards to south-eastern Queensland. Is also found throughout much of the central desert drainage systems of the Northern Territory, Queensland and South Australia. In south-eastern Australia has been found in the Paroo River system, possibly also in the Condamine and Warrego Rivers (Darling River system), and in artesian bore tanks in the Bulloo internal drainage system (P. Brown, personal communication).

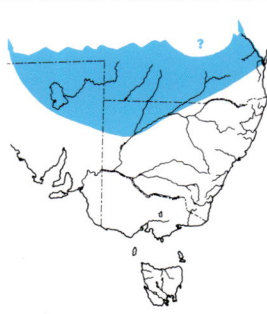

Conservation status Remains common and widespread.

Natural history Inhabits a wide range of habitats throughout its distribution, ranging from clear, flowing streams to turbid, stagnant pools. Spawning behaviour is stimulated by flooding, at least in northern Queensland. Diet comprises a wide variety of insects, crustaceans, molluscs and worms.

Utility Grows to an edible size in parts of its range, though is not often sought after as a food or sport fish. Makes a hardy aquarium fish, though its spines are venomous and can inflict painful wounds.

Similar species A number of similar species of *Neosilurus* with which this species could be confused occur in the drainages of northern and central Australia. Several other undescribed species of *Neosilurus* occur together with this species and *N. argenteus* in various parts of the Lake Eyre internal drainage system in north-eastern South Australia. Small freshwater catfish *Tandanus tandanus* might also cause confusion in coastal drainages of north-eastern Queensland.

Other names Common: mottled tandan, white tandan, yellowfin tandan, glencoe tandan, silver moonfish. Scientific: *Neosilurus glencoensis*.

Literature Allen, 1989; Brown, 1992; Glover, 1989, 1990; Grant, 1991; Lake, 1966; Merrick & Schmida, 1984; Orr & Milward, 1984; Pollard, 1974; Scott *et al.*, 1974; Taylor, 1964; Whitley, 1957.

18
Family Cobitidae
Loaches

M. LINTERMANS AND J. BURCHMORE

This large family is native to Europe and Asia, attaining its greatest diversity in southern Asia and the Malay Archipelago. Feral populations of cobitids have become established in three mainland states of the USA, as well as in Hawaii, Palau, Mexico and the Philippines. Australia has no native cobitids, but a single species has recently become established in south-eastern Australia.

ORIENTAL WEATHERLOACH *Misgurnus anguillicaudatus* Cantor

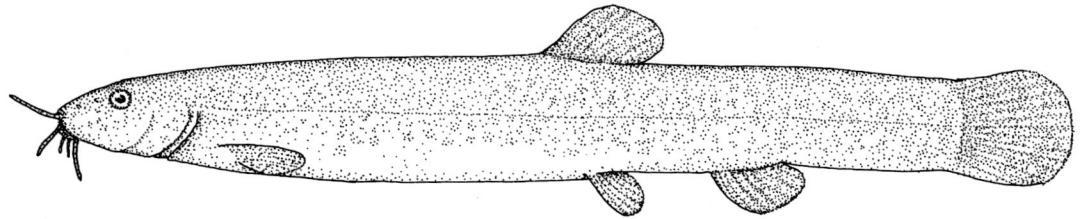

18.1 Oriental weatherloach *Misgurnus anguillicaudatus*. (R.M. McDowall)

Description Body elongate, cylindrical and sturdy. A single short-based dorsal fin (II-IV, 5-7) located about midway along back; anal fin (II-V, 5-6) short-based and rounded. Pectoral fins (I, 9-10) low on body, their shape varying according to sex; pelvic fins (I, 5-8) inserted about one third of length behind dorsal and extend to anus; tail rounded. Five pairs of conspicuous barbels around mouth. Body covered in mucus, which makes the species extremely difficult to handle.

Sexual dimorphism Mature males can be distinguished by the size and shape of the 2nd ray of the pectoral fins. This is conspicuously longer in males and about three times as broad as the 1st ray; dorsally it has an enlarged bony plate, the lamina circularis, that is thickly covered with skin at the base. The longer 2nd ray gives the pectoral fin a distinctly triangular shape, whereas in the female it is more rounded (Fig. 18.2).

Colour Generally a mottled brownish yellow, with numerous black spots on back and sides; underside pale and unmarked. A prominent black spot at base of caudal fin.

Size Maximum length about 250 mm but in Australia usually 200 mm or less.

Introduction and distribution Popularly kept as an aquarium species. It is thought that the discarding of unwanted specimens by aquarists or escape from ponds were the source of early feral populations in Australia. The first of these was recorded from the Yarra River, Victoria, in 1984. This population expanded to occupy almost the full length of the river. Also recorded from a small tributary of the Ovens River, near

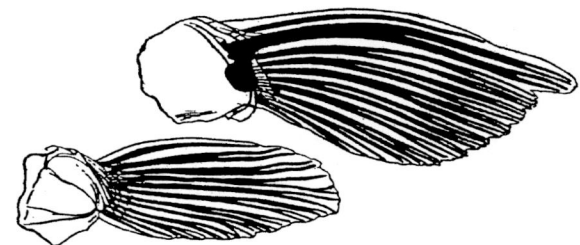

18.2 Pectoral fin of oriental weatherloach *Misgurnus anguillicaudatus*:, male (top) showing enlarged fin ray, and female (below)

LOACHES

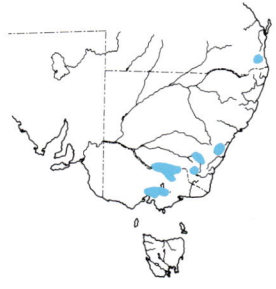

Wangaratta, Victoria, in 1985; this population now occupies a substantial portion of the Ovens River and the Murray River downstream to at least Barmah Forest. There are at least three other weatherloach populations in the Melbourne metropolitan area, and there has been a recent record from the Latrobe River (T. Raadik, personal communication). A population was recorded from the Australian Capital Territory in 1986 in Ginninderra Creek. This colonised most of the creek and its tributaries and spread into New South Wales waters, including the Murrumbidgee River down to Lake Burrinjuck. A population recorded from the Wingecarribee River in 1989 has colonised the Wollondilly River to Lake Burragorang and also the Coxs River. Further populations were recorded from Tuggeranong Creek and the Cotter/Paddys River areas in the ACT in 1991 and 1992 respectively. In late 1992, weatherloach were discovered in a small tributary of Lake Eucumbene in the Snowy Mountains in New South Wales. They have become established in a small suburban creek near Brisbane (R. McKay, personal communication), and there are unconfirmed reports of weatherloach from South Australia and Western Australia.

Natural history So-named because of its reported restlessness during changes in barometric pressure. Its preferred habitat is slack water with sand or mud substrates into which it can burrow to escape predators and aestivate. Tolerates waters with low dissolved oxygen content as it can utilise atmospheric oxygen by swallowing air and passing it through a highly vascularised hind gut. It is also eurythermal, thriving in water temperatures of 2–30°C, and has been recorded from a number of hot springs at temperatures of up to 38°C. It can move overland to disperse and colonise new water bodies. Sexually mature at about 100 mm total length. Is a multiple spawner, laying 4-8000 eggs per spawning on aquatic vegetation or mud in summer. Lifespan in captivity is 13+ years.

The weatherloach is omnivorous, with aquatic insect larvae, crustaceans, rotifers, algae and detritus recorded in the diet. Prey is sensed through a combination of chemical and tactile cues.

Utility A popular aquarium species in Australia, although its importation has been prohibited since 1986. Spread in Australia has been greatly facilitated by illegal use as live bait by anglers. Widely eaten in eastern Asia and has been introduced into Mexico and the Philippines for aquaculture.

Similar species Sometimes confused with small eels but can readily be differentiated by barbels around mouth and the short dorsal fin.

Other names Common: Japanese weatherloach, Japanese loach, Japanese weatherfish, loach, mud loach, weatherfish. Scientific: *Misgurnus fossilis anguillicaudatus, Cobitis fossilis, C. anguillicaudatus*.

Literature Allen, 1984; Burchmore *et al.*, 1990; Lintermans, 1993; Lintermans *et al.*, 1990a, b; McMahon & Burggren, 1987; Suzuki, 1983; Watanabe & Hidaka, 1983; Welcomme, 1988.

Oriental weatherloach *Misgurnus anguillicaudatus*. (R.H. Kuiter)

19
Family Poeciliidae
Livebearers

R.M. McDOWALL

The poeciliid fishes are popular aquarium fishes that are well known for their habit of giving birth to live young rather than laying eggs—hence the popular name 'livebearers' for the family. It is a large and diverse family with many species in tropical Central America, spreading north into the southern United States and south into northern areas of South America. Poeciliids mostly live in fresh water, but some species enter brackish water or the sea. The best known species in the family are probably the guppy (*Poecilia reticulata*) and various mollies (species of *Poecilia* formerly included in the genus *Mollienisia*), including the sailfin molly. These are widely kept as aquarium fishes. Several species have become established in the wild in south-eastern Australia, mostly as deliberate releases or accidental escapees from captivity. However, by far the most widespread and abundant is the eastern gambusia, *Gambusia holbrooki*, which was introduced and released for the control of troublesome aquatic insects. The family is characterised by the protrusible, upturned and terminal mouth, a single, soft-rayed dorsal fin, large scales, no lateral line, and a gonopodium in the males (consisting of a highly modified anal fin to facilitate internal fertilisation of the female).

Key to livebearers

1 Dorsal fin origin obviously well behind anal fin origin . . **Gambusia holbrooki** p. 116; Fig. 19.1
   Dorsal fin origin not obviously behind anal fin origin. 2

2 Dorsal fin origin well in front of pelvic fin bases. . . . . **Poecilia latipinna** p. 119; Fig. 19.3
   Dorsal fin origin more or less above anal fin origin 3

3 A large oval dark blotch on caudal peduncle, followed by a pair of smaller round dark blotches at tail base, one each at upper and lower margins of fin . . . . **Xiphophorus maculatus** p. 122; Fig. 19.5
   No obvious dark oval blotch or paired smaller blotches on caudal peduncle and tail base. . . . 4

4 Dorsal fin short, 7–9 rays, high and rounded, basal length shorter than height; gonopodium longer than head; no elongated sword at lower margin of tail in male . . . . **Poecilia reticulata** p. 118; Fig. 19.2
   Dorsal fin longer, 11–12 rays, lower and angular, base longer than fin is high; gonopodium shorter than head; lower margin of tail extended as a sword in male . . . . **Xiphophorus helleri** p. 120; Fig. 19.4

EASTERN GAMBUSIA
Gambusia holbrooki (Girard)

Description A tiny, stout fish with a deep rounded belly (particularly in females, and more so when carrying a brood of young); upper surface rather flattened, especially the head; mouth small, upturned and protrusible, lower jaw a little longer than upper; eyes large. One soft-rayed dorsal fin (6–8 rays, usually 7), well back on trunk,

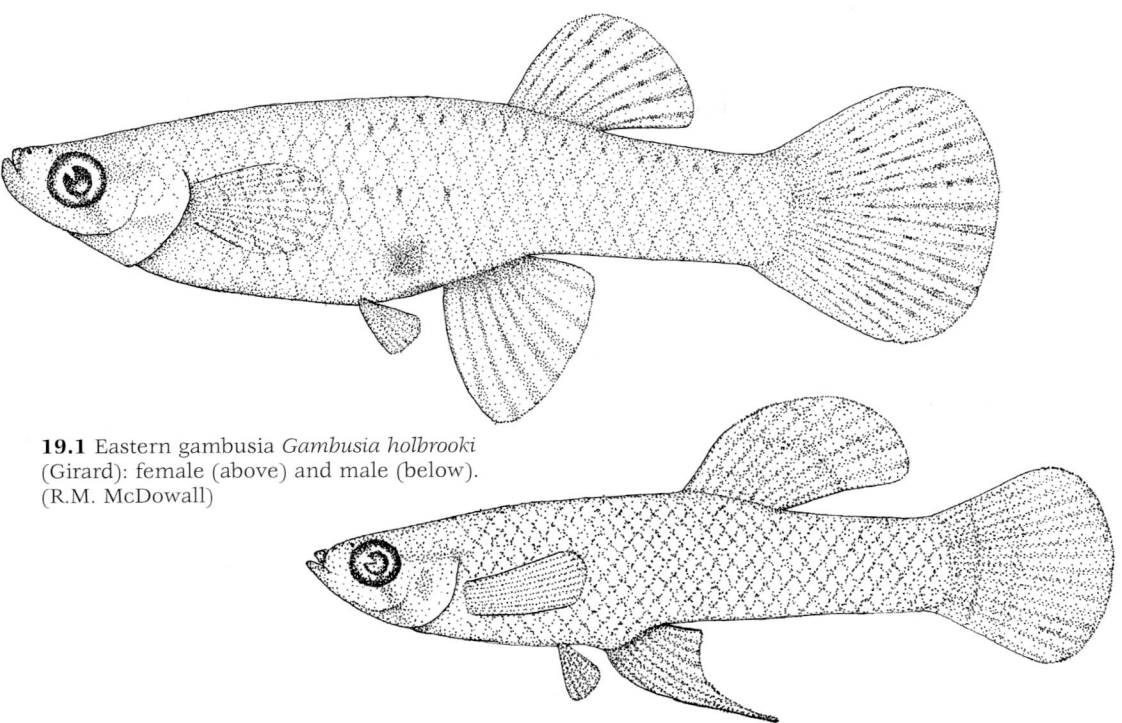

19.1 Eastern gambusia *Gambusia holbrooki* (Girard): female (above) and male (below). (R.M. McDowall)

short-based, high and rounded. Anal fin (9–11 rays, usually 10) also small and rounded; tail rounded. Large scales cover head and trunk (28–32, usually 30–31 along side); no lateral line; 31–33 vertebrae; 13–15 stout gill rakers of moderate length.

Sexual dimorphism Female much larger than male; male has anterior rays of anal fin much thickened and elongated, with very small hooks at tip. These form the gonopodium, which the male uses to facilitate internal fertilisation of eggs in the female.

Colour Generally greenish olive to brownish on back, sides grey with a bluish sheen, belly silvery white. In females a large black blotch surrounded by a golden patch occurs just above vent.

Size A tiny species, the female growing to about 60 mm but the male reaching only about 35 mm.

Introduction and distribution Very widespread and common throughout New South Wales, South Australia and Victoria in both inland and coastal drainages, and in coastal drainages of Queensland; not recorded from Tasmania. Also present in parts of the Northern Territory and Western Australia. Occurs especially in warmer waters that are still or gently flowing.

Native to rivers draining into the Gulf of Mexico. Because it reproduces rapidly and has a reputation for eating mosquitoes (hence the widely used common name mosquitofish), it has been introduced into many countries. Introduction to Australia has been poorly documented, but apparently it was first brought to the country in the 1920s for use in aquaria and was subsequently released into the wild for mosquito control. Original introductions came from the USA via Italy.

Natural history Is most abundant in warm and gently flowing or still waters, mostly around the margins and along the edges of aquatic vegetation beds; tolerates a very wide range of temperatures and other habitat conditions; prefers water between about 25°C and 38°C but can survive under ice and in temperatures up to about 44°C. Tolerates a wide range of salinities, from pure fresh water to full marine salinities. Sexes about equal in abundance. Eggs are fertilised within the females by sperm deposited at the mouth of her genital opening by the male's gonopodium. The young take 3–4 weeks to develop and are produced throughout the warm months, numbering about 50 on average but may exceed 100, with more than 300 having been reported. Up to 9 broods a year. Peak reproductive activity in spring (October).

New-born are just a few millimetres long but grow rapidly and may mature in under two months (females at only about 21 mm long). May breed several times a year, and this, with rapid maturation, means that populations may increase rapidly. Breeding season probably much longer in northern, warmer waters—lasted from October to April in a southern Queensland population. Feeds on a wide range of both terrestrial and aquatic organisms, especially terrestrial insects such as ants and flies, as well as aquatic bugs and beetles. The diet varies with the fauna available; this is an adaptable generalist predator.

Utility Although released into the wild to control malarial and other mosquitoes, this species is probably of no greater value than other small insectivorous fishes native to Australian waters. Its high reproductive rate may give it some advantage over them, and it may be a threat to other small fishes into whose habitat it has been introduced or spread. The decline of the softspined rainbow-fish *Rhadinocentrus ornatus* in Queensland is attributed to impacts of eastern gambusia. Is regarded by some as aggressive and a pest in Australian waters.

Is very easy to keep in aquaria, but because it is pugnacious and attacks other fish, nipping their fins, is best kept alone. Apart from the ease with which it is kept and breeds, offers little of interest to aquarists.

Similar species Females resemble female guppies; also small atherinids.

Other names Common: is widely called mosquitofish, though this name is not preferred as it gives a misleading impression of its value for mosquito control. Scientific: was formerly referred to as *G. affinis* or sometimes *G. affinis holbrooki*.

Literature Arthington, 1989; Arthington and Lloyd, 1989; Arthington *et al.*, 1986; Krumholz, 1948; Lloyd, 1984, 1990; Lloyd *et al.*, 1986; Lloyd and Tomasov, 1985; Milton and Arthington, 1983b; Myers, 1965; Rosen and Bailey, 1963.

GUPPY *Poecilia reticulata* Peters

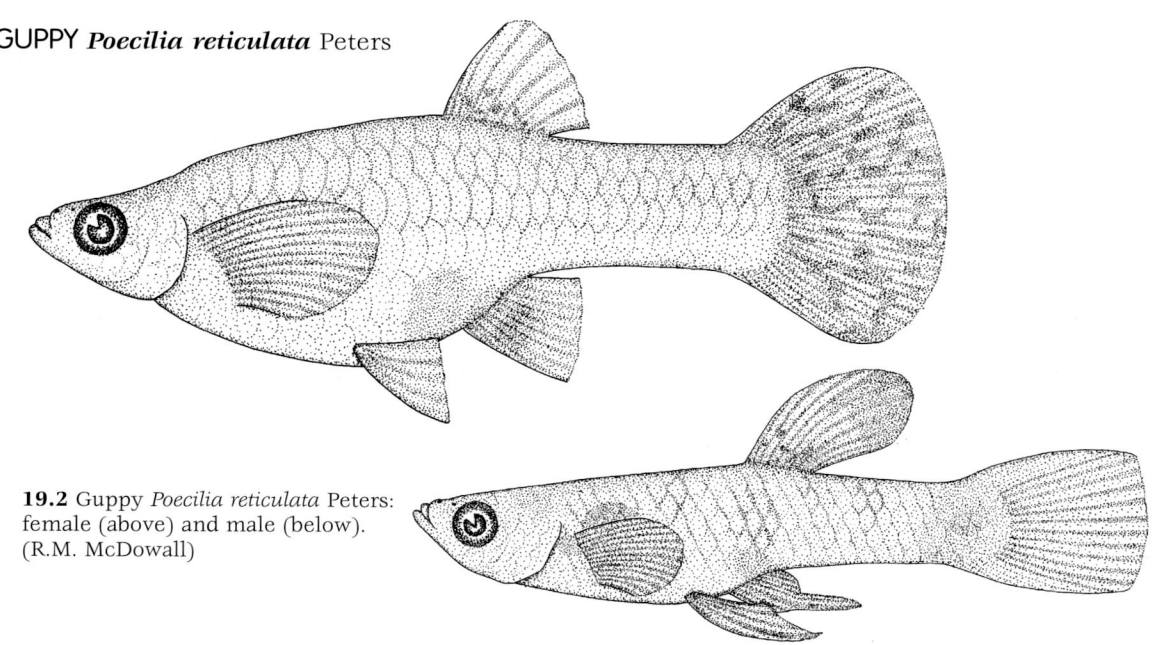

19.2 Guppy *Poecilia reticulata* Peters: female (above) and male (below). (R.M. McDowall)

Description A stout little fish, compressed towards tail; back arches in front of dorsal fin and belly deepens before vent, especially in females. Mouth small, upturned and protrusible, lower jaw a little longer than upper, eyes large. A single, soft-rayed dorsal fin (7–8 rays) a little behind middle of body, small and angular in female, but longer, almost flag-like in males. Tail rounded and may also be enlarged and flag-like in males. Anal fin (8–10 rays) originates below dorsal. Large scales on head and trunk (25–29 along midlateral); no lateral line; 13–16 short, well-spaced and slender gill rakers.

Sexual dimorphism Female may be much larger than male but lacks the enlarged dorsal fin and tail, though these features are less prominent in wild populations than in carefully selected captive stocks. The female also has no gonopodium and

lacks the bright and varied blotches of colouration characteristic of the male (see below).
Colour Female rather drab, silvery grey with scale outlines darker, giving the sides of the fish a hatched appearance; back a little darker, belly tending to silvery white; fins colourless to greyish, sometimes finely speckled; a darker grey-black patch on lower abdomen before vent. Male highly variable, brightly coloured with irregular and varied markings of green, turquoise, blue, red, orange and yellow and also pearly iridescence and black spots.
Size A tiny species, the female reaching about 60 mm but males only 30 mm.

Introduction and distribution
Introduction is not documented, but the species has been present and kept by aquarists for many decades. Establishment in the wild is also not well documented, but certainly took place before the 1970s. Occurs in the wild widely along coastal Queensland, around towns such as Cairns, Mackay, Rockhampton and also Brisbane, bringing it within the margins of the range of this book.

Native to northern South America (Brazil, Guyana, Venezuela) and some islands of the Caribbean such as Trinidad and Barbados. Is now present in many countries owing to its value to aquarists.

Natural history Inhabits only still or gently flowing waters, where it may be found in loose aggregations along sheltered margins or among vegetation. Is a warm-water species that does not tolerate water temperatures below about 15°C. Young guppies mature in just a few weeks. Mature females are constantly harassed by males seeking to deposit sperm at the mouth of urino-genital aperture; one copulation may fertilise several sequential broods of young. Gestation takes 3–4 weeks; pregnant females have a black blotch on the lower rear abdomen. If populations lack males, one or more females may become males. Feeds on diverse small terrestrial and aquatic animals (particularly ants, in a study of a Queensland population).

Utility Apart from its well-known popularity as an aquarium fish, has little use, though it appears that its spread into wild habitats may have been to assist mosquito control. Very easily kept and bred in captivity.

Similar species Resembles other poeciliids, such as eastern gambusia, particularly females of both species; may also be confused with small atherinids.

Other names Common: millionsfish. Scientific: was long known as *Lebistes reticulatus*.

Literature Arthington, 1989; Arthington and Lloyd, 1989; Arthington et al., 1986; Rosen and Bailey, 1963; Sterba, 1962.

SAILFIN MOLLY
Poecilia latipinna (Le Sueur)

Description Deep-bodied and highly compressed, body arching to dorsal fin; mouth small, upturned, lower jaw a little longer than upper; eyes large. Dorsal fin well forward of mid-body length, long-based (15–17 rays) and quite low in female but greatly enlarged in male; anal fin (8–10 rays) about mid-body length, modified to form gonopodium in male; tail slightly rounded to truncate. About 26–28 large scales along side; 24–28 moderately long gill rakers.

Sexual dimorphism Male has dorsal fin greatly enlarged, hence name 'sailfin', the fin being especially well developed in wild populations.

Colour Olive on back, paling to iridescent pale violet on sides, with longitudinal rows of orange spots; belly with a bluish sheen, silvery white below. Dorsal fin olive with a pale orange margin, stronger in male, which also has a band of vertically elongate blotches on outer quarter of dorsal fin.

Size Known to grow to 120 mm but more commonly reaches only 60–70 mm.

Introduction and distribution Another popular aquarium fish brought to Australia decades ago for the aquarium trade; established in the wild only since the late 1960s. Occurs in only a few creeks and drains a little north of Brisbane.

Native to the southern and eastern states of the United States, from North Carolina south and west to areas of Mexico bordering the Gulf of Mexico.

Natural history A warm-water species likely to find congenial habitats in northern parts of Australia, occurring in fresh to brackish or even fully marine waters, usually still or gently flowing.

FAMILY POECILIIDAE

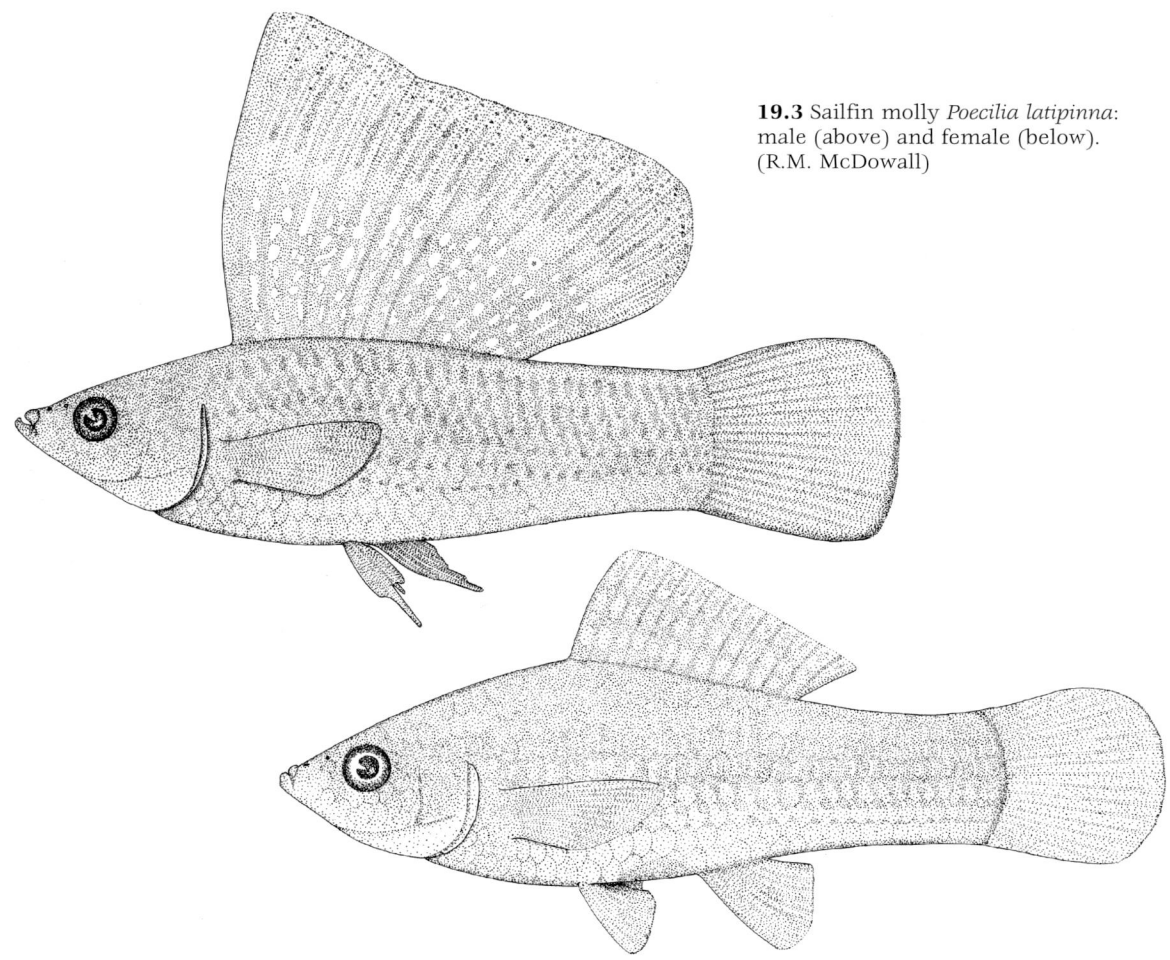

19.3 Sailfin molly *Poecilia latipinna*: male (above) and female (below). (R.M. McDowall)

Fertilisation is internal, with up to 120 young, about 12 mm long, being born per brood. Known to be omnivorous, though diet is not reported in Australian waters.
Utility Other than its value in aquaria, is of no known utility.
Similar species Small sailfin mollies may be confused with other poeciliids, such as guppy and eastern gambusia.
Other names Common: none. Scientific: formerly placed in genus *Mollienisia*, thus known as *M. latipinna*.
Literature Arthington, 1989; Arthington and Lloyd, 1989; Arthington *et al.,* 1986; Rosen and Bailey, 1963; Sterba, 1962.

SWORDTAIL
Xiphophorus helleri Heckel

Description Of modest size, deep-bodied and compressed, with dorsal fin high on arching back; mouth upturned and protrusible, lower jaw a little longer than upper; eyes large. Dorsal fin angular (11–12 rays), at about mid-point of body length, hind rays of male fin elongated; anal fin smaller (8–9 rays), below dorsal in female, forward of dorsal origin in male and modified to form gonopodium. Tail truncated, lower margin elongated to form a long sword in male. Scales cover head and trunk, 26–27 along midlateral, no lateral line; 18–20 slender gill rakers.
Sexual dimorphism The male can be distinguished by elongated rays at rear of dorsal fin, the well-developed sword at the lower edge of the tail and the anal fin modified to form the gonopodium.
Colour Highly variable, depending on recency of aquarium ancestry. Aquarium fish are typically bright orange on both body and fins, with alternating silvery and orange lines along trunk; may

revert to wild colouration of olive brown, with sides greenish, an orange-red midlateral stripe that continues along the tail sword.
Size Males may reach 80 mm and females 120 mm; size in Australian waters not well reported.
Introduction and distribution Found only in a few rivers around Brisbane. Occurs naturally in waters of the eastern (Atlantic) drainages of Central America, from Mexico to Belize. Introduced to Australia many years ago for aquarists, and wild populations presumably result from escapes from an aquarist's pond caused by flooding, or deliberate releases of aquarium-reared fish, perhaps by children. Wild populations date back to the mid-1960s.

Natural history Not much studied in Australian waters. Occurs naturally in gently flowing streams with sparse vegetation over gravelly substrates, but is reported in disturbed and degraded habitats in weed-infested creeks around Brisbane, where it occurs around the margins and along the edges of weed beds. Needs water at 10°C and preferably warmer and can tolerate up to 50 per cent sea water if acclimatised. Can also tolerate low oxygen levels, surviving by gulping air at the surface. May have large broods of over 200 young. Sex-reversal, with females becoming males, is well known. Females outnumber males in southern Queensland populations, in which females were found to be reproductively active in all months except June and reproductively most active in spring (October–December), with up to 9 broods a year. These may take 5–9 weeks to develop. Females may mature when only 23 mm long. Known as an omnivore, eating small insects and other aquatic animals, but feeds mostly on algal growth and diatoms from weeds and substrate. Limited Australian studies confirm this.
Utility Only use is as an aquarium species. As such it is easily kept and bred.
Similar species Small females may be confused with females of other poeciliids, though sword of male is highly distinctive.
Other names Common: none. Scientific: none.
Literature Arthington, 1989; Arthington and Lloyd, 1989; Arthington *et al.*, 1986; Milton and Arthington, 1983b; Rosen, and Bailey, 1963; Sterba, 1962.

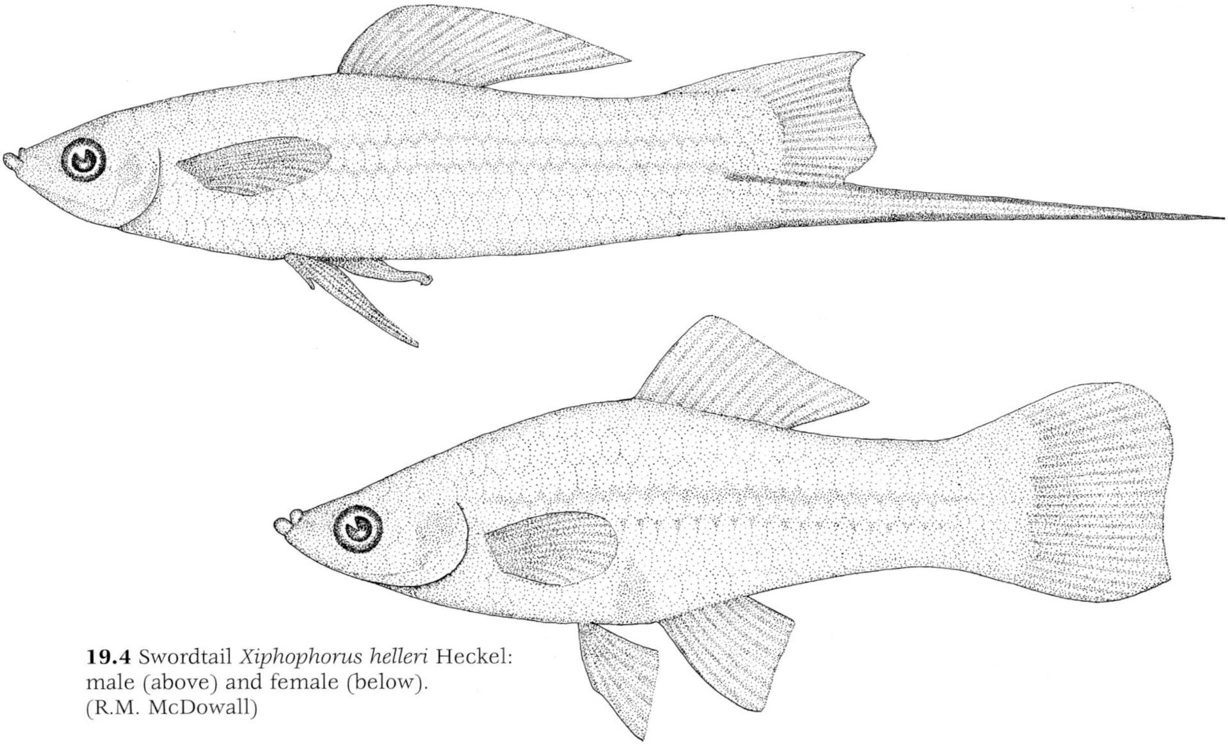

19.4 Swordtail *Xiphophorus helleri* Heckel: male (above) and female (below). (R.M. McDowall)

FAMILY POECILIIDAE

PLATY
Xiphophorus maculatus (Günther)

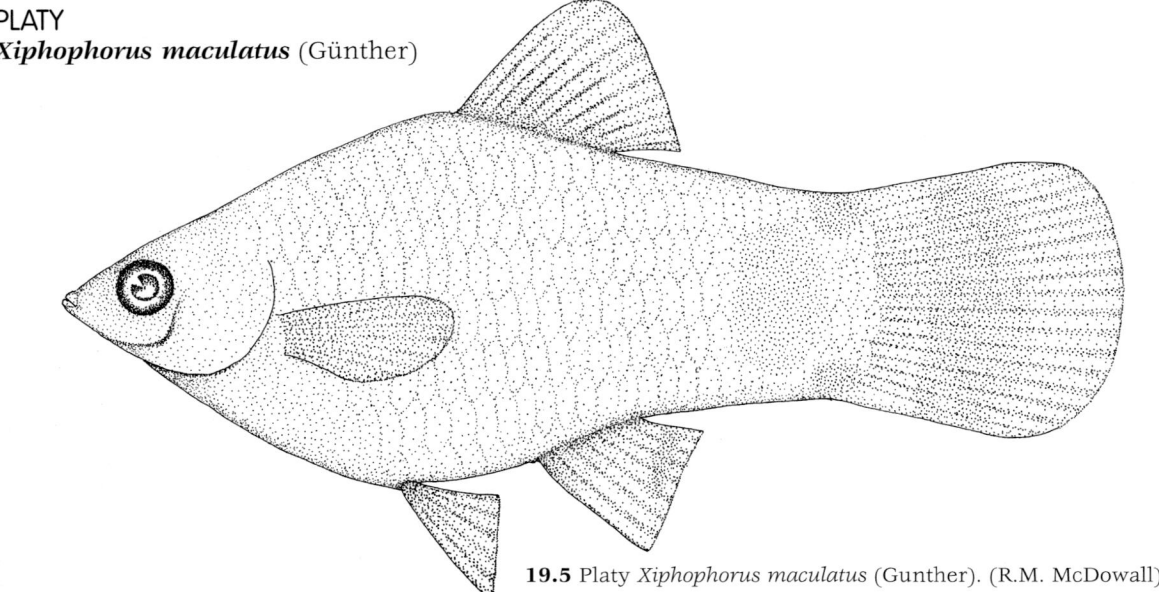

19.5 Platy *Xiphophorus maculatus* (Gunther). (R.M. McDowall)

Description A small fish, very deep-bodied and much compressed, with dorsal fin high on arching back; caudal peduncle deep and strongly compressed. Head small, mouth upturned and protrusible, lower jaw longer than upper; eyes large. Dorsal fin angular (10–11 rays), at about mid-point of body length, high and somewhat rounded; anal fin smaller (8–9 rays), origin a little behind dorsal origin; modified to form gonopodium in male. Tail slightly rounded to truncate. Scales cover head and trunk, 25–27 along midlateral, no lateral line.
Sexual dimorphism Sexes similar, though male is smaller and has anal fin modified to form a gonopodium.
Colour Highly variable, from irregularly blotched in black and red (especially in captivity or recent escapees) to brownish olive on the back, sides with a bluish iridescence, belly greenish yellow; caudal peduncle has a large oval blotch followed by two round black blotches at upper and lower margins of tail base; fins mostly colourless.
Size Females reach 60 mm but males much smaller, to 40 mm.
Introduction and distribution Another popular aquarium species long present in Australia but known from the wild since the late 1960s; present there only in a few swamps and creeks around and north of Brisbane.
Natural history Is largely unstudied in Australian waters but is another still-water species that needs warm waters to thrive, preferably above 20°C. Breeds from spring to autumn (September–March), perhaps with broods every 8–10 weeks; produces up to 100 live young. Is omnivorous but prefers animal foods, including terrestrial and aquatic insects as well as aquatic crustaceans.
Utility Apart from value as an aquarium fish, is of no use.
Similar species Resembles other poeciliids, especially females when small.
Other names Common: none. Scientific: none.
Literature Arthington, 1989; Arthington and Lloyd, 1989; Milton and Arthington, 1983b; Rosen and Bailey, 1963; Sterba, 1962.

20
Family Atherinidae
Silversides or Hardyheads

W. IVANTSOFF AND L.E.L.M. CROWLEY

The family Atherinidae is a large, widespread group of fishes present in most tropical and temperate shallow coastal waters, bays and estuaries, usually in large schools. In Australia, a number of species have entered fresh waters and are now found in the streams, rivers and lakes of most drainages. In the river systems of eastern Australia, including the Lake Eyre, Murray-Darling, South and North Coastal Drainages, seven species of the freshwater atherinid genus *Craterocephalus* are found. These fishes, commonly known as hardyheads, can be divided into three distinct groups. Freshwater species belong to the *C. eyresii* and *C. stercusmuscarum* groups, while a third group is restricted to estuarine and shallow marine habitats. Sympatry occurs between some members of the two freshwater groups in two of the eastern drainages.

The atherinid genus *Atherinosoma*, though not strictly a freshwater fish, is often found in areas of low to zero salinity in estuaries and enclosed coastal lakes of the east coast.

NOTE: Total length (TL, measured from the tip of the snout to the longest point of the tail fin) is given for maximum length of the fish throughout this section. In the key and in some parts of the text, proportions are based on standard length (SL, measured from the tip of the snout to the base of the tail fin). This measurement is frequently used as fins may be damaged by fungal disease or by being nipped by other fish.

Key to silversides and hardyheads

1 Gill rakers moderately short but never tuberculate, always more than half diameter of pupil; anal fin always 7–12 rays; dorsal process of premaxilla short and broad .... ***Atherinosoma microstoma*** p. 132; Fig. 20.8
   Gill rakers usually stumpy or tuberculate, never more than half diameter of pupil; anal fin always 9 or fewer rays; dorsal process of premaxilla long and slender 2

2 Midlateral scale count always 38 or more; transverse scale count 14–18; body scales small, circular, barely overlapping; scales on top of head, if present, small, circular, never large or irregular in shape .... ***Craterocephalus amniculus*** p. 124; Fig. 20.1
   Midlateral scale count rarely more than 38; body scales not usually small, circular or barely overlapping; scales on top of head always present, large and irregular in shape, never small or circular 3

3 Transverse scale count always 10 or more 4
   Transverse scale count never more than 8 5

4 Origin of ventral fin always in front of tips of pectoral fins. Body scales strong, rarely deciduous; opercular scales moderately large, irregular in shape; minimum body depth 8.3–12.0 (10.1) in standard length (SL) .... ***Craterocephalus eyresii*** p. 126; Fig. 20.3

Origin of ventral fin either in front or behind tips of pectoral fins; body scales thin and deciduous; opercular scales smallish and rounded, not large or irregular; minimum body depth 10.6–12.6 (11.6) in SL . . . . ***Craterocephalus fluviatilis*** p. 125; Fig. 20.2

5   Midlateral scale count never more than 30; transverse scales never more than 5.5–6.5; 5th ceratobranchial bones never fused in midline; lateral ramus of premaxilla never highly elevated; dusky or dark stripe through snout and eye to caudal peduncle never present . . . . ***Craterocephalus marjoriae*** p. 127; Fig. 20.4
Midlateral scale count always more than 29; transverse scales always more than 6; 5th ceratobranchial bones fused in midline; lateral ramus of premaxilla always highly elevated; dusky or dark stripe through snout and eye to caudal peduncle always present 6

6   Midlateral scales 29–31; transverse scale count 6–8; dark spots in even rows along sides of body forming checkerboard pattern present. ***Craterocephalus dalhousiensis*** p. 128; Fig. 20.5
Midlateral scales always 30 or more; transverse scale count always 7–7.5; dark spots in even rows along sides of body forming checkerboard pattern never present 7

7   Gill rakers in lower ramus of 1st gill arch, including those in angle, 10–13; minimum body depth 10.2–15.0 (12.5); distance origin of pectoral fin to anus 2.9–3.7 (3.3), both in SL. . . . . ***Craterocephalus stercusmuscarum fulvus*** p. 131; Fig. 20.7
Gill rakers in lower ramus of 1st gill arch, including those in angle, 7–10; minimum body depth 9.7–11.1 (10.3); distance origin of pectoral fin to anus 2.7–3.1 (3.0), both in SL . . . . ***Craterocephalus gloveri*** p. 129; Fig. 20.6

DARLING RIVER HARDYHEAD *Craterocephalus amniculus* Crowley and Ivantsoff

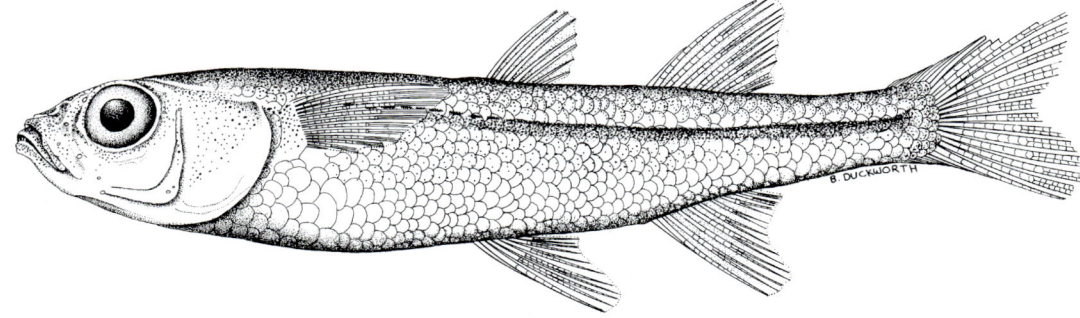

20.1 Darling River hardyhead *Craterocephalus amniculus*. (B. Duckworth)

Description   A small, robust fish with a protrusible mouth and thin lips; teeth, which are sparse but moderately long and pointed, are restricted to medial third of upper and lower jaws. First dorsal fin (V–VII) originates behind tips of pectorals (i, 11–14); 2nd dorsal (I, i, 5–7) originates above anal fin (I, i, 5–8) origin. Superficially, similar to *C. fluviatilis*, but differs in the midlateral (37–48) and transverse scale count (14–18). The scales are almost circular and barely overlapping, a character unique in this genus. There are usually no scales covering top of head but if present, are very small, circular and well spaced. Gill rakers (9–11) moderately short, less than half diameter of pupil, with the first 3–4 tuberculate.
Colour   Varies between localities; specimens from the Cockburn River at Nemingah are dusky gold above a dark silvery midlateral stripe, with silvery gold below, abdomen, opercles, eyes and chin silvery; top of head, snout and lower jaw very dark. Populations from other streams are usually paler.
Size   Maximum known size is about 55 mm TL.
Distribution   Found in upper tributaries of the Darling River, including Condamine River, Peel River, Namoi River, Warialda Creek, Macintyre River, Cockburn River and Boiling Down Creek.
Conservation status   Appears abundant at times but is more

usually found singly or in small schools of about 10–15. Concern for its status results in this species being listed as 'restricted' by the Australian Society for Fish Biology.

Natural history Usually found in gently flowing, shallow, clear water or in weed at the edges of such waters. Life history is not known, though small (subadult) specimens have been collected from Warialda Creek in September.

Utility Has no commercial uses but is probably eaten by larger fish and by birds. Being carnivorous, it probably contributes to the regulation of local insect populations. Is not generally known as an aquarium fish but survives well in captivity.

Similar species For many years this species and *C. fluviatilis* were confused with *C. eyresii* and had been placed in the synonymy of the latter species. Its specific status was recognised through osteological and electrophoretic work. Small specimens may also be confused with eastern gambusia or retropinnids.

Other names Common: none. Scientific: *C. eyresii* and *C. fluviatilis*.

Literature Crowley and Ivantsoff, 1990a.

MURRAY HARDYHEAD *Craterocephalus fluviatilis* McCulloch

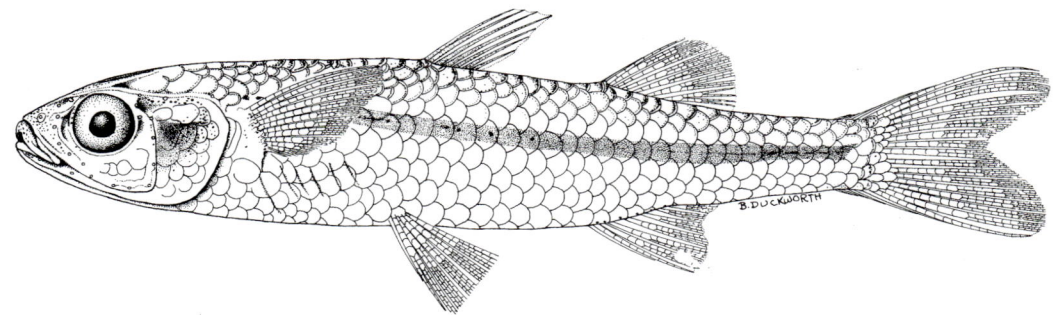

20.2 Murray hardyhead *Craterocephalus fluviatilis*. (B. Duckworth)

Description A moderately deep-bodied species; mouth small and protrusible but restricted by a labial ligament from $1/3$ to halfway along the thin lips. A single row of small teeth is restricted to anterior part of both jaws. First dorsal fin (IV–VII) with origin behind tips of pectorals (I, i, 11–13); 2nd dorsal fin (I, i, 5–8) originates above origin of anal fin (I, i, 6–9). Body scales thin, deciduous and almost circular (midlateral scale count 31–35; transverse scale count 10–12). Scales on top of head robust and large, with a single large interorbital scale reaching as far as the anterior margin of orbit. Gill rakers (10–12) short, often tuberculate, with last 3–4 in angle of gill arch, slightly longer.

Colour Varies from silver to dark golden dorsally, with a silver midlateral stripe; abdomen always pale, with a silvery iridescent sheen. Opercles and eye bright silver; top of head and snout slightly darker to dusky.

Size Known to about 72 mm.

Distribution Previously reported to have a wide distribution from the Murray and Murrumbidgee Rivers in southern New South Wales to northern tributaries of the Darling River. It is now considered that the latter fish were probably *C. amniculus*.

Conservation status While once abundant in southern waters of New South Wales, now appears to be restricted to a few lakes in Victoria associated with the Murray River. In many of the former localities from which this species was known, there are now large numbers of introduced European carp and eastern gambusia. The Australian Society for Fish Biology lists this species as 'potentially threatened'.

Natural history Previous descriptions of the development of fish named *C. fluviatilis* were probably of *C. stercusmuscarum fulvus*. Like other hardyheads, probably spawns eggs with adhesive filaments among weeds.

Utility Like most hardyheads, appears to eat insects as well as plant material; probably helps control local insect populations. Has no known commercial uses and is probably eaten by piscivorous fish and birds.

Similar species On external morphology and meristic counts, is similar to *C. eyresii* but differs

FAMILY ATHERINIDAE

in shape of nasal bone and is quite distinct from that species genetically. Small specimens may be confused with other species of hardyheads, retropinnids or eastern gambusia.

Other names Common: none. Scientific: was described in 1913 but for many years it and *C. amniculus* had been confused with *C. eyresii* and occasionally also with the Michellian hardyhead *C. stercusmuscarum fulvus*. This confusion has now been resolved and the fish is again recognised as distinct.

Literature Crowley and Ivantsoff, 1990a; Llewellyn, 1979.

LAKE EYRE HARDYHEAD *Craterocephalus eyresii* (Steindachner)

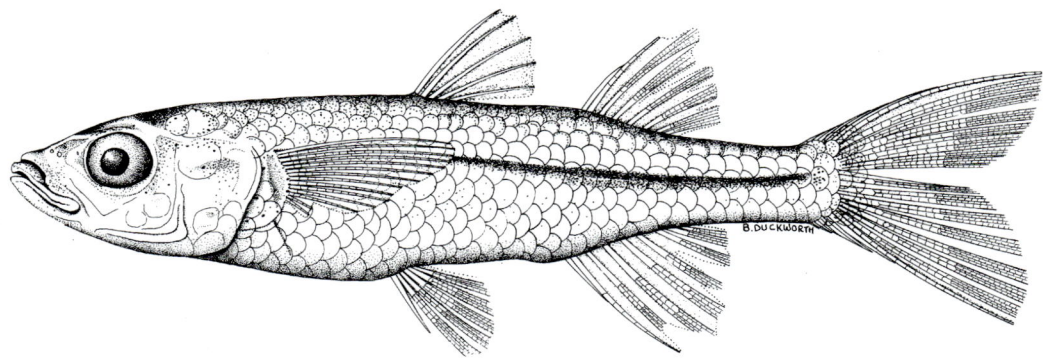

20.3 Lake Eyre hardyhead *Craterocephalus eyresii*. (B. Duckworth)

Description A small robust fish that appears to vary in body depth with age. Mouth protrusible, with fleshy lips; gape restricted by labial ligament about $1/2$ to $2/3$ of the way along lips. Teeth small, in a single row and restricted to medial $1/3$ of upper and lower jaws. First dorsal fin (IV–VII) origin in front of tips of pectorals (II, i, 12–14); 2nd dorsal fin (I, i, 5–7) origin above or slightly behind origin of anal fin (I, i, 5–8). Gill rakers (10–12) short, less than $1/2$ diameter of the pupil of eye, rarely tuberculate. Body scales small, strong and not always in even rows above or below mid-

Lake Eyre hardyhead *Craterocephalus eyresii*. (W. Ivantsoff)

lateral stripe (midlateral scale count 30–34; transverse scale count 11–14). Scales on top of head large and irregularly shaped.
Colour Usually a bright yellowish colour, with a distinct silvery midlateral stripe; opercles and abdomen iridescent green or silver; top of head and lips slightly dusky.
Size Generally about 60–70 mm but may reach 96 mm.
Distribution Found in the Lake Eyre Drainage west of the Flinders Ranges; also in the Frome River and in man-made bores and springs.
Conservation status Isolated collections indicate that this fish is not usually abundant but it appears that sufficient numbers survive dry seasons in refuge areas. During rainy seasons, or in times of flooding of Lake Eyre, they disperse and breed rapidly to become very abundant in the favourable conditions, only to largely die out again owing to evaporation and increasing salinity.

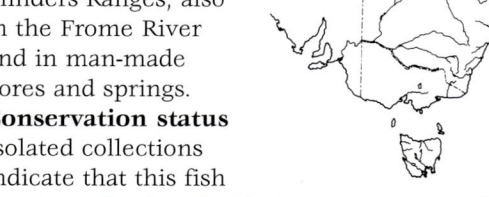

Natural history Although a freshwater species, can withstand salinities of up to 100 parts per 1000 (about 3 times the salinity of sea water). Is often found in weeds over gravel beds in slow-flowing sections of streams, rivers and bores. Little is known of its life cycle or breeding season, though breeding may be opportunistic according to water and food availability. Eggs are about 1 mm in diameter and develop between January and March.
Utility Has no commercial value but is apparently an important food for pelicans, cormorants and whiskered terns when Lake Eyre is in flood. Is not known as an aquarium fish as it is fairly inconspicuous, staying close to bottom of tank, but survives for a long time and is easy to maintain in captivity.
Similar species As pointed out above, several hardyhead species are superficially similar and have been placed in synonymy. However, some of these similar forms are distinct genetically and in transverse scale row counts. Small specimens could be confused with sand mullet and eastern gambusia.
Other names Common: occasionally called smelt, minnow or whitebait. Scientific: none.
Literature Crowley and Ivantsoff, 1990a.

MARJORIE'S HARDYHEAD *Craterocephalus marjoriae* Whitley

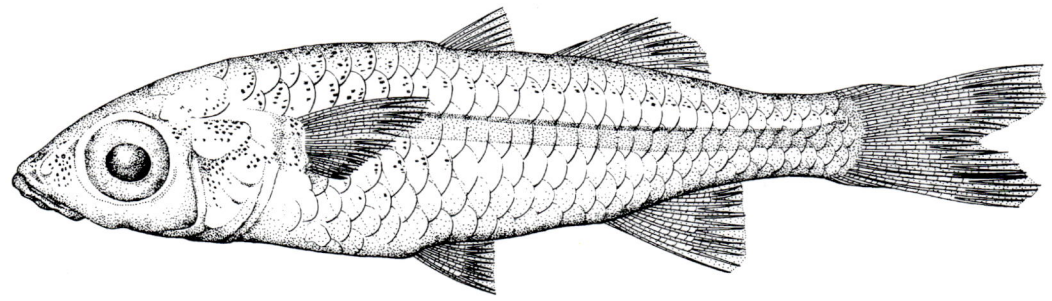

20.4 Marjorie's hardyhead *Craterocephalus marjoriae*. (B. Duckworth)

Description A small, moderately deep-bodied, robust fish, the head usually blunt, slightly flattened and sloping towards snout. Mouth protrusible, with small, sharp, inwardly pointing teeth restricted to medial third of jaws. Upper jaw usually overhangs lower when mouth is closed. First dorsal fin (IV–VII) origin behind tips of pectorals (I, i, 11–15); 2nd dorsal fin (I, i, 5–7) originates directly above or slightly behind origin of anal fin (I, i, 5–8). Gill rakers (10–13) short and tuberculate. Body scales (midlateral 28–30, transverse 5.5–6.5) large, sturdy and dorsoventrally elongated, with circuli prominent. Top of head covered by large, irregularly shaped scales.
Colour Body golden to sandy yellow, with a dark midlateral stripe, flecked with golden iridescence; abdomen and opercles silvery; top of head dusky, darker around snout. A dark, triangular blotch, lateral to vent, present but not always very obvious. Northern populations may have a row of dusky spots on scales directly above midlateral stripe.
Size Known to reach 97 mm.
Distribution Occurs from the Mary River south through coastal rivers to the Brisbane and Logan Rivers in south-eastern Queensland and also in

FAMILY ATHERINIDAE

Marjorie's hardyhead *Craterocephalus marjoriae*. (W. Ivantsoff)

the Clarence River in north-eastern New South Wales.

Conservation status Abundant in both the Clarence and Mary Rivers but has not been collected in great numbers from other rivers. Had previously been collected from the Burnett and Burdekin Rivers, both north of the Mary River, but extensive collecting in recent years has failed to find any there.

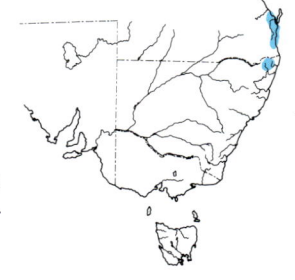

Natural history Is often found in large schools in shallow water with a gravelly or sandy bottom, though also frequents weedy sections of streams, making collection and assessment of numbers difficult. Little is known of natural history but it has been bred in captivity. Frequent sampling in the wild indicates that it may spawn in /September in the Mary River. The eggs of hardyheads are usually about 1 mm in diameter, with adhesive threads that hold the eggs firmly to weeds. The young are well developed at hatching and take fine food readily. Protrusion of upper jaw over lower when the mouth is closed is associated with the feeding habit of these fish as they 'pick' food as well as particles of sand and gravel from the bottom, leaving small depressions.

Utility Has no commercial value and few aquarists keep it, though is easy to keep and has its own particular charm. Appears to be carnivorous and so probably plays a role in keeping insect populations in check. It may also form part of the diet of larger piscivorous fish and birds.

Similar species May be difficult to distinguish from eastern gambusia, sand mullet, smelt and some other hardyheads, particularly when small.

Other names Common: Mary River hardyhead (local name). Scientific: none.

Literature Crowley and Ivantsoff, 1992; Merrick and Schmida, 1984; Milton and Arthington, 1983.

DALHOUSIE SPRINGS HARDYHEAD
Craterocephalus dalhousiensis Ivantsoff and Glover

Description A deep-bodied, robust fish. Mouth protrusible, lips thick, the teeth small, in several rows in medial third of jaws and visible on the outside of the lips when mouth is shut. First dorsal fin (IV–VI), origin well behind tips of pectoral fins (I, i, 11–13); 2nd dorsal fin (I, i, 4–6), origin behind origin of anal fin (I, i, 6–8). Body scales (midlateral 29–31; transverse 6–8) strong, rather rounded in juveniles but more dorsoventrally elongated in adults. Scales on top of head large and irregular to rectangular in shape. Gill rakers (7–8) short and tuberculate, longer in angle of gill arch but always

SILVERSIDES OR HARDYHEADS

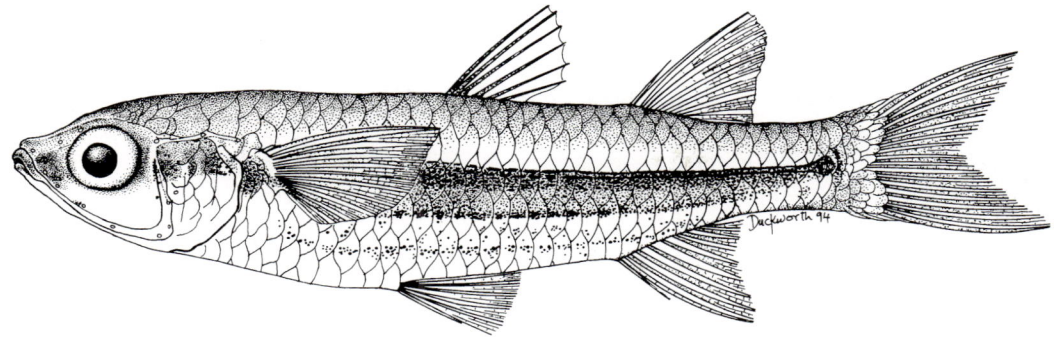

20.5 Dalhousie Springs hardyhead *Craterocephalus dalhousiensis.* (B. Duckworth)

less than 1/2 diameter of pupil.
Sexual dimorphism Is the only species of *Craterocephalus* that displays sexual dimorphism. Adult males are slightly smaller and have dorsal surface almost horizontal from snout to origin of 1st dorsal fin; interorbital space flat. In females dorsal surface is curved, tapering downwards towards snout; interorbital space dished.
Colour Bright golden brown, with a very dark-to-black midlateral stripe running from snout, through eye to base of caudal fin. There are continuous rows of dark spots, one on each scale, along sides of body, giving a striped or checkerboard effect. Eyes dark, and top of head, lips, and snout dusky. The golden colour may fade in captivity, but spotting does not fade (as it often does in *C. s. stercusmuscarum*).
Size Grows to about 82 mm but generally slightly smaller.
Distribution Restricted to several pools in the Dalhousie Springs complex.

Conservation status Abundant in the pools at Dalhousie Springs, but the limited number of habitats results in this species being listed as 'potentially threatened' by the Australian Society for Fish Biology.
Natural history Natural history has not been studied in detail. Experiments indicate that it can withstand quite high temperatures (to about 39°C) but not extremely low temperatures. Live fry collected at Dalhousie Springs in July indicate breeding at that time of year; this is supported by the size range of fish collected then. Fry transported to laboratory survived and grew in normal Sydney tap water (pH about 7). Gut contents indicate that they eat algae, insects and some molluscs.
Utility There is no known use for these fish, though they may be eaten by water birds or spangled perch. Are difficult to breed in captivity and so are not prized as aquarium fish.
Similar species Might be mistaken for *C. gloveri* or *C. s. stercusmuscarum*, but colouration along the sides is quite distinctive, particularly in larger specimens.
Other names Common: none. Scientific: none.
Literature Ivantsoff and Glover, 1974; Crowley and Ivantsoff, 1989.

GLOVER'S HARDYHEAD
Craterocephalus gloveri Crowley and Ivantsoff

Description A slender species with a small mouth, lips moderately thin and mouth restricted by a labial ligament about 1/2 to 1/3 of way along mouth. Teeth small, pointed and in 3–4 rows in the medial 1/3 of upper and lower jaws, not visible unless mouth is open. First dorsal fin spines weak (V–VI), fin origin behind tips of pectoral fin (I, i, 10–13); origin of 2nd dorsal fin (I, i, 5–6) always behind origin of anal fin (I, i, 6–7). Body scales (midlateral 30–33, transverse 7–7.5) strong, more rounded than dorsoventrally elongated and with circuli distinct posteriorly. Scales on top of head large and irregularly shaped. Gill rakers (7–10) on 1st lower gill arch short and stumpy, slightly longer in angle of arch.
Colour Pale silvery yellow, slightly darker along dorsal surface; abdomen a pale silvery colour. Snout, lips and top of head dusky; eye pale gold and opercles iridescent silver. Unlike *C. dalhousiensis* or *C. s. stercusmuscarum*, it lacks distinct spots along sides of body, but as in those species, the midlateral stripe running from snout, through eye and opercle and back to caudal peduncle is a

FAMILY ATHERINIDAE

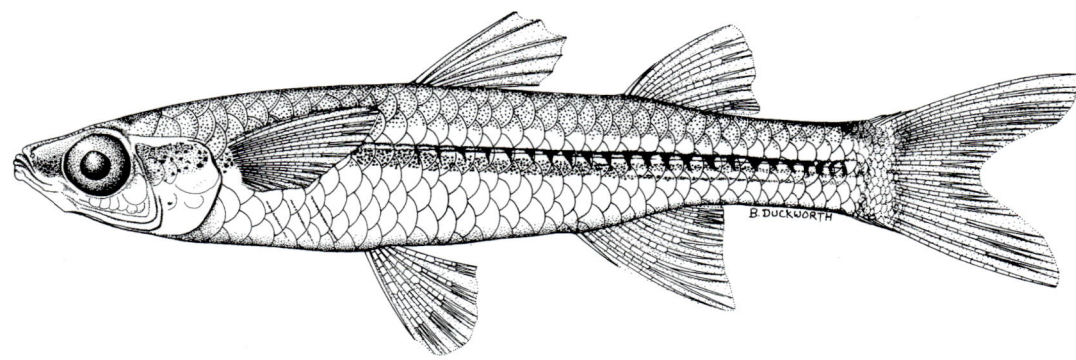

20.6 Glover's hardyhead *Craterocephalus gloveri*. (B. Duckworth)

dusky or silvery in colour.
Size Smaller than *C. dalhousiensis* (which is also endemic to Dalhousie Springs), with the largest only to about 50–52 mm.
Distribution Like *C. dalhousiensis*, is restricted to a few pools at Dalhousie Springs, though it does not co-occur with the Dalhousie hardyhead.
Conservation status Does not appear to be as abundant as *C. dalhousiensis*.

Natural history Restricted to thermal pools at Dalhousie Springs and is not found in any of the seeps away from the pools where water becomes cold at night in winter. Nothing is known of breeding times or habits. Diet consists of algae and small insects.
Utility Like other species of hardyhead, is not attractive enough to be prized as an aquarium fish; is probably eaten by aquatic birds.
Similar species Has been confused with *C. fluviatilis* and *C. stercusmuscarum*.
Other names Common: none. Scientific: *Craterocephalus stercusmuscarum*; *C. fluviatilis*.
Literature Crowley and Ivantsoff, 1989, 1990b.

Glover's hardyhead *Craterocephalus gloveri*. (W. Ivantsoff)

FLYSPECKED HARDYHEAD *Craterocephalus stercusmuscarum fulvus* Ivantsoff, Crowley and Allen

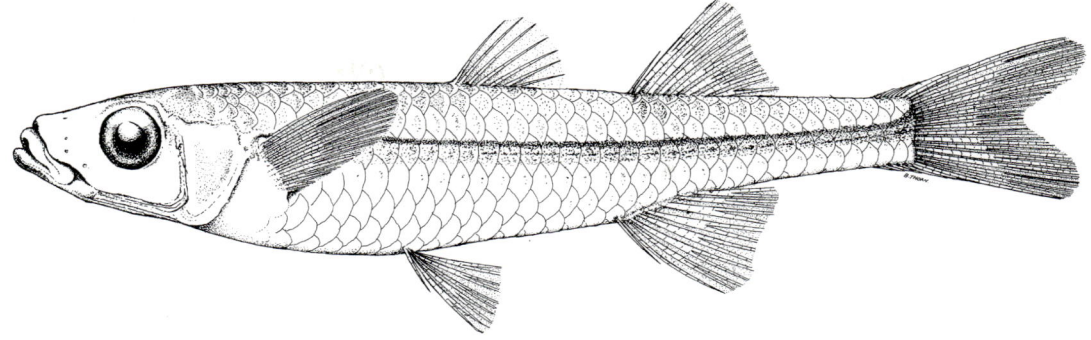

20.7 Flyspecked hardyhead *Craterocephalus stercusmuscarum fulvus*. (B. Thorn)

Description A small fish, more slender than most other hardyheads; head in larger specimens tending to slope downwards towards snout. Lips moderately thick; a small, protrusible mouth. Small teeth in 2–3 rows, restricted to the medial $1/3$ of both jaws. First dorsal fin (IV–VIII) origin behind tips of pectorals (I, i, 11–14); 2nd dorsal fin (I, i, 5–9) origin directly above origin of anal fin (I, i, 6–9). Body scales (midlateral 32–35) relatively large and dorsoventrally oval, in regular rows (7–7.5) along sides of body. Gill rakers (10–13) short and tuberculate.

Colour Varies from one locality to another; those from the Namoi River (NSW) are bright golden yellow, those from Fraser Island (Queensland) greenish gold. There is a dusky stripe from snout, through eye and across opercle, extending to tail base; along the body this may vary from black to bright golden or silvery. There is a dark blotch around the vent of females when running ripe, while males have a golden abdomen during the breeding season. Colour fades in captivity, becoming a dull, pale yellow or silvery colour.

Flyspecked hardyhead *Craterocephalus stercusmuscarum fulvus*. (R.H. Kuiter)

Size Is not known to exceed 78 mm, though the northern subspecies *C. stercusmuscarum stercusmuscarum* may reach about 108 mm.

Distribution Previously present in most parts of the Murray-Darling drainage system in New South Wales and extending into Victoria. Occurs in northern tributaries of the Darling River and east of the Great Dividing Range in south-eastern Queensland, including the Mary and Brisbane Rivers.

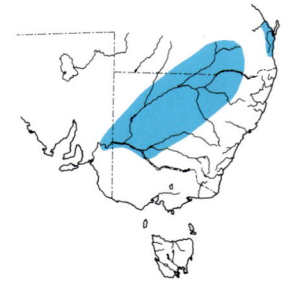

Conservation status Once abundant throughout the Murray-Darling in New South Wales and Victoria but now considered rare, if present at all, in southern regions. Remains abundant in northern parts of range.

Natural history Usually schools in still or gently flowing water over sand, gravel or mud. In the Mary River is also found among marginal weeds. Has an extended breeding season (mid-October to mid-February), eggs large (up to 1.46 mm diameter), with adhesive filaments over the surface. The larvae are well developed at hatching and take food immediately. Small specimens are found from September to about March.

Utility Is carnivorous and so may contribute to regulating local insect populations and may provide food for larger piscivorous fish and for birds. Is not usually kept by aquarists as it is small and not showy, but is easy to keep in captivity.

Similar species Similar to the more northern *C. s stercusmuscarum* but less strongly coloured. May also be confused with the eastern gambusia, sand mullet, retropinnids and other small hardyheads.

Other names Common: Mitchellian hardyhead, western freshwater hardyhead, freshwater silverside or line-eye. Scientific: *C. fluviatilis*; *C. stercusmuscarum*.

Literature Ivantsoff *et al*, 1989; Milton and Arthington, 1983.

SMALLMOUTHED HARDYHEAD *Atherinosoma microstoma* (Günther)

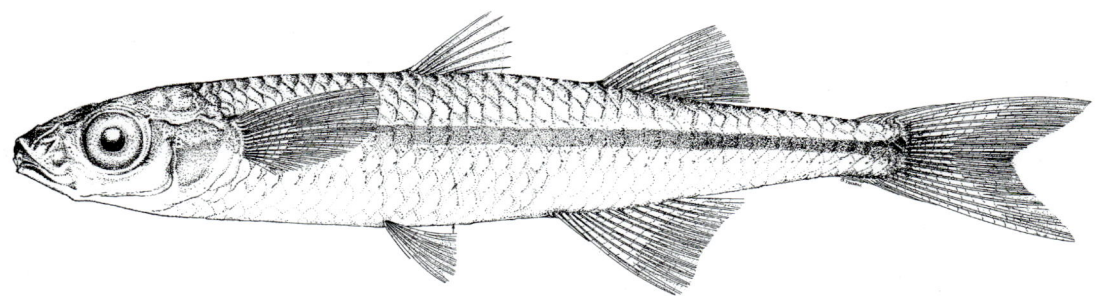

20.8 Smallmouthed hardyhead *Atherinosoma microstoma*. (B. Thorn)

Description A small, slender fish with a barely protrusible mouth; gape unrestricted by labial ligament. Teeth in jaws in several rows and not restricted to the medial $1/3$; teeth also present on vomer and mesopterygoid. Body scales (midlateral 36–41; transverse 7) small, oval to rectangular in shape, in even rows along sides of body. First dorsal fin (V–IX) origin well behind tips of pectoral fins (I, i, 11–15); origin of 2nd dorsal fin (I, i, 7–10) directly above or slightly behind origin of anal fin (I, i, 7–12). Gill rakers (12–15) longer than in *Craterocephalus* species, being more than half diameter of pupil.

Colour Variable, from a deep coppery red in some to the more usual green or green-brown above the midlateral stripe, and light green, yellowish, white or silvery below. Abdomen, opercles and eyes silvery, with or without blue-green iridescence; body translucent.

Size In New South Wales rarely grows to more than 83 mm, but in Tasmania it may reach 107 mm.

Distribution Occurs in coastal waters from Tuggerah Lakes (NSW) to Hobart (Tasmania) and Woodwell (SA, near the Victorian border).

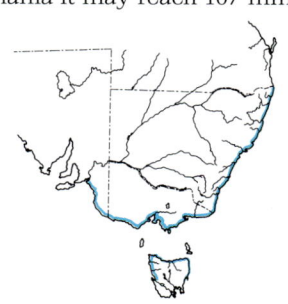

Conservation status One of the most common and abundant atherinids of the south-east coast of

Smallmouthed hardyhead *Atherinosoma microstoma*. (R.H. Kuiter)

Australia. Can be collected in thousands from lakes and lagoons.

Natural history Not strictly a freshwater species but has been collected in areas with very low salinities (3–0 parts per thousand); favours eelgrass (*Zostera*) beds. New recruits about 11–20 mm (SL) first appear in October. By the following September the size range is 50–54 mm, but numbers decline rapidly after November, indicating that this species probably has a life cycle of about one year. The colour change to copper shades along body and through eye does not appear to be related to sexual maturity. Fish tend to school by size, and females predominate. Feeds on small crustaceans, other small animals commonly found around eelgrass beds, and also on insects.

Utility Is probably eaten by birds and piscivorous fishes and appears occasionally to be used as a bait fish; does not survive well in aquaria.

Similar species Other hardyheads, small mullet.

Other names Common: greyback. Scientific: *Taeniomembras microstoma*; *T. endorae*; *Pranesella endorae*; *Atherina microstoma*.

Literature Ivantsoff, 1994; Potter *et al.*, 1986.

21
Family Melanotaeniidae
Rainbowfishes

G.R. ALLEN

The family Melanotaeniidae is closely related to the hardyheads (family Atherinidae, Chapter 20). It is restricted to freshwater habitats in continental Australia, New Guinea and adjacent islands. The family contains 55 species in six genera. In Australia it is represented by four genera and 13 species, most of which inhabit the tropical north. Three species in two genera are found in south-eastern Australia. Rainbowfishes are extremely abundant in tropical inland waters. They are an important food source for waterfowl and predatory fishes such as barramundi. Rainbows, in turn, feed on a variety of insects and their aquatic larvae and are useful in the control of mosquitos. They also consume algae and tiny crustaceans. Spawning may occur over much of the year in the tropics but tends to be seasonal in southern populations. Males exhibit bright courtship colouration and engage in vigorous chasing activity before spawning. Females deposit small numbers of eggs each day that are simultaneously fertilised by the male. The eggs have adhesive filaments that become attached to vegetation. Hatching requires about 7–10 days in most species. Rainbows are the only group of Australian native fishes that have made an impact on the international aquarium trade. Many species are now available in pet shops throughout Japan, Europe and North America.

The family is characterised by two separate dorsal fins (the 1st small and composed of relatively few rays, the 2nd long-based), a long-based anal fin, large scales, no lateral line, and several rows of fine teeth in the jaws, the outer row often extending forward onto the lips.

Key to rainbowfishes

1 Spines of 1st dorsal fin very thin and flexible; no stiff spines at beginning of 2nd dorsal, anal and pelvic fins; origin of anal fin distinctly in advance of dorsal fin origin . . . . ***Rhadinocentrus ornatus*** p. 135; Fig. 21.1
   Spines of 1st dorsal fin relatively rigid and stiff; also a stiff spine present at beginning of 2nd dorsal, anal and pelvic fins; origin of anal fin about even with that of 1st dorsal fin . . 2

2 Vertical scale rows usually 33–35; cheek scales usually 8–11; inhabits eastern or coastal drainage systems of northern New South Wales and southern Queensland . . . . ***Melanotaenia duboulayi*** p. 137; Fig. 21.2
   Vertical scale rows usually 35–37 (occasionally 34 or 38); cheek scales usually 11–16; inhabits inland Murray-Darling system . . . . ***Melanotaenia fluviatilis*** p. 138; Fig. 21.3

SOFTSPINED RAINBOWFISH *Rhadinocentrus ornatus* Regan

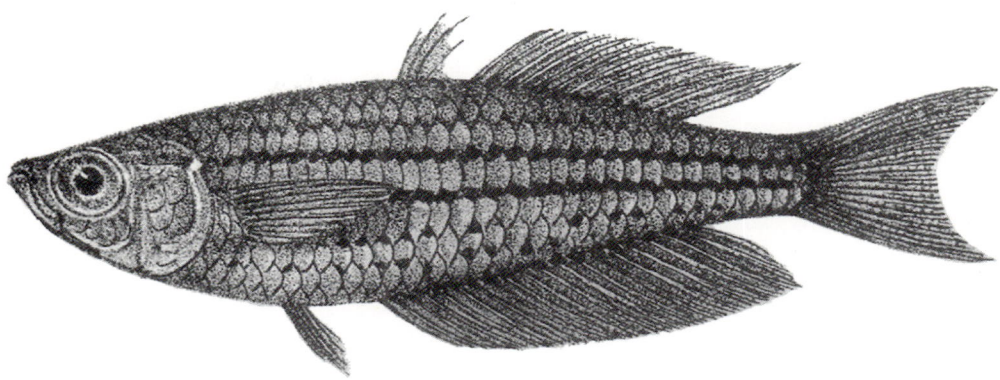

21.1 Softspined rainbowfish *Rhadinocentrus ornatus*. (C.T. Regan)

Description A small species that is slender and compressed, with a moderate-sized head and relatively large eye; upper edge of eye close to upper profile of head; mouth very oblique, upper jaw protruding slightly. Jaw teeth conical to canine-like, one or more rows extending outside of mouth; vomer and palatines toothless. Two dorsal fins (III–V, 11–15) separated by a small gap. First dorsal fin origin about midway between snout tip and tail base, above about 5th–6th ray of anal fin (I, 18–22). Anal fin originating on anterior $^1/_2$ of body; usually only last few anal and dorsal fin rays branched. Tail slightly forked, with 10–13 branched rays; pectoral rays 11–13. Scales relatively large, 31–37 in midlateral series; about 8 horizontal rows on side of body.

Sexual dimorphism Adult males have longer fin rays in the 2nd dorsal and anal fins. In addition, the rear tips of the dorsal and anal fins are pointed in males and rounded in females. Males tend to be deeper-bodied and exhibit brighter body and fin colours, particularly during courtship and spawning

Colour Pattern and colouration variable according to locality and environment, but ground colour usually translucent, light brown to bluish,

Softspined rainbowfish *Rhadinocentrus ornatus*. (G. Schmida)

FAMILY MELANOTAENIIDAE

Softspined rainbowfish *Rhadinocentrus ornatus*. (N. Armstrong)

or occasionally pink. Scales on side of body have dark brown to blackish margins; sometimes the heavier pigmentation on the upper and lower margins of the midlateral scale row forms a pair of parallel dark stripes. Scales on the upper portion of sides are frequently iridescent blue. Fin colouration variable, often orange to reddish with blackish to iridescent blue-green margins. There are two distinct fin colour varieties that appear to be unrelated to geography as they sometimes occur in the same stream. One variety is characterised by a wine-red hue on the fins, the other by bluish fins.

Size Males reach a maximum total length of about 80 mm and females 55–65 mm.

Distribution Inhabits coastal areas east of the Great Dividing Range between Coffs Harbour in northern New South Wales and Byfield in southern Queensland; also occurs on some low sandy islands of southern Queensland, including North Stradbroke, Moreton and Fraser Islands. Fraser Island was formerly believed to be the northern limit of distribution, but specimens were recently collected from Sandy Creek, near Byfield (22°49'S, 150°39'E), representing a northward extension of about 320 km.

Conservation status Relatively abundant on the Queensland islands but has a patchy localised distribution on the mainland. Habitat is apparently under threat in some areas.

Natural history Inhabits coastal streams in sand-soil or wallum swamp country, where it is frequently found in stagnant pools or streams with little or no flow. The water is often strongly stained and tea-like in appearance and usually very soft and acidic. Temperature range is generally 20–28°C and pH value 5.4–6.5. During the breeding season a few eggs are laid among plants each day. Hatching takes 7–10 days. Food includes insects and their aquatic larvae, microcrustaceans and algae. Is an active swimmer that congregates in small schools. The life span is about 3–4 years.

Utility Is well known to aquarists but is slightly more difficult to maintain than most other rainbowfishes. Will accept live, frozen and dried aquarium foods. The fry grow quickly and may reach 20 mm in a few months. Natural populations are probably useful for mosquito control.

Similar species Could possibly be confused with small crimsonspotted rainbowfishes (*Melanotaenia duboulayi*), especially females with low, bluntly rounded fins.

Other names Common: ornate rainbowfish, southern softspined rainbowfish, sunfish or jewelfish, Fraser or Moreton Island sunfish, neon sunfish, porthole fish. Scientific: none.

Literature Allen, 1982; Allen & Cross, 1982; Hansen, 1992; Leggett, 1983; Marshall, 1988; Merrick & Schmida, 1984.

DUBOULAY'S RAINBOWFISH *Melanotaenia duboulayi* (Castelnau)

21.2 Duboulay's rainbowfish *Melanotaenia duboulayi*, adult male. (B. Thorn)

Description A small species that is slender and compressed when young but increasing in depth with age. Head is moderate in size, with a relatively large eye; upper edge of eye close to upper profile of head. Mouth very oblique, upper jaw protruding strongly. Jaw teeth conical to canine-like, several rows extending outside of mouth; teeth present on vomer and palatines. Two dorsal fins (V–VIII, I, 8–13), separated by a small gap. First dorsal fin origin closer to snout tip than to tail base, slightly in front of origin of anal fin (I, 15–21). Tail slightly forked, usually with 15 branched rays; pectoral rays 12–15. Scales relatively large, usually 33–35 (rarely 36) in midlateral series; about 11–13 horizontal rows on side of body; usually 8–11 (occasionally 7 or 12) scales occur on cheek.

Sexual dimorphism Mature males have a higher 1st dorsal fin (it overlaps 2nd dorsal fin when depressed, whereas that of females falls short of 2nd dorsal). In addition, the rear tips of the dorsal and anal fins are pointed in males and rounded in females. Males have a deeper body than females and exhibit brighter body and fin colours, particularly during courtship and spawning.

Colour Ground colour olive-brown on the back, silvery to greenish on sides and whitish on belly and underside of head. Most of the body scales have a dusky brown margin, and there is a narrow reddish stripe (faint or absent in females) between each row of scales. Frequently a diffuse, blackish longitudinal band along middle of side and usually a bright red spot on upper portion of gill cover. Fins mainly clear in juveniles and females, but males usually have red spotting on dorsal, anal and caudal fins. Courting males usually exhibit a blackish margin on all fins.

Size Males may reach at least 130 mm, but most do not exceed 90 mm in the wild; females usually under 75 mm.

Distribution Inhabits coastal drainages east of the Great Dividing Range between the Macleay River in northern New South Wales and the Burnett River in southern Queensland; also occurs on Fraser Island.

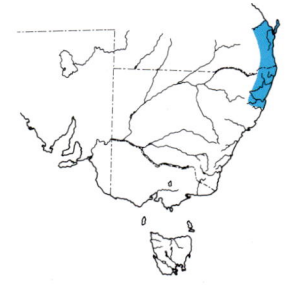

Conservation status Relatively common over much of range.

Natural history Inhabits a variety of environments, including rivers, creeks, drains, ponds, dune lakes and reservoirs. Is found in both coastal sand-soil habitats and forested areas farther inland, and sometimes co-occurs with the softspined rainbowfish (see above). Water conditions vary from very soft and acidic (coastal) to relatively alkaline (inland). Temperature range generally 16–28°C and pH value 5.4–7.8. During the breeding season a small batch of eggs is laid among plants each day. These hatch in about a week at 25–29°C. Food includes insects and their aquatic larvae, microcrustaceans and algae. Is an active swimmer that congregates in small to large schools, often near aquatic vegetation, log snags or other debris. Life span in the wild is probably about 3–4 years, but captive specimens may live twice as long.

Utility A popular aquarium fish, both in Australia and overseas, and was probably the first Australian

FAMILY MELANOTAENIIDAE

Duboulay's rainbowfish *Melanotaenia duboulayi*. (N. Armstrong)

fish introduced to the overseas aquarium trade, in the 1930s. Will accept live, frozen and dried aquarium foods. Captive specimens are readily bred and the fry are easy to raise. Natural populations are probably useful for mosquito control.
Similar species *Melanotaenia duboulayi* was formerly considered a synonym of *M. fluviatilis* (treated below). However, recent study indicates that the species are distinct. Although morphological differences are slight, they can be genetically separated by electrophoretic analysis. In addition, there are clear differences in egg characteristics (e.g. size, number of oil droplets, colour, etc.) and larval development.
Other names Common: common or spotted sunfish, crimsonspotted rainbowfish, crimsonspotted jewelfish, pinkear, Australian rainbowfish. Scientific: *Melanotaenia fluviatilis*, *Melanotaenia splendida fluviatilis*.
Literature Allen, 1989; Allen & Cross, 1982; Crowley *et al.*, 1986.

CRIMSONSPOTTED RAINBOWFISH
Melanotaenia fluviatilis (Castelnau)

Description A small species that is slender and compressed when young but increasing in depth with age. Head moderate in size, with a relatively large eye; upper edge of eye close to upper profile of head. Mouth very oblique, and upper jaw protrudes strongly. Jaw teeth conical to canine-like, several rows extending outside of mouth; teeth present on vomer and palatines. Two dorsal fins (V–VIII, I, 10–13), separated by a small gap. First dorsal fin origin closer to snout tip than to caudal base, slightly in front of origin of anal fin (I, 17–21). Tail slightly forked, usually with 15 branched rays; pectoral rays 11–14. Scales relatively large, usually 35–37 (occasionally 34 or 38) in midlateral series; about 10–13 horizontal rows on side of body; usually 11–16 (rarely 10) scales on cheek. Adult fish from Victoria frequently develop a blunt, rounded snout and very broad interorbital region.
Sexual dimorphism Mature males have a higher 1st dorsal fin (it overlaps 2nd dorsal fin when depressed; that of females falls short of 2nd dorsal). In addition, the rear tips of the dorsal and anal fins are pointed in males and rounded in females. Males tend to exhibit brighter body and fin colours, particularly during courtship and spawning. Females are generally more slender than males.
Colour Ground colour silvery on sides, with a

RAINBOWFISHES

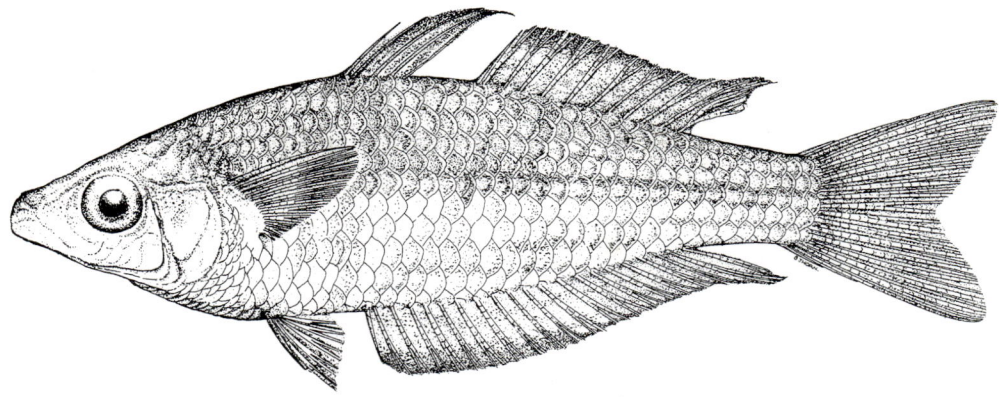

21.3 Crimsonspotted rainbowfish *Melanotaenia fluviatilis*, adult male. (B. Thorn)

greenish iridescence, and whitish on belly and lower portion of head. Most body scales have a dusky brown margin, and there is sometimes a narrow reddish stripe (faint or absent in females) between each row of scales. Usually a reddish spot on upper portion of gill cover. Fins mainly clear in juveniles and females, but males usually have red spotting on dorsal, anal and caudal fins. Courting males usually exhibit a blackish margin on all fins.

Size Males may reach a length of about 90 mm, but females are usually under 60–70 mm.

Distribution Inhabits the inland Murray-Darling system of South Australia, northern Victoria, New South Wales and southern Queensland.

Conservation status Relatively common over much of its range, although population numbers in Victoria have fluctuated greatly in recent years. Sampling at several Victorian sites failed to produce specimens after a drought in 1982–83.

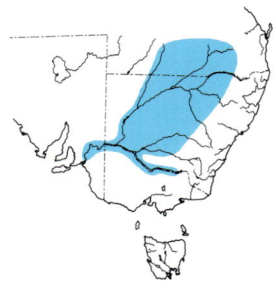

Crimsonspotted rainbowfish *Melanotaenia fluviatilis*. (G. Schmida)

Natural history Habitats includes rivers, creeks, drains, billabongs, ponds and reservoirs. Victorian populations exist in a marginal habitat situation as far as melanotaeniids are concerned. Cold winter temperatures must certainly be the dominant factor in limiting the size of populations. Experiments indicate that Victorian fish can survive a period of several days at 7°C but are prone to a high incidence of protozoan and bacterial infection at low temperatures. Spawning is seasonal, at least over the southern part of range. In Victoria it occurs mainly between October and January. Each female lays several eggs a day and these are summarily fertilised by the male. The eggs attach to aquatic plants by means of adhesive threads. Hatching occurs in about one week at 25–29°C. Food includes insects and their aquatic larvae, microcrustaceans and algae. This rainbowfish is an active swimmer that congregates in small to large schools, often near the surface.

Utility Useful as an aquarium fish. Care and feeding are similar to those of Duboulay's rainbowfish. Wild fish feed on mosquito larvae, thus helping to control this pest.

Similar species See this section in *Melanotaenia duboulayi*, above.

Other names Common: Murray River rainbow, inland rainbow. Scientific: *Melanotaenia fluviatilis*, *Melanotaenia splendida fluviatilis*.

Literature Allen, 1989; Allen & Cross, 1982; Crowley *et al.*, 1986; Merrick & Schmida, 1984; Romanowski, 1986.

22
Family Pseudomugilidae
Blue-eyes

W. IVANTSOFF AND L.E.L.M. CROWLEY

Blue-eyes are all small, colourful fishes found mainly in coastal freshwater, brackish or estuarine habitats. There are 12 species found throughout eastern and northern drainages of Australia as well as in Papua New Guinea, Irian Jaya and some islands of eastern Indonesia, but only two are found in south-eastern Australia. Members of this family have formerly been included in the family Atherinidae, but studies have shown that they are a separate family.

Key to blue-eyes

Head pores large, mandibular pores present.......... ***Pseudomugil signifer*** p. 141; Fig. 22.1
Head pores small, mandibular pores absent .... ***Pseudomugil mellis*** p. 143; Fig. 22.2

SOUTHERN BLUE-EYE *Pseudomugil signifer* Kner

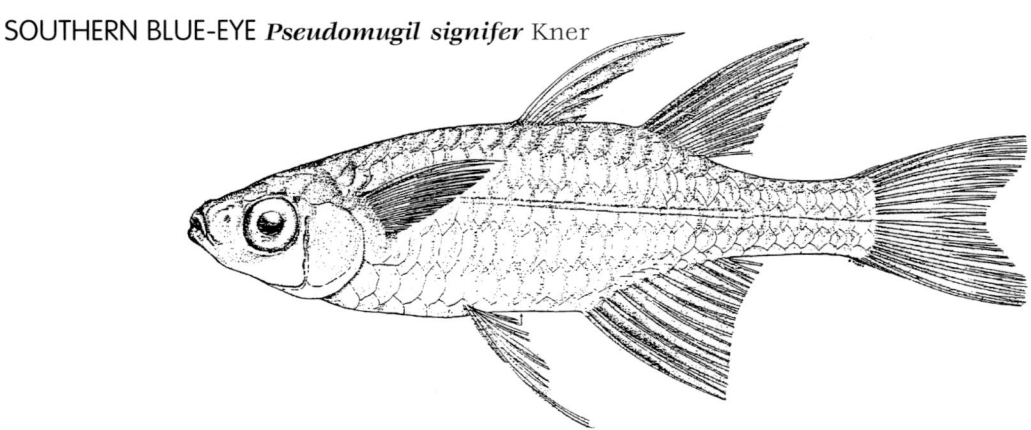

22.1 Southern blue-eye *Pseudomugil signifer.* (B. Thorn)

Description A small, moderately compressed, elongate fish with a very large eye (about 2.8 in head); large sensory pores on dorsal surface of head, and also large mandibular pores. Mouth protrusible and subvertical, with thick, rubbery lips. Posterior, premaxillary teeth enlarged and exposed when mouth is shut. First dorsal fin (III–VI) origin about midway along dorsal surface, about level or just behind pectoral (i, 8–13) tips; 2nd dorsal fin (i, 6–10) origin just behind origin of anal fin (i, 9–12). Scales along sides of body (midlateral 25–31) relatively large, dorsoventrally elongated and in 5–6 even rows. There are 9–12 gill rakers in the 1st lower gill arch, including those in angle of arch.

Sexual dimorphism Male differs from female in having the dorsal, anal, and pelvic fins with extended filaments. This is particularly pronounced in the 1st dorsal, where the filaments lie well past origin of 2nd dorsal fin; pelvic fin rays of males are extended to lie past origin of anal fin. In females, 1st dorsal fin just reaches origin of 2nd, and pelvic fins do not reach anal fin. Colouration also varies between sexes. In males the fins are

FAMILY PSEUDOMUGILIDAE

Southern blue-eye *Pseudomugil signifer.* (R.H. Kuiter)

spectacular, particularly at breeding, but in females they are clear to yellowish, without markings.

Colour The most notable features of these fishes are the brilliantly coloured fins of males, though this is not constant. First dorsal fin has a dark blotch at base, with white extended filaments at the front, followed by yellowish to dark spines. Second dorsal and anal fins have an extensive black marking at front, followed by white tips to the extended rays. The rest of the fin is yellowish. Caudal fin has white along dorsal and ventral edges, then several dark to black rays, with a dusky middle. Pectoral fins dusky, and pelvic fins white or yellowish. The scales are outlined by melanophores, forming a reticulate pattern over whole body. Eye bright blue and opercle iridescent, with a midlateral iridescent sheen along body.

Size Greatest known length of males is 88 mm, females usually smaller, only to about 62–63 mm.

Distribution Widespread throughout eastern drainages, from northern Queensland as far south as Narooma in New South Wales (South-East Drainage), and also found in lakes and streams of

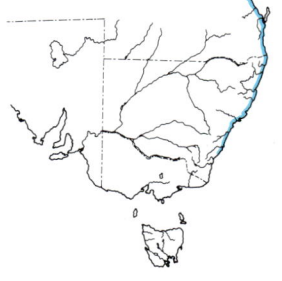

the sand islands of south-eastern Queensland. May occasionally be found in rainforest streams on the escarpment.

Conservation status Abundant in fresh or brackish coastal waters but does not usually penetrate far inland and is not usually abundant in rainforest streams.

Natural history The natural history of these fishes has been studied in captivity. The female sheds 1–2 eggs at each spawning, and up to 9 eggs may be laid a day at peak of breeding season. Eggs (about 1.8 mm in diameter) have adhesive filaments that attach to vegetation at the spawning site. Development takes about 18–21 days at 22–24°C; the larvae are well developed at hatching and take food immediately. Under aquarium conditions they grow rapidly and reach sexual maturity at about 6 months.

Utility Is a very popular aquarium fish as it is very colourful and easy to keep and breed. In the wild is a voracious feeder on mosquito larvae and other insects and so probably assists in reducing numbers of small biting insects.

Similar species *Pseudomugil mellis.*

Other names Common: Pacific blue-eye; common blue-eye. Scientific: *Pseudomugil signatus.*

Literature Howe, 1987; Saeed *et al.*, 1989; Semple, 1986; Merrick & Schmida, 1984.

HONEY BLUE-EYE *Pseudomugil mellis* Allen and Ivantsoff

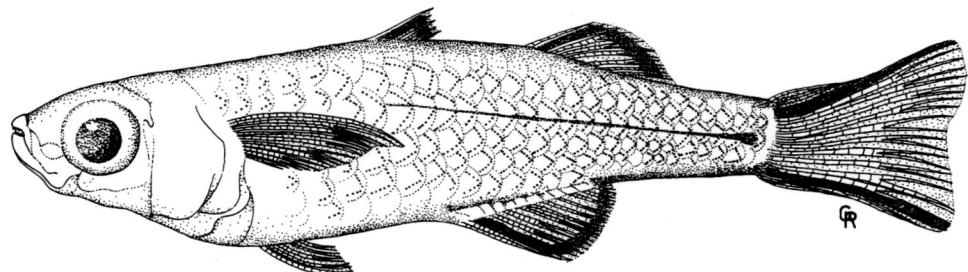

22.2 Honey blue-eye *Pseudomugil mellis*. (G. Rankin)

Description A small fish with head tapering towards snout, eye (as in most blue-eyes) large. A few small sensory pores are present on head above and below eyes but no large mandibular pores (as in *P. signifer*). Mouth small, with small villiform teeth on jaws; posterior maxillary teeth not visible when mouth is closed. First dorsal fin (IV–V) origin in front of tips of pectoral fin (i, 8–11); 2nd dorsal (i, 5–8) origin well behind origin of anal fin (ii, 8–11), which is in line with, or just behind, last ray of 1st dorsal; pectoral fin placed high on side of body. Scales relatively large (midlateral scale count 26–29, transverse 5–6) and in even rows along sides of body; 10–13 gill rakers in 1st lower gill arch, including those in angle of arch.

Sexual dimorphism Males have extended pelvic, 1st and 2nd dorsal and anal fins, which are much larger than in females, particularly 1st dorsal. Body and fin colour of males is also more distinct than in females.

Colour Body honey- to copper-coloured when in breeding condition, particularly in males. Top of head dusky, with melanophores looking like fine grains of pepper. A faint reticulate pattern is present at edges of scales. A single dark midlateral line runs along side of body from pectoral base to base of tail. Eyes blue and cheeks iridescent blue. First dorsal fin principally black, with a white leading edge; 2nd dorsal and anal fins of males are clear to honey coloured and edged with a broad black band with white outside. In females fins are clear to pale honey coloured.

Size Not known to exceed 38 mm.

Distribution Range very restricted; found only in wallum country in south-eastern Queensland from about Brisbane north to Bundaberg, and also on Fraser Island.

Conservation status Has never been considered abundant and, with the development for housing and tourism of much of the area where it occurs, habitat is rapidly being destroyed. The Australian Society for Fish Biology lists this species as 'vulnerable'.

Natural history Reproductive behaviour has been studied in captivity; males appear to be territorial, guarding vegetated spawning sites. In small aquaria, aggressive behaviour occasionally led to death of one of males. Males were sexually active from about 20 mm (SL), with females slightly larger at onset of breeding. Females released 1–16 eggs at each spawning. Water-hardened eggs were 1.3–1.6 mm in diameter. Larvae hatched about 112 hours after spawning, the larvae (from 3.6 mm SL) retained the yolk sac and took food 3–80 hours after hatching. Growth to maturity took about 3 months in aquarium conditions.

Utility This small, attractive fish is easy to keep and so is sought by aquarists, particularly those interested in Australian fish. Probably provides part of the diet of birds and larger fish.

Similar species Southern blue-eye, juvenile sand mullet.

Other names Common: none. Scientific: *Pseudomugil signifer; P. signatus*.

Literature Allen & Ivantsoff, 1982; Saeed *et al.*, 1989; Semple, 1991; Merrick & Schmida, 1984.

23

Family Scorpaenidae

Scorpionfishes

D.A. POLLARD AND P. PARKER

Members of this large family are distributed worldwide, with about 40 representatives in Australian waters, most of them marine. They are characterised by numerous sharp and often venomous spines on the head and fins. One species occurs in fresh to brackish waters of south-eastern Australia.

BULLROUT *Notesthes robusta* (Günther)

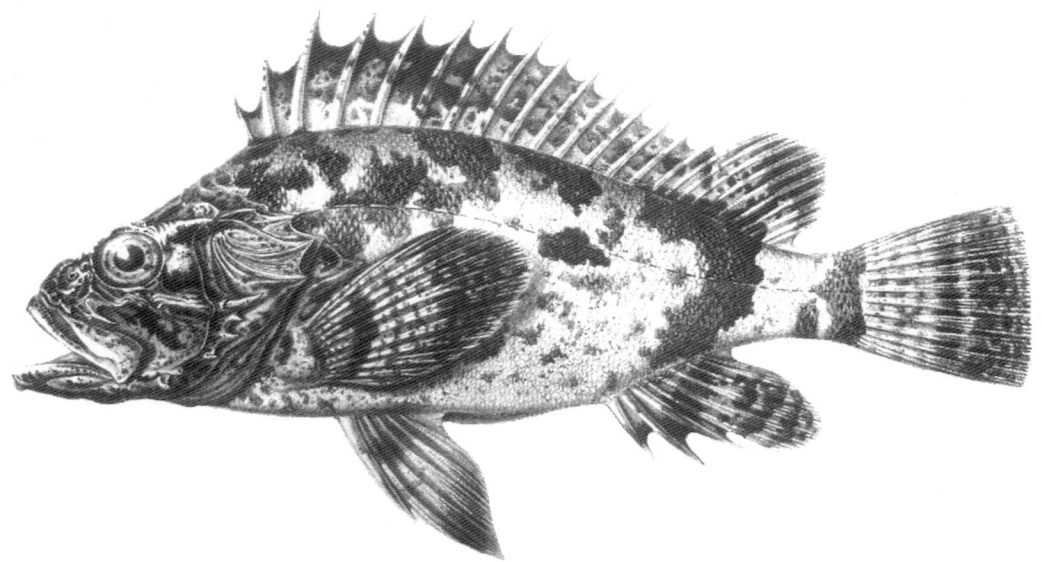

23.1 Bullrout *Notesthes robusta*. (F. Steindachner)

Description A moderate-sized, stocky fish with a large broad head that has numerous very sharp, stout spines; trunk somewhat compressed and elliptical. Mouth large, with lower jaw longer than upper. There is a single dorsal fin (XV, 9–10), originating well forward above back of head, the fin evenly rounded in front, becoming concave towards rear. Dorsal fin spines of moderate strength and pungent; last dorsal ray attached to caudal peduncle by a fine membrane. Anal fin (III, 5) small. Pectoral fins rounded and symmetrical, with no free ray ventrally; pelvic fins inserted a little behind pectoral bases. Scales small (about 85 in lateral line), with none on head. There are 16 gill rakers.

Colour Colour variable, allowing the bullrout to remain inconspicuous among rocks and weeds; generally brown with irregular black marbling or blotching that may form broad transverse bands. The fins are mottled with blue-grey to yellow and black, and there is usually a large black-mottled area in front of dorsal fin.

Size Reaches about 300 mm in length, and is commonly found up to about 200 mm.

Distribution Occurs from tidal estuaries upstream into fresh waters in rivers flowing

Bullrout *Notesthes robusta*. (N. Armstrong)

eastwards from the Great Dividing Range, from about the Daintree River in northern Queensland to the Clyde River in southern New South Wales; has also occasionally been taken from the sea.

Conservation status Though not particularly common anywhere, is not considered threatened.

Natural history Usually occurs in slow-flowing rivers, where it leads a sedentary life among weeds and stones. When aroused, it erects venomous spines on the head and dorsal, anal and pelvic fins; for this reason it should be handled with great care, even when dead. (Immediate pain develops when the skin is punctured by a spine and may become unbearable. Rapid local irrigation through puncture with 2–5 ml of 1–2% lignocaine provides relief.)

Little is known about breeding, although it has sometimes been thought to be catadromous. The occurrence of tiny juveniles 20–30 mm long many kilometres upstream and of populations above dams and weirs, however, suggests that the whole life cycle can be completed in fresh water. There is little information on diet, though it feeds in captivity on shrimps and small fish and is probably a generalised carnivore.

Utility Like many scorpaenids, is a good table fish; because of its small size and low abundance, however, it is seldom eaten. Is an impressive aquarium fish, having bold and striking colouration, and is easily kept in captivity.

Similar species None in fresh waters.

Other names Common: kroki. Scientific: *Centropogon robustus*.

Literature Allen, 1989; Harris & Pearn, 1987; Merrick & Schmida, 1984; Ogilby, 1903.

24
Family Chandidae
Glassfishes, chanda perches

G.R. ALLEN

The family Chandidae contains small, partly transparent fishes (mainly under 100 mm long) that inhabit coastal seas and inland waters in the Indo-Pacific region. About half of the approximately 40 species are restricted to fresh water, and six are native to northern and eastern Australia. Another seven species also occur in Australia but are confined mainly to brackish estuaries or coastal seas, sometimes venturing into the lower reaches of freshwater streams. Glassfishes are apparently nocturnal. During the day they usually form stationary aggregations among vegetation. Family characters include two dorsal fins (or a single fin that is very deeply notched), the presence of tiny spines (serrae) on various head bones, and a variable lateral line (complete in some species, interrupted in others, or nearly obscure in a few species). The names Chandidae and Ambassidae have been used by previous authors for this family. However, in view of its chronological priority and more widespread usage, Chandidae is the appropriate designation. Two species occur in the region covered by this book.

Key to glassfishes

Supraorbital spines 2–4; soft anal rays usually 10 (rarely 11); gill rakers on lower limb of 1st gill arch 22–24 . . . . *Ambassis marianus* p. 148; Fig. 24.2
A single supraorbital spine; soft anal rays 7–9; gill rakers on lower limb of 1st gill arch 15–18 . . . . *Ambassis agassizii* p. 146; Fig. 24.1

OLIVE PERCHLET
Ambassis agassizii Steindachner

Description A small species with an oval, laterally compressed body. The moderately large and oblique mouth reaches just beyond anterior edge of very large eye. Dorsal fin (VII, I, 7–9) divided by a deep notch; anal fin (III, 7–9) similar in shape to soft portion of dorsal fin and directly below it; tail forked. There are two rows of scales on the cheek just below eye. Lateral line variable, occasionally absent, but usually consisting of 1–15 tubed scales that terminate below the spinous dorsal fin or anterior to it; occasionally it is continued in the form of a few weakly tubed or pitted scales on middle of caudal peduncle. Tiny spines (serrae) present on edge (or on bony ridges near edge) of preorbital and preopercle bones. There is also a single tiny spine above upper rim of eye.
Colour Generally semitransparent, with dusky scale edges imparting a network pattern on sides. Fins generally clear, except that there is often a broad, blackish band along margins of pelvic and anal fins.
Size Maximum length about 70–80 mm, but most specimens are under 60 mm.
Distribution Inhabits the Murray-Darling system of South Australia, Victoria, western New South Wales and southern Queensland. Is also present in coastal drainages of eastern Australia between

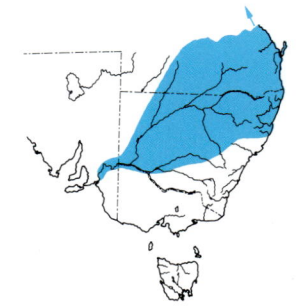

GLASSFISHES, CHANDA PERCHES

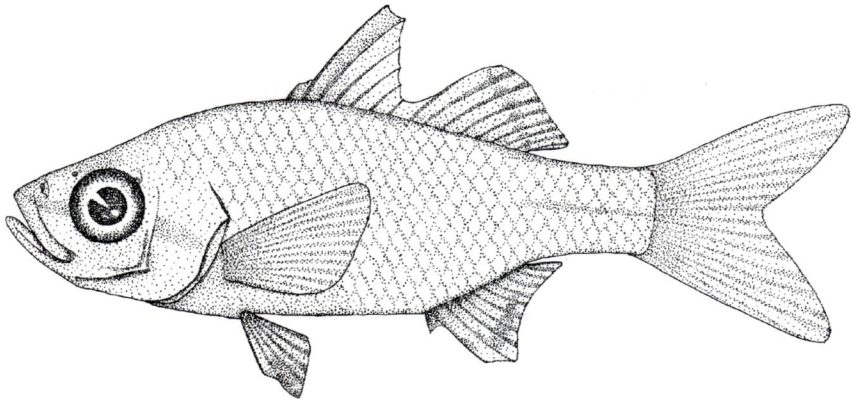

24.1 Olive perchlet *Ambassis agassizii*. (R.M. McDowall)

Lake Hiawatha in New South Wales and the Mowbray River in north Queensland.
Conservation status Relatively common throughout most of its range.
Natural history Habitat consists of rivers, creeks, drainage ditches, ponds and swamps. Small to large aggregations are common among log snags and aquatic vegetation. Spawning occurs in November and December, or when water temperature reaches 23°C. Eggs are adhesive, about 0.7 mm in diameter and are apparently scattered among vegetation. A female 49 mm long contained 2350 eggs. The diet consists largely of microcrustaceans and insects (including larvae).

Utility Suitable for captivity, though not as popular as other native fishes such as rainbows and gudgeons. In the wild it helps to control mosquito populations and is also an important food for larger predators such as grunters.
Similar species Estuary perchlet, juveniles of golden perch and redfin.
Other names Common: silver spray, doody, Agassiz's glassfish, western chanda perch. Scientific: *Ambassis castelnaui, Ambassis nigripinnis, Pseudambassis pallidus, Priops olivaceus*.
Literature Allen, 1989; Allen & Burgess, 1990; Merrick & Schmida, 1984.

Olive perchlet *Ambassis agassizii*. (G.R. Allen)

FAMILY CHANDIDAE

ESTUARY PERCHLET *Ambassis marianus* Günther

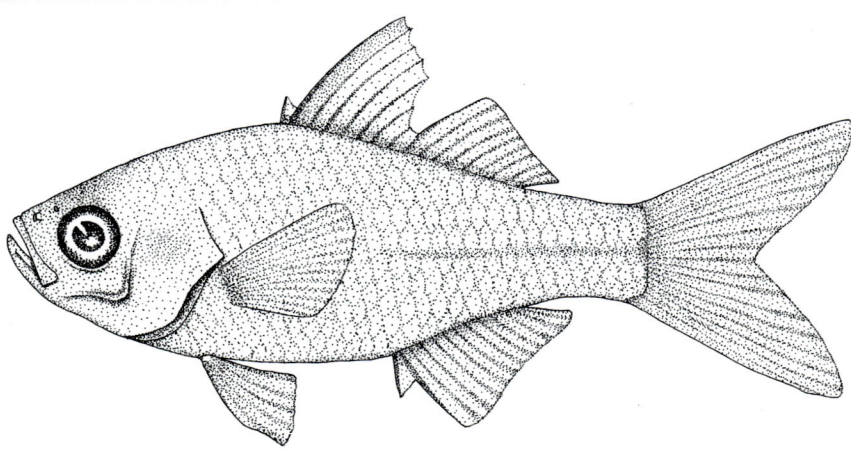

24.2 Estuary perchlet *Ambassis marianus*. (R.M. McDowall)

Description A small species with an oval, laterally compressed body. The moderately large and oblique mouth reaches just beyond anterior edge of very large eye. Dorsal fin (VII, I, 9–10) divided by a deep notch. Anal fin (III, 10–11) similar in shape to soft portion of dorsal fin and directly below it; tail forked. There are 2–3 rows of scales on cheek, just below eye. Lateral line interrupted by 3–8 tubeless scales in its middle portion. Tiny spines (serrae) are present on edge (or on bony ridges near edge) of preorbital and preopercle bones; also 2–4 serrae above upper rim of eye.

Colour Generally semitransparent, with dusky scale edges. Breast, belly and gill cover silvery, with usually a silvery midlateral band running along sides. Fins are clear, except that the membrane between 2nd and 3rd dorsal spines is blackish.

Estuary perchlet *Ambassis marianus*. (G.R. Allen)

GLASSFISHES, CHANDA PERCHES

Size Maximum known length about 90 mm, but most specimens are under 70 mm.

Distribution Known only from the south-eastern coast of Australia between Narooma, southern New South Wales and Maryborough, southern Queensland.

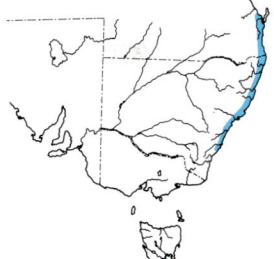

Conservation status Generally abundant throughout its range.

Natural history Habitat consists of brackish mangrove estuaries, tidal creeks and lower reaches of freshwater streams. Small to large aggregations are common among mangrove roots, log snags and aquatic vegetation. There is no information on the life history.

Utility Suitable for captivity, although is seldom maintained. Owing to its abundance in many areas, is probably an important food for larger predators.

Similar species Estuary perchlet, juveniles of golden perch and redfin.

Other names Common: silver perchlet, convex perchlet. Scientific: *Pseudambassis ramsayi*, *Pseudambassis convexus*.

Literature Allen, 1989; Allen & Burgess, 1990; Merrick & Schmida, 1984.

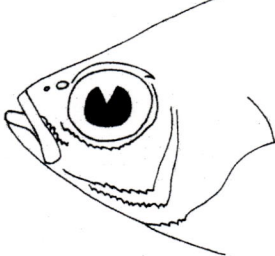

24.3
Supraorbital spines in olive perchlet *Ambassis agassizii* (left) and estuary perchlet *A. marianus* (right). (G.R. Allen)

25
Family Percichthyidae
Australian freshwater cods and basses

J.H. HARRIS AND S. J. ROWLAND

The percichthyids are a diverse, generalised percoid family closely related to the Serranidae—the gropers and their relatives. There are about 25 genera, including freshwater, estuarine, catadromous and oceanic species. Of the seven species and one subspecies in two genera found in Australian fresh waters, two are more or less catadromous, spawning in marine environments. Percichthyids are characterised by their single, long dorsal fin with a long spinous portion in front, strong fin spines, smallish ctenoid scales, a lateral line, and the 1st ray of the pelvic fins usually elongated and finger-like. Australian freshwater species have two opercular spines and all are carnivorous.

Key to cods and basses

1. Caudal fin forked or emarginate 2
   Caudal fin usually distinctly rounded, sometimes almost truncate 3

2. Gill rakers on lower arch of 1st gill 14–18; dorsal head profile concave; outer rays of pelvic fin uniformly greyish. .... ***Macquaria colonorum*** p. 157; Fig. 25.4
   Gill rakers on lower arch 12–15; dorsal head profile straight or only slightly concave; outermost ray of pelvic fin whitish .... ***Macquaria novemaculeata*** p. 155; Fig. 25.3

3. Head deep and laterally compressed, with a tapered snout; conspicuous open pores in lower jaw region; 42–63 scales in lateral line. 4
   Head broad and depressed, with a rounded snout; no conspicuous open pores; 63–82 scales in lateral line 5

4. Lower jaw protrudes in larger specimens; vertebrae 26; 40–56 scales transversely between front of dorsal fin and anal fin. .... ***Macquaria ambigua*** p. 151; Fig. 25.1
   Lower and upper jaws equal in larger specimens; 29–30 vertebrae; 24–33 scales transversely between front of dorsal fin and anal fin. ..... ***Macquaria australasica*** p. 153; Fig. 25.2

5. Lateral head profile straight; upper jaw longer than lower; blue-grey to dark brown dorsally, with speckled pattern of dark grey to black spots and small bars dorsally and laterally; 14 precaudal vertebrae .... ***Maccullochella macquariensis*** p. 162; Fig. 25.7
   Lateral head profile concave; jaws equal or lower jaw protruding; distinct mottled pattern dorsally and laterally; 15 precaudal vertebrae 6

6. Cream to olive to yellow green, with pale to dark green reticulated mottling dorsally and laterally, sometimes with small dark spots between reticulations; ventral surface white to creamy white; pelvic fins whitish, relatively long, 13.2–16.4% of standard length; otolith 2.5–2.9 total length, has a prominent antirostrum ....
   .... ***Maccullochella peelii peelii*** p. 158; Fig. 25.5
   Pale green to yellow green to golden, with black to very dark green, heavily reticulated mottling dorsally and laterally, this sometimes extending onto ventral surface, which is dark grey to whitish; pelvic fins clear, relatively long, 16.4–20.7% of standard length; otolith 2.9–3.5% total length, has a small antirostrum 7

FRESHWATER CODS AND BASSES

7   Depth of caudal peduncle 12.3–14.2% of standard length; orbit length 10.0–13.8% and interorbital width 25.1–30.4% in head length; 30–34 scale rows below lateral line; 5th/6th dorsal spine 6.1–9.8% of standard length; 12–13 anal rays; length of 1st anal pterygiophore 79–86% of distance from its base to 1st caudal vertebra; otolith length 2.9–3.3% of total length .... ***Maccullochella peelii mariensis*** p. 160
Depth of caudal peduncle 10.7–12.5% of standard length; orbit length 12.0–19.0% and interorbital width 20.8–27.5% in head length; 32–52 scale rows below lateral line; 5th/6th dorsal spine 8.5–13.0% of standard length; 11–12 anal rays; length of 1st anal pterygiophore 89–97% of distance from its base to 1st caudal vertebra; otolith length 3.2–3.5% of total length. .... ***Maccullochella ikei*** p.161; Fig. 25.6

GOLDEN PERCH *Macquaria ambigua* (Richardson)

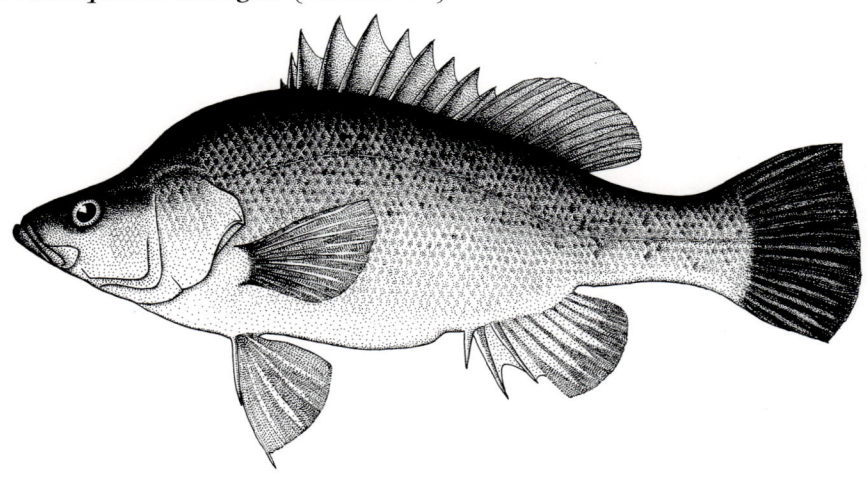

25.1 Golden perch *Macquaria ambigua*. (J. Hannan)

Description A moderate-to-large fish, elongate-oval and laterally compressed. Eyes large and dorsal profile of head strongly convex, with a markedly arched nape, particularly in large fish; lower profile of body weakly convex. Mouth large, extending to below middle of eyes, lower jaw protruding slightly. Margin of preoperculum finely serrate, operculum with two flat spines, lower largest. Large open pores present on snout, lower jaw and lower preoperculum. Scales of moderate size (50–63 along lateral line); snout naked but cheeks and gill covers with small ctenoid scales. Dorsal fin continuous, without a distinct notch between spiny and soft-rayed portions (VIII–XI, 11–13), 4th–5th spines longest. Tail rounded. Anal fin III, 7–10; pectorals 15–18. First ray of pelvic fin elongated into a filament, fin inserted below pectoral bases. Vertebrae 24–28; gill rakers 18–24.
Colour Varies from olive green through bronze and yellow to white, with lower surface normally lighter. Colouration appears to be related to water colour. Fins range from yellow with red margins to white.
Size Recorded up to 760 mm and 23 kg, but fish up to 5 kg are generally considered of good size. Fish stocked in impoundments are often larger than those in rivers.
Distribution Lives throughout the Murray-Darling system except at higher altitudes, and also in the Lake Eyre and Bulloo internal drainage systems of Queensland, New South Wales and South Australia; also occur in the Dawson River system of coastal central Queensland. Clear genetic differences indicate that several population groups have been isolated from each other for considerable, and different, lengths of time. Fish of the central Australian drainages (Lake Eyre and Bulloo River) have recently been identified as a separate species, with significant differences between the two populations. Similar genetic differences were also seen between the Dawson and Murray-Darling populations, and relatively isolated stocks,

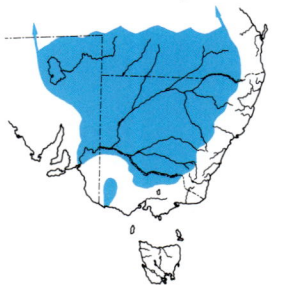

FAMILY PERCICHTHYIDAE

Golden perch *Macquaria ambigua*. (R.H. Kuiter)

with limited gene-flow, occur within the Murray-Darling basin.

Are occasionally found as escapees from farm dams and impoundments in coastal rivers where they have been introduced in Queensland and New South Wales.

Conservation status Like other native species, is now rare or has disappeared from large areas in Murray-Darling tributaries and higher reaches of the main channels. The main cause of this decline has been habitat degradation, especially through numerous dams and weirs obstructing migration and altering the natural regimes of streamflow and water temperature. The movements of juveniles during small river rises are likely to be impeded by even low weirs as these remain effective barriers at such times.

Natural history Is mainly a fish of warm, turbid, slow-flowing inland rivers and their floodplain lakes and anabranches and is well adapted to the widely varying streamflow conditions. Juveniles are known to migrate upstream during small rises in river levels at all times of day and night, though especially at dawn and dusk. Females are usually around 400 mm (1.4 kg) and 4 years old before maturing, but stunted males may mature at 190 mm and 2–3 years. Fish have been aged to 26 years. They spawn in floods during spring and summer when water temperature exceeds about 23°C. The female can hold her ova at an advanced stage of development until conditions are suitable for spawning; if these do not occur, the ovaries regress until the following season. Fish of 2.5 kg produce well over 500,000 eggs. When water-hardened after being shed, eggs are about 4 mm in diameter, have a thin, smooth chorion and are transparent, colourless, non-adhesive and semi-buoyant. Development is pelagic in floodwaters and takes 24–33 hours at 20–31°C. Newly hatched larvae are 3.2 mm long, unpigmented, buoyant and float upside-down. They are positively phototactic, swim actively and start to feed on zooplankton after about 5 days. Metamorphosis begins at 25 days when 12 mm long. Young fish reach about 150 mm in their first year and about 430 mm in their fifth year in the Murrumbidgee River. In earthen ponds their food is mainly crustaceans of suitable size and small larval insects are also eaten. Young fish can be handled readily and survive temperatures of 4–37°C.

Are migratory, adults making long-distance upstream migrations during high flows, probably to colonise habitats that are only occasionally accessible. Juveniles and sub-adults make strong upstream dispersal migrations during spring and summer, being stimulated to move by rises in stream flow, with peaks in activity in the evening and early morning.

Utility Remains sufficiently abundant in the lower Murray-Darling rivers to provide roughly $1/3$–$1/2$ the catch in a small, multi-species commercial fishery, which averages about 200 tonnes a year for all finfish species. Is taken commercially in drum nets in rivers and gill nets in still waters. Is also one of the most significant species in the extensive recreational fisheries in inland rivers, lakes and impoundments. Is readily caught by angling with bait or lures, but bites spasmodically.

Techniques are well established for large-scale artificial propagation using hormone induction of spawning, and rearing in earthen ponds. Large numbers of juvenile fish are produced in government and private hatcheries for stocking of impoundments and farm dams. Though they survive and grow in many diverse dams, they only rarely breed there, apparently when substantial flooding occurs. Government controls have been placed on the translocation and introduction of golden perch into stream systems. These are necessary because of the risks to genetically distinct wild populations of damage resulting from mass stocking, as well as the risk of disease transfers.

Similar species Macquarie perch; small fish may be confused with small silver perch, spangled perch, goldfish, carp, pigmy perch, Murray cod and trout cod.

Other names Common: yellowbelly, callop, perch, Murray perch, white perch. Scientific: *Plectroplites ambiguus*.

Literature Anderson *et al*, 1992b; Koehn & O'Connor, 1990; Lake, 1967a, b, c; McDonald, 1976; MacKay, 1973; Mallen-Cooper, 1994; Merrick & Midgley, 1985; Merrick & Schmida, 1984; Musyl & Keenan, 1992; Rowland, 1983b; Stephenson & Grant, 1957.

MACQUARIE PERCH *Macquaria australasica* Cuvier

25.2 Macquarie perch *Macquaria australasica*

Description A moderate-sized fish, elongate-oval and laterally compressed; snout profile convex and nape arched. Eyes large and jaws of equal length, with mouth reaching to below front of eyes. There are large open pores around eyes and on preoperculum, and the two opercular spines are serrated. Dorsal fin divided by a moderate notch (VIII–XII, I, 11–14), 4th and 5th spines longest. Anal fin III, 8–11, 2nd spine longest. Pelvic fins inserted slightly behind base of pectoral fins (14–17 pectoral rays). Tail rounded. Scales of moderate size (42–60 along lateral line), snout naked. Vertebrae 28–31; gill rakers 17–21.

Colour In the Murray-Darling basin varies from almost black or dark silvery grey to bluish grey or even green-brown above, paler to off-white below, often with a yellowish tinge. Pelvic fins often rosy, with black margins; pectorals yellowish, other fins often with a purplish tinge. Fish from east-flowing Shoalhaven and Hawkesbury River systems are blotched with grey-brown, buff and dark greyish patches over head and body and can be pale grey-brown when living in shallow sandy streams.

Size In western drainages has been recorded up

to 460 mm and 3.5 kg but is not commonly more than 1.5 kg. Fish in eastern drainages are distinctly smaller in maximum size and size at maturity, generally reaching less than 180 mm.

Distribution It is likely that there are two species of Macquarie perch, living in western and eastern-flowing streams respectively, as well as other genetic groups within these species. Although clear genetic differences have been shown, separate species have not yet been described. The western form in the Murray-Darling system is typically found in cooler upper reaches of tributaries from the Lachlan River southward, but not extending to the sources of these streams. Is also recorded from lowland reaches, such as the Barmah Lakes area near Deniliquin. Persists in some impoundments, notably Dartmouth Dam on the Mitta Mitta, with relict populations in older storages where they have declined, such as Burrinjuck, Googong and Cotter dams (Murrumbidgee), Wyangala Dam (Lachlan) and Eildon Reservoir (Goulburn). These western Macquarie perch have been established in southern Victorian streams, including the Yarra, and may also have been the source of the Mongarlowe River population in the east-flowing Shoalhaven system. The eastern form of the Macquarie perch is reasoably abundant in the Hawkesbury and Shoalhaven systems, including the large impoundments, but lives mostly upstream of populations of Australian bass. As with golden perch, the identification of genetically distinct populations has prompted new controls over the translocation and introduction of this species into new waters.

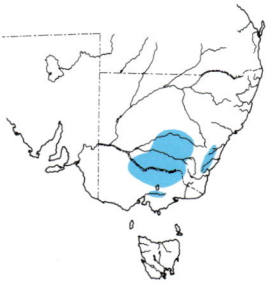

Conservation status Is a threatened species, formally classified as 'indeterminate' because of its limited distribution and long-term declines in abundance, though not enough is known to be certain of its status; is fully protected in New South Wales. Overfishing has contributed to past declines. Its survival is now threatened by habitat degradation, especially siltation and river regulation, and perhaps by interactions with trout, redfin perch and other alien fishes. Redfin perch also carry a viral disease, epizootic haematopoietic necrosis (EHN), which is fatal to Macquarie perch and was probably implicated in its decline in

Macquarie perch *Macquaria australasica*. (N. Armstrong)

areas where the two species' distributions overlap.
Natural history Occurs widely in riverine and lake habitats. Males breed at two years and 210 mm, females at three years and 300 mm. They spawn in shallow upland streams in October and November when the temperature reaches about 16°C. Fish in impoundments make an upstream spawning migration; whether this occurs in riverine populations is not certain. The spherical, demersal, adhesive eggs have been found among stones and gravel in riffle areas up to 50–75 cm deep, with a flow of about 1 metre per second. Fecundity is about 32,000 eggs per kg of fish weight. Ova in ripe females are opaque and pale creamy-amber in colour, 1–2 mm in diameter, swelling to 4 mm after spawning. When water-hardened, they become transparent and pale yellow, with a cluster of small oil globules. Embryos develop eyes about 6 days after fertilisation and hatch in 13–18 days at 11–18°C. They reach about 127 mm in their first year and up to 370 mm after 5 years. Larvae feed on zooplankton in hatchery tanks; older fish feed mainly on stream invertebrates, particularly aquatic insects.

Utility As a threatened species, the Macquarie perch is no longer exploited, except by anglers in the relatively abundant population in Dartmouth Dam. Artificial propagation has proved difficult, although ripe wild fish have been bred successfully using hormone induction.

Similar species Golden perch, juveniles similar to small silver perch, redfin perch and adult pigmy perch.

Other names Common: silvereye, white-eye, mountain perch, bream, black bream. Scientific: None.

Literature Cadwallader & Eden, 1979; Cadwallader & Rogan, 1977; Dufty, 1986; Gooley, 1986; Gooley & McDonald, 1988; Koehn & O'Connor, 1990; Lake, 1967a, b, c; McDonald, 1976; McKeown, 1934; Merrick & Schmida, 1984; Wharton, 1971.

AUSTRALIAN BASS *Macquaria novemaculeata* (Steindachner)

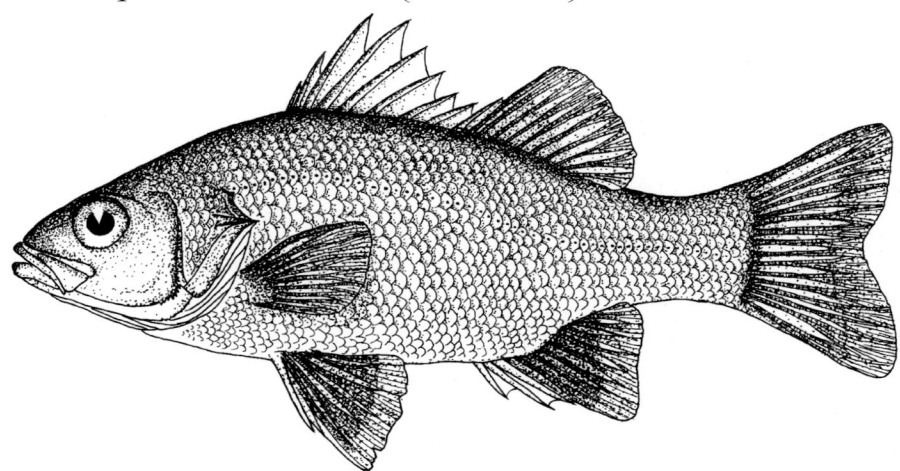

25.3 Australian bass *Macquaria novemaculeata*. (J.H. Harris)

Description Of moderate size, body elongate-oval and laterally compressed, slightly shallower than estuary perch, snout profile slightly concave to straight. Eyes moderately large. Profile of back evenly arched from above eyes to tail. Snout somewhat tapered, naked and of moderate length. Lower opercular spine largest, broad and very sharp. Lower jaw protrudes. Spinous and soft-rayed portions of dorsal fin separated by a moderate notch (VII–IX, I, 8–11); 4th dorsal spine longest. Anal fin III, 7–9. Pelvic fins inserted behind base of pectoral fins. Pectorals have 12–16 rays; tail moderately forked. Scales are mostly ctenoid and of moderate size, with 48–55 along lateral line; 25 vertebrae.

Colour Dark olive-grey-green on back fading to off-white or yellowish white below, generally less glossy or silvery than estuary perch. Anal and pelvic fins have white tips, the most lateral ray of each pelvic fin whitish, lighter coloured than others. Juvenile fish generally have five dark vertical bands across head, back, and dorsal and pelvic fins. These persist longest on fins.

Size Recorded to about 600 mm and 3.8 kg and

FAMILY PERCICHTHYIDAE

Australian bass *Macquaria novemaculeata*. (R.H. Kuiter)

possibly larger, but fish of 1 kg are of good size.
Distribution Lives in coastal rivers from the Mary River and Fraser Island in Queensland south to tributaries of Gippsland Lakes in Victoria, and there are reports of their occurrence south to Wilsons Promontory. Reports of distribution, habitats and size of bass are often dubious owing to confusion with the very similar estuary perch. The bass travels extensively upstream, historically reaching altitudes of about 600 m in the Hawkesbury River system.

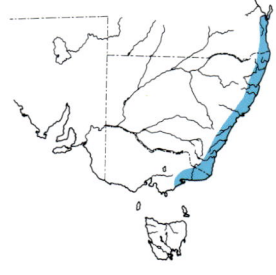

Conservation status Has declined severely as access to about half of the total potentially available habitat has been obstructed by dams and weirs blocking migration paths; river regulation also interferes with spawning cues provided by flooding and with subsequent population recruitment. Recruitment may also be prevented by acidification of streams whose catchments are affected by artificial drainage schemes in potential acid–sulphate soils.
Natural history Is catadromous, migrating downstream into estuaries to breed from May to August before their return homing migration. Males mature at about 180 mm (2–4 years), females at about 280 mm (5–6 years). Highly fecund, producing an average of 440,000 eggs (up to 1.5 million in large fish), and may spawn repeatedly in a season. Spawns in brackish waters about $1/3$–$1/2$ sea water when temperature is 11–18°C, although artificial propagation is done at higher salinity and temperature. Eggs small, transparent, non-adhesive and demersal in the spawning salinities, and about 1 mm diameter when water-hardened. Larvae hatch 2–3 days after fertilisation and are about 2.5 mm long. Feeding begins about 3 days after hatching when 4 mm long. Metamorphosis takes place at about 3 months when the fish are about 25–30 mm long. Young fish reach about 100 mm in their first year. There is marked sexual size dimorphism, with females being much larger than males, although this difference is less marked in smaller rivers in north of range. Growth varies greatly between different habitat types, being fastest in floodplain lagoons and slowest in tidal reaches. Juvenile fish migrate upstream through spring and summer; most males remain in tidal waters while females travel further upstream, so that the population is sexually segregated in the non-breeding season. Larvae feed on zooplankton and chironomid larvae; older fish are generalised carnivores, eating a wide range of fish, crustaceans and other

invertebrates, especially insects. About half the diet in summer can be from terrestrial sources, mostly associated with riparian vegetation.

Utility Is a highly sought-after angling species that supports a major recreational fishery, although overfishing has damaged the quality of many stocks by reducing their average age and size. Is readily caught on a range of baits, artificial flies and lures, often at the water's surface. Is a good table fish, but the increasing popularity of catch-and-release fishing is resulting in smaller proportions of fish being killed. Artificial propagation in hatcheries supplies a growing demand for stocking farm dams and impoundments. As with other native angling species, the genetic and disease aspects of this practice are causing concern. In particular, loss of gene-pool diversity, swamping of specifically adapted, isolated, relict populations, and introductions of new parasites or disease organisms are seen as major hazards in the developing fish-stocking practices, especially when applied in rivers.

Similar species Estuary perch, Macquarie perch.

Other names Common: perch, freshwater perch. Scientific: formerly in genus *Percalates*.

Literature Battaglene *et al.*, 1989; Harris, 1985a, b; 1986; 1987a, 1988; Koehn & O'Connor, 1990; McCarraher, 1986; McDonald, 1976; Merrick & Schmida, 1984.

ESTUARY PERCH *Macquaria colonorum* (Günther)

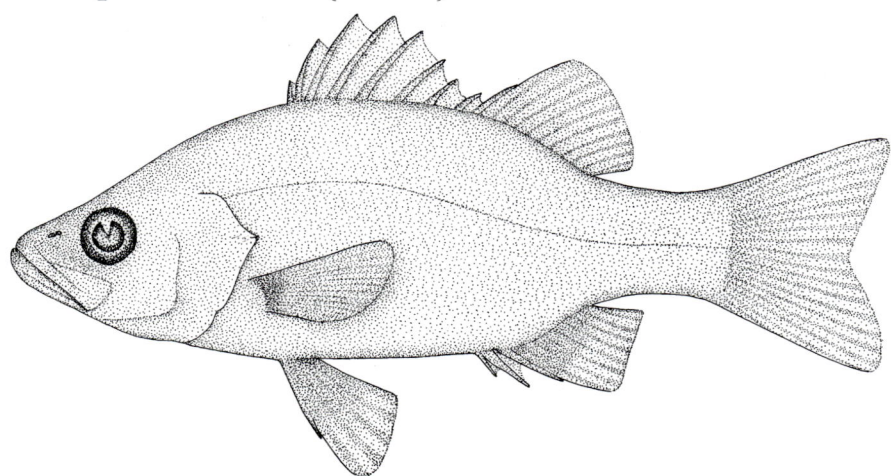

25.4 Estuary perch *Macquaria colonorum*. (R.M. McDowall)

Description Of moderate size, body elongate-oval and laterally compressed, slightly deeper than Australian bass, profile of back evenly arched from nape to tail. Snout tapered and fairly long (noticeably longer than in Australian bass), snout profile concave. Eyes moderately large. Post-orbital head length less and opercular spines also narrower. Number and form of fin rays and spines, serrations on operculum and most other characters are similar to those of Australian bass (dorsal VIII–IX, I, 8–11, anal III, 7–9, pectorals 12–16). Scales of moderate size, with 48–55 along lateral line; vertebrae 25; gill rakers 14–18.

Colour Dark grey and silvery on the back, becoming paler below; lower sides whitish, generally slightly paler than in Australian bass. All fins dark except outer half of last 2–3 rays of anal fin. Fish under 110 mm have a dark spot between the two opercular spines and on head behind eyes; tips of all fins except pectorals are darker.

Size Recorded to about 750 mm and 10 kg but fish over 3 kg are now uncommon.

Distribution Lives in coastal rivers and lakes from the Richmond River in northern New South Wales south and west as far as the mouth of the Murray River in South Australia; is also found in the Arthur and Ansons Rivers in northern Tasmania.

Conservation status Not clear; no commercial catch data or other stock-assessment methods are currently available. However, commercial and recreational fishing pressures have been followed by an apparent decline in abundance, although the species, together with bass, is now protected from commercial fishing and conservative bag limits are applied to angling catches.

Natural history Lives in tidal waters in the northern parts of its range but moves well upstream into flowing waters south of the main distribution of Australian bass; generally prefers deeper and more saline waters than bass but is commonly found in fresh or slightly brackish reaches of estuaries. Breeds in sea water at the mouths of estuaries in winter (around July and August) in temperatures between about 14 and 19°C. Males mature at about 220 mm, females at about 280 mm. Eggs are numerous, spherical, transparent, non-adhesive, planktonic in sea water and about 1 mm in diameter when water-hardened. Larvae hatch after about 2–4 days, are about 2.2 mm long and begin feeding at about 4.5 mm. Scales appear at 21 mm, when vertical bands also develop and metamorphosis begins. Diet is less variable than that of bass and includes shrimps, prawns and small fishes, other crustaceans to a lesser extent, bivalve molluscs and worms; feed closer to the bottom than bass.

Utility Is a popular angling species, readily caught using lures, bait and flies, and is a good eating fish. Is more susceptible to handling than bass, and this may limit its value in aquaculture.

Similar species Australian bass and Macquarie perch.

Other names Common: perch. Scientific: was previously placed in genus *Percalates*.

Literature Koehn & O'Connor, 1990; McCarraher & McKenzie, 1986; McDonald, 1976; Merrick & Schmida, 1984.

MURRAY COD *Maccullochella peelii peelii* (Mitchell)

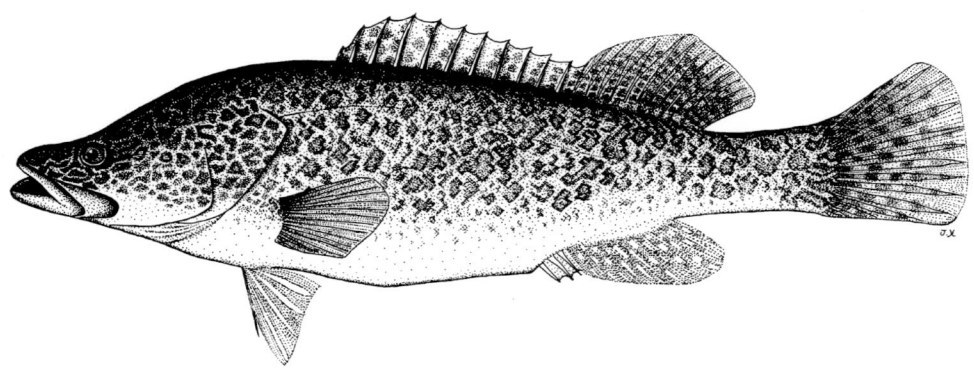

25.5 Murray cod *Maccullochella peelii peelii*. (J. Hannan)

Description A large, elongated, deep-bodied fish with small eyes and a short, rounded, depressed snout with a distinctly concave profile; lower jaw protruding, or jaws equal in some, usually small, specimens. Mouth large, reaching to behind eyes. Spiny and soft-rayed portions of dorsal fin partially separated by a notch (X–XII, 13–16, usually XI, 15), spines much shorter than rays, 5th–6th spines longest. Pelvic fins with an elongate filament, insertion anterior to rounded pectorals (18–21 rays). Caudal peduncle short, tail rounded. Anal fin rounded (III, 11–15, usually 12). Cheeks and opercula scaled but snout naked; scales small, mostly ctenoid, some cycloid (65–81 along lateral line). Operculum has a fleshy margin and two spines, the lower largest. Vertebrae 34–36, precaudal vertebrae 15; gill rakers 16–20. Otolith length 2.5–2.9% of total length.

Colour Cream to olive to yellow-green, with pale to dark green reticulated mottling dorsally and laterally, sometimes with small dark spots between reticulations, extending onto opercula and snout. Some very large specimens have a speckled, green-grey appearance. Ventral surface white to creamy white. Dorsal, caudal and anal fins olive to dark green, with varying amount of white on margins. Pelvic fins white; pectorals clear to whitish.

Size Is Australia's largest wholly freshwater fish, recorded to 1800 mm and 113.5 kg; occasional specimens 20–40 kg are caught regularly, but most are now under 10 kg.

Distribution Originally present throughout most of the Murray-Darling system except for

FRESHWATER CODS AND BASSES

Murray cod *Maccullochella peelii peelii*. (R.H. Kuiter)

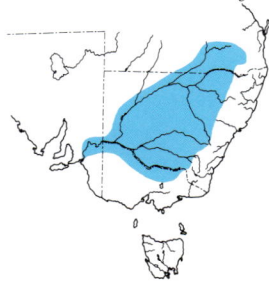

upper reaches of some southern tributaries. Because of its great popularity with early settlers, was stocked into many waters outside its natural range, including the Yarra River (Victoria), the east-flowing Coxs, Nepean and Wollondilly Rivers, Mulwarree Ponds and Lake George (New South Wales), as well as some Western Australian rivers. Apart from occasional specimens from the Nepean and Yarra Rivers, the Murray cod is no longer found at these locations. During the 20th century, was stocked successfully into Lakes Bathurst (NSW), Charlegrark, Green and Taylor and the Wimmera River (Victoria), and Cataract Dam, a Sydney water supply reservoir. Since the 1970s, has been reared in government and private, commercial hatcheries in New South Wales and Victoria and stocked into many farm impoundments and natural waters.

Conservation status Once very abundant but has declined dramatically in range and abundance and is now relatively uncommon in most areas and rare in most Victorian tributaries of the Murray. Overfishing by a relatively large commercial fishery caused a decline between the 1800s and the 1930s. Environmental changes to the Murray-Darling system, particularly altered flow and temperature regimes and reduced frequency, magnitude and duration of floods caused by construction of dams, high-level weirs and levee banks, have adversely affected juvenile recruitment, leading to a dramatic and lasting decline of stocks since the 1950s. Other factors, such as recreational fishing, competition with or predation by introduced redfin perch, siltation, desnagging and pollution, have also contributed to decline in many areas.

Natural history Found in habitats ranging from small, clear, rocky streams in the upper western slopes of New South Wales to the generally turbid, slow-flowing rivers and creeks of the western plains. Is generally found in or near deep holes and prefers habitats containing cover such as rocks, fallen trees, stumps, clay banks or overhanging vegetation. Growth rates of sexes are similar but vary greatly between populations in different types of habitat. Averages 236 mm (0.2 kg) at age 1, 348 mm (0.8 kg) at 2, 500 mm (2 kg) at 3, 580 mm (3.5 kg) at 4 and 640 mm (5 kg) at 5 in New South Wales rivers; in Lake Mulwala, averages 654 mm (5.8 kg) at 4 but only 427 mm (1.8 kg) in Lake Charlegrark. When older than 10 years, grows mostly by increase in weight.

Reaches maturity at 4–5 years; spawns in spring and early summer, when water temperature rises to about 20°C, usually in September–November, but spawning may be later and at lower temperatures in Victoria. Becomes very active before spawning, forming pairs. Fecundity ranges from

10,000 to 90,000 eggs in females 2.5–23 kg in weight. The eggs are large (3–3.5 mm diameter), adhesive, and are spawned onto solid objects such as logs, rocks, or clay banks in water as shallow as 300 mm. Overhead cover and hollow logs or pipes are not needed, though protected sites may be preferred. The male guards, and probably fans, the eggs during incubation. Hatching starts at 5–7 days at 20–27°C, and a batch of eggs takes 3–4 days to hatch. The larvae are 5–8 mm at hatching, with a large yolk sac, and may remain clumped for 8–10 days until the yolk sac is absorbed. They then disperse and begin feeding on copepods and cladocerans, shifting to chironomid larvae and other aquatic insects when 15–20 mm long. Is the top predator in inland waters and is at times a voracious feeder; diet consists of fish, crustaceans, molluscs, water birds, turtles, frogs, and some terrestrial animals, including snakes and mice.
Utility Is the most famous and highly valued inland fish and played a prominent part in mythology and life of the Aborigines, for whom it was *the* fish. Was used as fresh food by explorers and settlers and sustained a large commercial fishery from the mid-1880s to early 1900s. Is still keenly sought by anglers and supports a small commercial fishery in south-western New South Wales. There is a closed season in New South Wales and Victoria during September–November, with a recreational bag limit of 2 fish per day and 4 in possession. There is also a minimum size limit of 500 mm in Victoria. About 500,000 Murray cod fry are produced annually at Government and private hatcheries for stocking.
Similar species Mary River cod, eastern cod, trout cod.
Other names Common: cod, codfish. Scientific: *Maccullochella peeli, Maccullochella macquariensis*.
Literature Anderson *et al.*, 1992a; Berra & Weatherley, 1972; Cadwallader *et al.*, 1979; Dakin & Kesteven, 1938; Gooley, 1992; Koehn & O'Connor, 1990; Lake, 1967a, b, c; McDonald, 1976; Merrick & Schmida, 1984; Rowland, 1983, 1985, 1989.

MARY RIVER COD
Maccullochella peelii mariensis Rowland

Description Similar in general shape to the nominate subspecies *M. peeli peeli*, but with the following differences: Dorsal fin XI, 15; pelvic fin with relatively long tapering filament; anal fin III, 12–13; lateral scales 68–74; vertebrae 34–35, 15 precaudal; gill rakers 17–21; otolith length 2.9–3.3% of total length.
Colour Yellowish to pale green, with dark, heavily reticulated mottling dorsally and laterally, extending onto ventral surface in some specimens. Ventral surface grey-green to whitish. Dorsal, pectoral, caudal and anal fins clear to dark, with grey-green mottling on bases, fin margins whitish. Pelvic fins colourless to whitish, filaments white.
Size Recorded up to 23.5 kg, but uncommon over 5 kg and 700 mm.
Distribution Is a subspecies endemic to the Mary River system in south-eastern Queensland.
Conservation status Was once extremely abundant and was used as fresh food by the early settlers and as pig food. There has been a major decline in numbers since the early 1900s, and the natural population is restricted to 3–5 small tributaries. Loss of habitat following agricultural and urban development, together with overfishing, probably caused the decline. Is regarded as 'endangered' by the Australian Society for Fish Biology.
Natural history Is little known, but probably similar to related species. Type locality, Tinana Creek, is slow-flowing and clear to slightly turbid, with a mud/clay substrate and cover from overhanging vegetation and fallen trees. Spawns large (3 mm) adhesive eggs during spring after water temperature reaches 20°C; in hatchery ponds spawning is mainly in narrow concrete pipes. Larvae feed on zooplankton and aquatic insects, especially chironomid larvae; adults feed on fish and crustaceans.
Utility Was commonly used as food for humans and pigs by early settlers; is highly valued by anglers. Stock is produced in hatcheries for stocking farm dams and natural waters; has been stocked into farm dams and impoundments in the Mary, Brisbane, and possibly other river systems in south-eastern Queensland.
Similar species Murray cod, eastern cod, trout cod.
Other names Common: Murray cod, cod, codfish, eastern freshwater cod. Scientific: *Maccullochella peeli, M. macquariensis*.
Literature Rowland, 1993; Simpson 1994.

EASTERN COD *Maccullochella ikei* Rowland

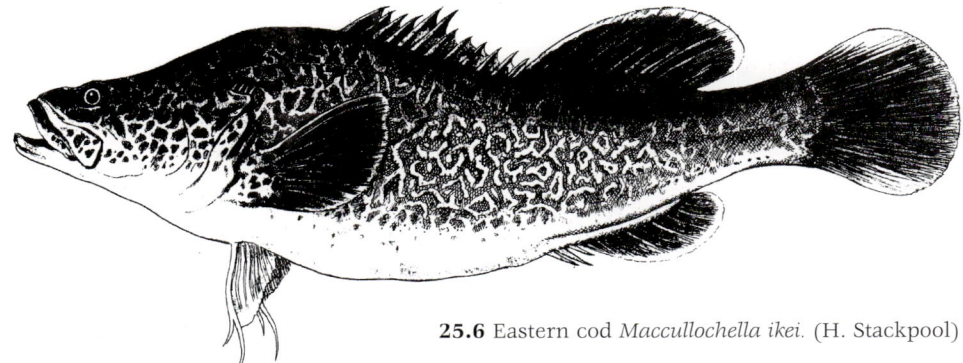

25.6 Eastern cod *Maccullochella ikei*. (H. Stackpool)

Description A large, elongate, deep-bodied fish with relatively small eyes and a short, rounded, depressed snout with a distinctly concave profile; lower jaw protrudes. Mouth large, extending to below posterior eye margin. Dorsal fin (X–XII, 13–16) has spines shorter than rays, 5th–6th spines longest. Pelvic fins have 1st ray elongated into 2 filaments, the 2nd long, tapering into a fine tip, fin inserted anterior to rounded pectoral fins (16–19 rays). Caudal peduncle short, tail rounded. Anal fin rounded (III, 11–12). Cheeks and opercula scaled, snout naked; scales mostly ctenoid. Lateral line scales 65–82. Operculum with fleshy margin and two spines, the lower largest. Vertebrae 35, 15 precaudal; gill rakers 18–20. Otolith length 3.2–3.5% of total length.

Colour Yellow-green to golden, with black to very dark green, heavily reticulated mottling dorsally and laterally, extending onto ventral surface in some fish; ventral surface otherwise grey, becoming grey-white in turbid water. Dorsal, pectoral, caudal and anal fins clearish to dark grey-green, with mottling on fin bases, pale grey to whitish margins on these fins (rayed portion of dorsal). Pelvic fins colourless, filaments white.

Size Recorded up to 41 kg, but mostly less than 5 kg and 660 mm since the 1960s.

Distribution Is now found only in the Clarence and Richmond Rivers, in northern New South Wales.

Conservation status Cod were once abundant, but by the 1930s were extinct in the Richmond and Brisbane Rivers and were found in only small numbers in isolated, pristine habitats in the Clarence. Factors contributing to decline are thought to be: numerous large fish kills caused by natural pollution following extensive bush fires and then heavy rain; pollution from gold and tin mines; dynamiting of tributaries during construction of railways; habitat degradation following agricultural and urban development; and gross overfishing until the late 1980s. Is classified as 'endangered' by the Australian Society for Fish Biology.

In 1988 and 1989 about 30,000 fry were produced by artificial spawning and larval rearing and were successfully stocked into the Richmond and Clarence Rivers. Recent captures suggest that the stocked fish may have spawned and wild populations are recovering following a ban on fishing.

Natural history Is little known, but in some aspects probably resembles related species. Type habitat in the Nymboida River is typically clear-flowing, with very rocky substrate and instream cover. Growth is slower than in Murray cod, reaching only 445 mm and 1.3 kg at 5 years. Becomes mature at 4–5 years of age, when 0.7–1.5 kg. Spawns large (3 mm), adhesive eggs during spring when temperatures exceed 16°C. Known diet includes fish, crustaceans, frogs and snakes.

Utility Was commonly taken for food by Aborigines and early settlers and is highly valued as an angling fish. Capture of *Maccullochella* species from eastern drainages north of the Macleay River as far as the Queensland border is currently prohibited because of the eastern cod's endangered status.

Similar species Murray cod, Mary River cod, trout cod.

Other names Common: Clarence River cod, cod, codfish, Murray cod, eastern freshwater cod. Scientific: *Maccullochella peeli, M. macquariensis*.

Literature Rowland, 1985, 1993.

FAMILY PERCICHTHYIDAE

TROUT COD *Maccullochella macquariensis* (Cuvier)

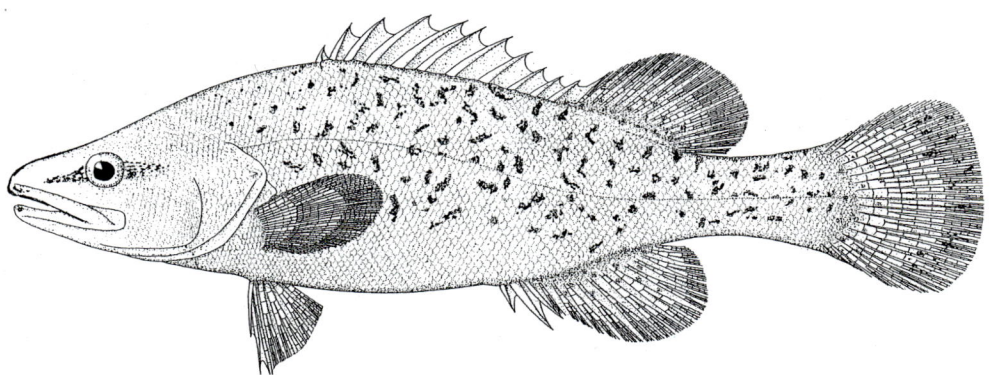

25.7 Trout cod *Maccullochella macquariensis*. (NSW Fisheries)

Description A large, elongated, deep-bodied fish with relatively large eyes and a long, broad, rounded snout with a straight profile. Upper jaw overhangs lower in fish of all sizes. Mouth large, reaching to below posterior border of eye. Dorsal fin XI (occasionally XII), 14–16, 5th–6th spines longest. Pelvic fins inserted below pectorals (19–20 rays). Anal fin III, 10–13; 63–82 scales in lateral line; vertebrae 34–36, 14 precaudal. Gill rakers 18–19.

Colour Bluish grey or sometimes dark to light brown, with small dark, irregular spots or bars extending onto lower sides and onto base of rayed dorsal, anal and caudal fins; markings fewer or absent on head. Ventral surface creamy white to grey. Fins clearish to grey, rayed dorsal, caudal and anal, with distinct, relatively broad, white or creamy margins, sometimes tinged with orange or yellow. A dark horizontal stripe extends from snout through nostrils and eyes and onto preoperculum.

Size Known to about 850 mm and 16 kg, but most fish caught are less than 5 kg.

Distribution Was once widespread throughout

Trout cod *Maccullochella macquariensis*. (R.H. Kuiter)

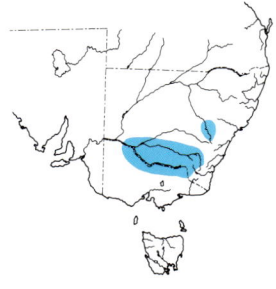

southern tributaries of the Murray-Darling system, where its distribution overlapped that of Murray cod; range extended down the Murray at least to Mildura and possibly to Mannum (South Australia), and to the north in the Macquarie River. There are no confirmed records from the Darling or Lachlan Rivers, and those from Glenbawn Dam are erroneous. Only two substantial self-sustaining populations are known: in the Murray between Yarrowanga Weir and the Barmah State Forest, and a small translocated population in the upper reaches of Seven Creeks (Goulburn River system, northern Victoria). Trout cod translocated to Lake Sambell, near Beechworth, in 1928 were lost after a fish kill in the late 1970s. Both trout cod and Murray cod were liberated in the Cataract Dam near Sydney in the early 1900s, and these persist there as viable fertile hybrids. Viable hybrids have also been reported from the Murray River recently.

Conservation status Trout cod, once abundant and widespread in the southern Murray-Darling, underwent a dramatic decline in both range and abundance in the past 50 years. Overfishing, altered flow and temperature regimes, other forms of habitat degradation, deterioration in water quality and competition with alien fish species have been implicated in their decline. Since 1986, more than 200,000 juveniles, bred at hatcheries in New South Wales and Victoria, have been liberated into sites within the trout cod's natural range: Turon River; upper reaches of Macquarie and Murray Rivers; Murrumbidgee River near Cooma, Adaminaby, and Gundagai; and Talbingo Dam (NSW); Bendora Dam (ACT); and in the Broken, Goulburn, Ovens, Campaspe and Mitta Mitta Rivers (Victoria). Releases into the Abercrombie River (tributary of the Lachlan River) and Ryans Creek (Broken River) may have been outside the historical range. It is difficult to determine whether these releases have generated self-maintaining populations, but liberated hatchery fish have survived and there is recent, preliminary evidence of spawning. Measures are also being taken to prevent further habitat degradation. Is listed as 'endangered' by the Australian Society for Fish Biology.

Natural history Is currently found in two distinct habitat types: Murray River with various substrates and abundant in-stream cover of snags and woody debris; and Seven Creeks, a narrow stream with rock, gravel and sand substrates, and pools interspersed with rapids and cascades. At both sites, fish are often found close to cover and in relatively fast currents.

Little is known of trout cod in the wild, and information on reproduction and early life history comes from captive fish. Sexual maturity is reached at 3–5 years and 0.75–1.5 kg; spawning is in spring, 2–3 weeks earlier and at lower temperatures than in Murray cod. The eggs, numbering between about 1200 and 11,000, are large (2.5–3.6 mm), adhesive and opaque and are probably deposited on hard substrates; they hatch in 5–10 days at 20°C; the larvae, 6.0–8.8 mm long, begin feeding on zooplankton in ponds at about 10 days. Is a carnivorous top predator, its diet including crustaceans, fish and aquatic insects.

Utility Is an excellent angling fish with good eating qualities and was once keenly sought and readily taken by anglers and commercial fishermen. However, its endangered status has led to the taking of trout cod being prohibited in New South Wales, Australian Capital Territory and Victoria.

Similar species Murray cod, Mary River cod, eastern cod.

Other names Common: bluenose cod, blue cod. Scientific: *Maccullochella mitchelli*.

Literature Berra, 1974; Berra & Weatherley, 1972; Douglas *et al.*, 1994; Ingram & Rimmer, 1992; Lake, 1967b; McDonald, 1976; Merrick & Schmida, 1984.

26
Family Terapontidae
Freshwater grunters or perches

J. R. MERRICK

This family contains about 45 small to medium-sized carnivorous or omnivorous species that are found in cool to warm waters (marine, brackish and fresh) of both hemispheres in the Indo-Pacific region. Many terapontids occur in Australia, especially in northern areas of the continent, which has been the site for a major radiation into fresh waters. Some species make grunting sounds, hence their common name. Terapontids are strongly built, perch-like fishes with a single, but notched, dorsal fin with stout spines forward of the notch, small to moderate-sized scales, and a lateral line.

Key to terapontids

Scales small, 70–90 in lateral line; tail weakly forked; dorsal fin XII, 11–12; sides without blotches; commonly over 300 mm . . . . ***Bidyanus bidyanus*** p. 164; Fig. 26.1
Scales moderate, 44–53 in lateral line; tail straight to weakly forked; dorsal fin X–XIII, 8–11; sides with rusty brownish to golden spots; commonly to 150 mm, rarely to 250 mm . . . . ***Leiopotherapon unicolor*** p. 166; Fig. 26.2

SILVER PERCH *Bidyanus bidyanus* (Mitchell)

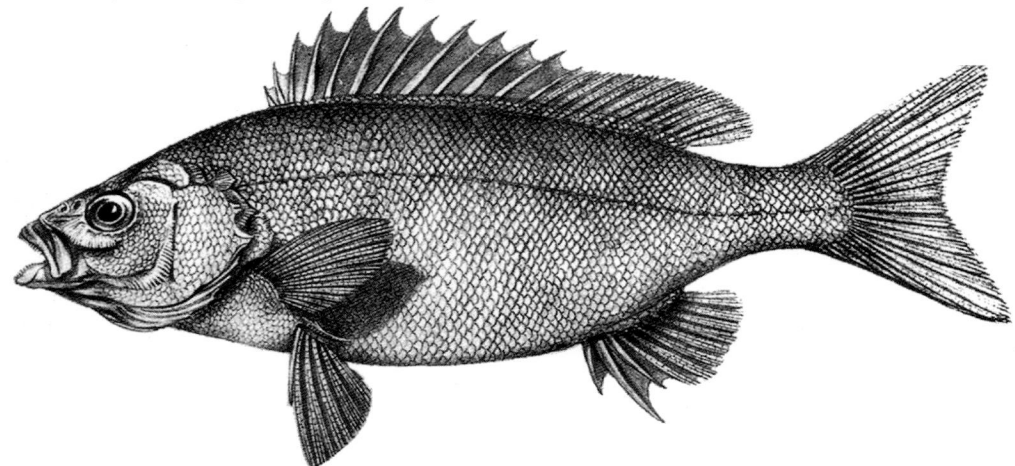

26.1 Silver perch *Bidyanus bidyanus*. (J.R. Richardson)

Description A moderate to large fish, body elongate and slender in the young, becoming deeper and compressed in adults. Head small, becoming more beak-like with increasing size, snout almost straight; jaws equal; eyes small. Dorsal fin (XII, 12) origin above pelvic fin bases, rayed portion beginning above origin of anal fin (III, 7–9), rays decreasing evenly backwards, posterior margin of dorsal fin rounded; tail weakly forked. Pectoral rays 14–17. Scales thin and smallish (70–90 along lateral line), lateral line following profile of back but often indistinct. Vertebrae 25.

FRESHWATER GRUNTERS OR PERCHES

Colour Varies with water conditions—an adult may be black, grey, olive greenish or gold on back, with grey to greenish or gold to silvery sides and white on belly. Dorsal, tail and anal fins grey; pelvic fins white. Juveniles may be mottled with vertical, dark bars.

Size Commonly reaches 300–400 mm and 0.5–1.5 kg, but is known to have attained 8 kg.

Distribution Natural range included most of the Murray-Darling drainage, excluding the cool, high, upper reaches of streams on the western side of the Great Dividing Range; not known upstream of Chinchilla on the Condamine River in the north, Bonshaw on the Dumaresq River in the north-east, or Albury on the Murray River in the south. Has been introduced into many eastern coastal river systems of New South Wales and south-eastern Queensland as well as south-western, Western Australia.

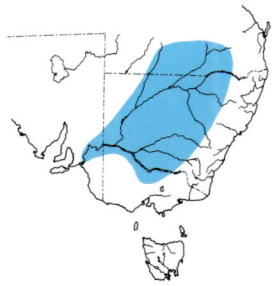

Conservation status Populations in most areas have declined and it is now reported as rare in the Brewarrina area. Known movements of juveniles, stimulated by small rises in river level, are likely to be impeded even by low weirs, since these remain effective barriers at such times. Migrations through the fishway at Euston on the Murray River declined 93% between 1940 and 1990. The Australian Society for Fish Biology lists this species as 'potentially threatened'.

Natural history Occurs mainly in fast-flowing waters, especially where there are rapids and races. Spawns in spring and summer after a long migration upstream to areas behind the peaks of floods. A female may shed 300,000 or more semi-buoyant eggs about 2.75 mm in diameter. These develop in a few days to free-feeding stages that drift downstream. Is a schooling fish that is taken in large numbers when moving upstream to spawn, and in summer, often congregating in large numbers below rapids, barrages and weirs. Movements during the rest of the year are little known, though immature fish are known to move upstream during the day after small rises in river level. Males mature at age 3 when about 250 mm long and females at 5 when about 290 mm. Silver perch have been recorded 27 years old. Feeds on small aquatic insects, molluscs, earthworms and green algae.

Utility Regarded as a good sporting fish with fair to medium eating qualities; well suited to chilling and smoking. Has now been demonstrated to have considerable potential for intensive fish farming.

Silver perch *Bidyanus bidyanus*. (L.C. Llewellyn)

FAMILY TERAPONTIDAE

Is now frequently sold as an aquarium fish, but may not be easy to induce to feed in captivity and is easily alarmed; successful maintenance in small tanks requires considerable care.
Similar species Closely resembles two other terapontids in drainages adjoining the Murray-Darling to the north and west: Welch's grunter (*Bidyanus welchi*) and Barcoo grunter (*Scortum barcoo*).
Other names Common: bidyan, black or silver bream. Scientific: *Terapon bidyanus*.
Literature Koehn & O'Connor, 1990; Lake, 1967a, b, c, d; Leggett & Merrick, 1987; Mallen-Cooper, 1992, 1993, 1994; Merrick & Schmida, 1984; Vari, 1978.

SPANGLED PERCH *Leiopotherapon unicolor* (Günther)

26.2 Spangled perch *Leiopotherapon unicolor*. (A.R. McCulloch)

Description A small to moderate-sized fish, body elongate and compressed in juveniles, becoming thicker and deeper in adults. Head large with small to moderate-sized eyes and a rounded snout. Mouth large, jaws equal. Dorsal fin (X–XII, 8–11) rather low anteriorly, origin just behind bases of large pelvic fins, beginning of rayed portion above beginning of anal fin, higher and shorter-based; anal fin II–III, 8–10; pectorals 12–15. Tail truncate to very weakly forked. Scales thin, of moderate size (44–53 along lateral line); lateral line follows profile of back. Vertebrae 23–25. Some variation in morphology noted in isolated, local populations.
Colour Varies with locality, but usually brown to steely blue on the back, with golden to silvery sides and white on the belly; characteristic marking is a pattern of rusty, brownish red spots that are nearly always present regardless of the base colour. The spotting can be random, resulting in a mottled appearance, or in rows, giving a banded appearance when the fish is alarmed. In specimens up to 90 mm long, the lower margin of the tail is black. Pelvic and anal fins normally white; pectoral fins light in colour, the outer margin of dorsal fin sometimes dark.
Size A small species commonly growing to 150 mm, rarely to 250 mm and 0.5 kg.
Distribution Range includes all the warmer fresh waters of Australia. It is one of the most widespread of our native freshwater fishes, inhabiting bore-drains, small pools, isolated billabongs, dams of all sizes, and natural streams. Occurs in the internal Lake Eyre and Bulloo drainages, the coastal South-Eastern and North-Eastern Slope drainages, Gulf of Carpentaria, Timor Sea and Indian Ocean drainages and in all areas of the Murray-Darling system north of Condobolin.

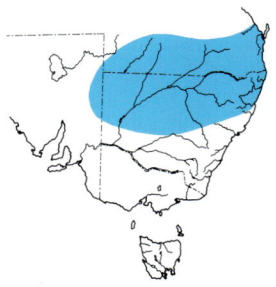

Conservation status Locally may be very abundant.
Natural history A remarkably hardy species with wide temperature tolerances, generally in largest numbers in warm, often turbid, still or slowly flowing waters. Is one of the few freshwater terapontids that will breed in dams. Spawns from November into summer; many thousands of tiny eggs about 0.75 mm in diameter are spread randomly over the bottom and hatch in only 2 days.

Feeds on a wide range of small aquatic insects, crustaceans, molluscs and some plant material. Is thought to be a schooling fish, especially when young, and is reported to aestivate (survive in bottom mud) when the water dries up. Aestivation reports are coupled with observations of large numbers of this species appearing suddenly a few days after rain in an area previously dry and with no connections to permanent water. No comprehensive studies of this ability have been undertaken.

Utility Although it bites avidly on a variety of baits, and fights well, has limited significance as a sports fish because of its small size. The flesh has excellent flavour but numerous bones are a disadvantage. When other angling species are uncommon or absent, larger specimens are retained for eating. In some areas where numbers are low, it is reported to grow unusually large. Spangled perch will spawn in ponds and small dams, and its potential for aquaculture is currently being investigated. Is an active and handsome species, but is not often kept in aquaria; although easily maintained, it is also easily alarmed, may jump and is aggressive.

Similar species Superficially resembles juvenile Murray cod and other native 'perches'.

Other names Common: trout cod (Queensland), jewel perch, bobby perch or cod, spangled grunter. Scientific: *Madigania unicolor*.

Literature Beumer, 1979a, b; Lake, 1967a; Leggett & Merrick, 1987; Llewellyn, 1973; Merrick & Schmida, 1984; Vari, 1978.

Spangled perch *Leiopotherapon unicolor*. (L.C. Llewellyn)

27
Family Nannopercidae
Pygmy perches

R.H. KUITER, P.A. HUMPHRIES AND A.H. ARTHINGTON

The classification of this small family, with only six species, has long been uncertain. Though they are sometimes placed in their own family Nannopercidae, most previous authors included them with the Kuhliidae, or occasionally Percichthyidae (the freshwater basses and cods of southern Australia and South America). The latest view is they should be in their own family Nannopercidae. It is not clear whether the latter is really distinct from Percichthyidae (see p. 150). Especially similar to the pygmy perches is the small Western Australian percichthyid *Bostockia porosa*, which in turn shows close affinities to *Gadopsis* spp., the blackfishes, which are also placed in their own family (Gadopsidae, see p. 186). Future studies may show that they are all percichthyids, with perhaps some groups, such as pygmy perches and blackfishes, warranting subfamilial status.

The pygmy perches occur in freshwater streams of coastal southern Queensland and all southern states. They are generally small, reaching about 85 mm; the body is distinctly scaled and, if present, the lateral line is divided and consists of irregularly spaced tubed scales; there is one deeply notched dorsal fin. Two genera are now recognised: *Nannatherina* is monotypic and restricted to south-western Australia, and all the others are placed in *Nannoperca*. The genus *Edelia*, used by some authors for two species (*obscura* in eastern Australia and *vittata* in the west) was based on minor differences that vary between populations and are often lost in large individuals. In addition, a large hybrid, *N. australis* x *obscura*, was collected from the wild in Lancefield, Victoria, perhaps further suggesting that these species belong in a single genus. Reproduction is usually in spring, and most species have been bred in captivity. Eggs are reported to be spherical and transparent, non- or slightly adhesive, and dropped in small numbers randomly on the bottom; over the breeding period they may accumulate to many hundreds. Hatching occurs in 3 or 4 days.

Because of the introduction of trout, many populations have suffered greatly and, with additional habitat loss, some species are just surviving. Four species occur in south-eastern waters and are determined with the following key.

Key to pygmy perches

1. Lateral line absent; 7 dorsal spines . . . . **Nannoperca oxleyana** p. 174; Fig. 27.4
   Lateral line present; 8 or more dorsal spines. 2

2. Third dorsal spine longest; 14–15 pectoral rays . . . . . . **Nannoperca variegata** p. 171; Fig. 27.2
   Second dorsal spine longest or equal to 3rd; 11–13 (rarely 14) pectoral rays 3

3. Lower edge of preorbital bone hidden under skin, rounded and smooth . . . .
   . . . . **Nannoperca australis** p. 169; Fig. 27.1
   Lower edge of preorbital bone free, straight to rounded and serrate to smooth in small
   to large specimens respectively . . . . **Nannoperca obscura** p. 172; Fig. 27.3

SOUTHERN PYGMY PERCH *Nannoperca australis* Günther

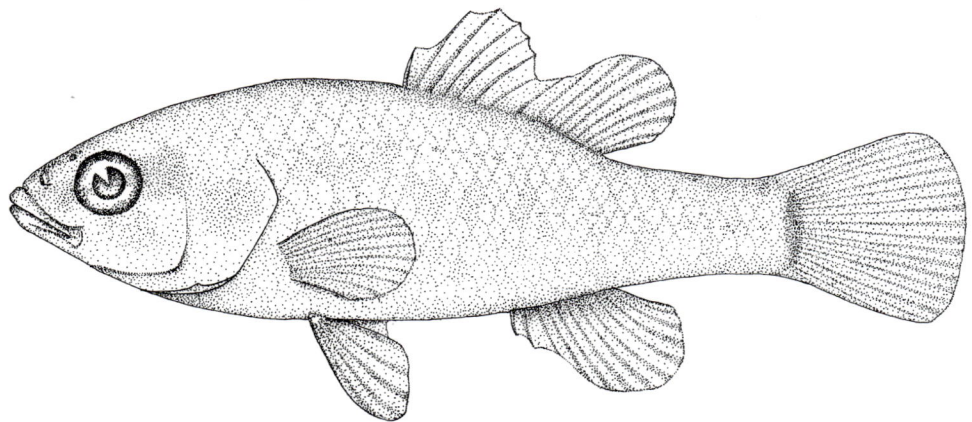

27.1 Southern pygmy perch *Nannoperca australis*. (R.M. McDowall)

Description Body moderately compressed, depth highly variable, usually more than 30% in standard length when small, but sometimes less in adults. Mouth small, just reaching below eye. Preorbital lower edge hidden by skin. Dorsal fin VI–IX, 7–10 (usually VIII, 9); anal fin III, 6–9 (usually 7–8); pectoral fin 11–14 (usually 12–13); gill rakers 3–4 + 6–9, total 10–12; lateral line consisting of interrupted series of tubed scales, anteriorly following contour of back and posteriorly straight along midlateral body. Body covered by ctenoid scales in 28–32 midlateral rows and over most of head. Meristic variation appears to have no relation to distribution or population: e.g. the dorsal fin count may vary little in one area but differ from that of another population nearby or may show a complete range of counts in one population.

Colour Highly variable, geographically and between clear or turbid water. Base colour varies from pale creamish to green-brown, darkest on top and grading to almost white on belly. Spots or blotches on sides, but extremely variable, from almost plain to dark blotches, these sometimes forming longitudinal banding. Colour brightens in breeding males, dorsal and anal fins becoming bright red. Outer edge of anterior part of dorsal fin and leading edges of posterior part, as well as pectoral and anal fins, and edges of caudal peduncle, region around vent, and breast between pectoral fins and chin, turn black. Females never as brightly coloured as males during breeding.

Size Largest size known is 85 mm, but normally reaches about 65 mm total length.

Distribution Common in southern Victoria and abundant in some tributaries of the Yarra River. Occurs in the Murray and Murrumbidgee River systems. Is found widely in lakes and wetlands in northern Tasmania, in rivers draining north, and on King and Flinders Islands.

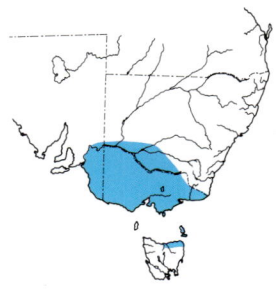

Conservation status Large-scale reductions in range of this species since European settlement, especially in the Murray-Darling system (South Australia, New South Wales and Victoria), are of major concern. Habitat and population fragmentation are occurring rapidly and may hinder recolonisation of habitats where individual populations are wiped out.

Natural history Commonly found in small, slow-flowing systems and still, vegetated habitats in streams, lakes, billabongs, irrigation ditches and other types of wetlands. Prefers cover to open water habitats and is rarely found in fast-flowing parts of streams unless displaced by floods. Juveniles and adults seem to form loose aggregations.

Most fish in populations aged 1+ or 2+, but can live for more than 5 years. Females grow larger than males, though males grow faster in first few years of life. Sex ratios are usually equal. Both sexes mature in their first year, males from about 30 mm and females from about 33 mm. Is a protracted or multiple spawner, breeding usually occurring between September and January, when temperatures are over 16°C. Males are territorial during breeding. Eggs transparent, spherical and non-adhesive, scattered over vegetation or the benthos, 1.2–1.4 mm in diameter when fertilised and fully hydrated. Fecundity variable, with reports of less than 100 maturing oocytes in small

FAMILY NANNOPERCIDAE

Southern pygmy perch *Nannoperca australis*. (R.H. Kuiter)

0+ fish of 37 mm, to over 600 in a 3+ fish 63 mm long, and there may be as many as 2000. In some individuals or populations, only a small proportion of these oocytes may mature, whereas in others most do. Eggs hatch after 2–4 days, and larvae are 3–4 mm long; juveniles recruit into the adult habitats at about 12 mm.

Diet carnivorous, consisting mainly of such crustaceans as amphipods, cladocerans, ostracods and copepods, insects such as larval chironomids, mayflies, adult hemipterans and other terrestrial insects. Diet changes with size and age, juveniles tending to consume more planktonic crustaceans, larger fish taking more terrestrial insects and other larger food items.

Occurs in the same habitat as several introduced species, such as eastern gambusia and juvenile tench, and native species such as dwarf galaxias. Also shares similar foods at times. Predation by perch and brown trout can be significant where their distributions overlap. In one Tasmanian river more than 6% and 20% of the gut contents of these two predators were southern pygmy perch. Is dependent on cover for refuge from piscivores.

Changes in water regimes during recent decades may have enhanced habitats, but further reductions in flows may dewater preferred habitats, making the fish more vulnerable to predation. In addition, reversed or unnaturally large or aseasonal fluctuations in water levels may be detrimental to successful colonisation by aquatic plants. Ensuring flow regimes that protect habitats is an integral step in ensuring the perpetuation of populations.

Utility Plays an important role in mosquito control in Victoria and is an ideal fish to put in ponds or dams for this purpose. Makes an attractive aquarium fish that can look quite spectacular when in its breeding colours.

Similar species Other pygmy perches and juveniles of percichthyids such as Murray cod or golden perch.

Other names Common: none. Scientific: *Paradules leetus, Nannoperca riverinae, Microperca tasmaniae, Nannoperca australis flindersi*.

Literature Koehn & O'Connor, 1990; Kuiter & Allen, 1986; Lake, 1967a; Leggett & Merrick, 1987; Llewellyn, 1974; Mitchell, 1976; Sanger, 1978; Unmack, 1992.

EWEN PYGMY PERCH *Nannoperca variegata* Kuiter and Allen

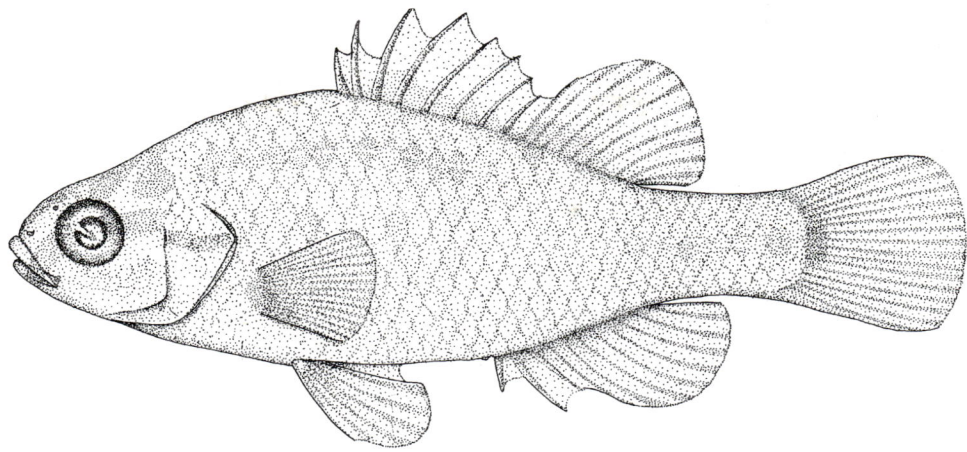

27.2 Ewen pygmy perch *Nannoperca variegata*. (R.M. McDowall)

Description Moderately compressed, body depth about 32% in standard length. Mouth small and reaching to just below eye. Dorsal fin VIII to IX, 9–10; anal fin III, 8–9; pectoral fin 14 (rarely 15); gill rakers 1–3 + 2–5, total 5–8. Body covered by ctenoid scales in about 30 midlateral rows, but scales on head mostly cycloid. Lateral line consists of an interrupted series of tubed scales, anteriorly following contour of back and posteriorly along midlateral body.

Colour Highly variable, base colour from pale creamish white to bright green above and orange below. Usually a dark spot on the base of the caudal fin and a series of dark blotches midlaterally, forming a line in juveniles. Median fins clear to pale orange. Males intermittently flash a combination of green and orange colours on body and fins during courtship.

Size Largest reported specimen was 62 mm.

Ewen pygmy perch *Nannoperca variegata*. (R.H. Kuiter)

Distribution The least known species in family and only recently discovered. Known only from Ewens Ponds in South Australia and some tributaries of the Glenelg River in south-western Victoria.

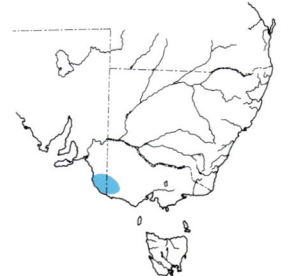

Conservation status Locally abundant but presently known over a very limited geographical area. The South Australian Museum has specimens from Piccaninnie Ponds in the collection but none have been found there in recent times. Trout co-occur in Victoria and could threaten the population there. Regarded as 'vulnerable' by the Australian Society for Fish Biology and a recovery plan has been prepared for it.

Natural history Can be found in fresh and slightly brackish waters, associated with dense aquatic vegetation, but unlike southern pygmy perch, is commonest in habitats with high water flows. Occurs mostly over substrates of gravel, cobble or boulder in the absence of silt, although at Ewens Ponds in South Australia is associated with large amounts of detritus.

Is thought to live for 3 or 4 years. May be a protracted or multiple spawner, as indicated by the presence of about 5000 ova at varying stages of development in ripe females in mid-September; adults appear to migrate upstream for spawning, which occurs between mid-July and mid-November. Aquarium observations indicate that there is no pre-spawning behaviour until just before the eggs are scattered among vegetation. Is probably entirely carnivorous, with benthic insect larvae and crustaceans in the diet.

Several piscivorous fishes in the Glenelg River may be major predators—redfin perch, brown and rainbow trout and golden perch, all of which have been introduced there. The species' preference for vegetated habitats with rapid flow indicates the importance of conserving this habitat. Its absence from silty areas suggests that land or water management activities that increase silt loads may harm populations.

Utility Is an interesting species for the aquarium and has potential species for mosquito control. Captive specimens need well-oxygenated water. Small shrimps and mosquito larvae are favourite foods.

Similar species Other pygmy perches; some records of *N. obscura* are based in this species.

Other names Common: None. Scientific: None.

Literature Koehn & O'Connor, 1990; Kuiter & Allen, 1986; Saddlier, 1992.

YARRA PYGMY PERCH *Nannoperca obscura* (Klunzinger)

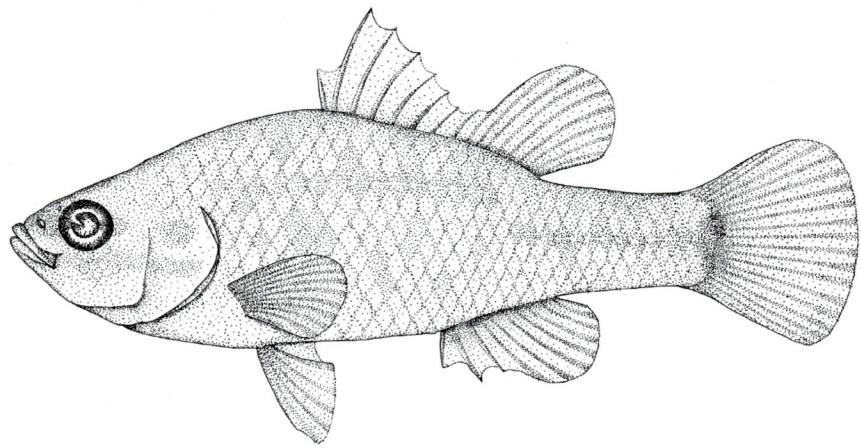

27.3 Yarra pygmy perch *Nannoperca obscura*. (R.M. McDowall)

Description Body moderately compressed, depth to about 36% in standard length when small, but large adults resemble the shape of *N. australis* and as a result, apart from colouration, can be difficult to distinguish. Preorbital free edge about straight and in young strongly serrate; however it varies in populations and in large adults becomes smooth and more rounded. Mouth small, usually not reaching to below eye. Dorsal fin VIII–IX, 7–9 (usually IX, 8); anal fin III, 6–7 (usually 7); pectoral fin 11–13 (usually 12); gill rakers 2–4 + 7–9, total 10–13. Body covered by ctenoid scales in 28–30 midlateral rows; lateral line consists of an interrupted series of tubed scales, anteriorly following contour of back and posteriorly along midlateral body.

Colour Dusky pale brownish grey, sometimes greenish, with a pale belly. Spots mainly along midline. Indistinct pale barring on caudal peduncle. Median fins variously clear, faint yellow to black, and spines usually dusky. There is some sexual dimorphism similar to that of southern pygmy perch, with males having brighter colouration. Breeding males best recognised when median fins turn black.

Size Largest size known is 75 mm, but normally to about 65 mm.

Distribution Recorded from all coastal systems in Victoria west of Frankston to just over the border in Bool Lagoon, South Australia.

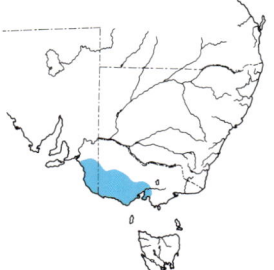

Conservation status Is a common species in a few places in Victoria but not known to be abundant anywhere; most records are of less than 10 fish. The Australian Society for Fish Biology lists it as 'potentially threatened' owing to the effects of habitat alteration.

Natural history Inhabits well-vegetated streams. Usually found in small groups, often mixed with *Nannoperca australis*, but seems to prefer slightly stronger flows. Aquarium-kept specimens readily take to mosquito larvae and other invertebrates, with insect larvae as favourite. Has bred in ponds in September and October, at temperatures between 16° and 24°C, when males were 35 mm long and females 40 mm.

Utility Is an interesting species for the aquarium and a potential species for mosquito control.

Similar species Other pygmy perches and juveniles of other perch-like species.

Other names Common: none. Scientific: *Edelia obscura, Paradules obscurus, Microperca yarrae.*

Literature Koehn & O'Connor, 1990; Kuiter & Allen, 1986; Leggett & Merrick, 1987; Sanger, 1978; Unmack, 1992.

Yarra pygmy perch *Nannoperca obscura.* (R.H. Kuiter)

FAMILY NANNOPERCIDAE

OXLEYAN PYGMY PERCH *Nannoperca oxleyana* Whitley

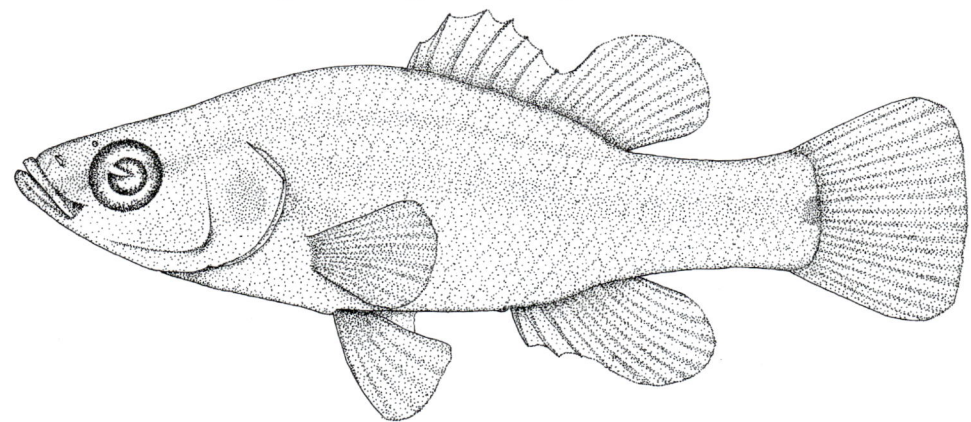

27.4 Oxleyan pygmy perch *Nannoperca oxleyana*. (R.M. McDowall)

Description Moderately compressed, body depth to about 32% in standard length. Mouth small, reaching to just below eye; teeth in lower jaw enlarged. Preorbital lower edge hidden by skin. Dorsal fin VI–VIII, 7–9 (usually VII, 8); anal fin III, 7–9 (usually 8); pectoral fin 11–13 (usually 12); gill rakers 2–4 + 6–8, total 9–12. Body covered by ctenoid scales in 26–28 midlateral rows; lateral line lacking. Tail truncated.

Colour Light brown to olive, darkest on back, sides paler, with three to four patchy, dark brown lines extending from head to tail; belly whitish. Opercular flap has a blue iridescence and there is a conspicuous dark spot with orange margins at base of tail. A blue ring around eye. During breeding the dorsal, pelvic and anal fins darken and the lateral stripes and tail turn red.

Size Attains about 75 mm, but commonly to only 45 mm.

Distribution Known only from south-eastern Queensland in small coastal and swampy drainages on the mainland and on Fraser and

Oxleyan pygmy perch *Nannoperca oxleyana*. (N. Armstrong)

Moreton Islands; also throughout the Noosa River; also known from one lake in Bundjalung National Park, south of the Richmond River in northern New South Wales; apparently absent from Lake Hiawatha near Grafton, and status in Richmond River at Coraki unknown.

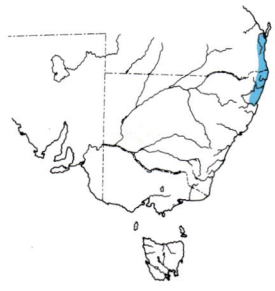

Conservation status Owing to habitat loss, this species has much more restricted geographical range than formerly and is known from only 18 localities, all populations apparently very small. Threats include habitat destruction, water pollution, and the presence of the exotic eastern gambusia. Is classified as 'endangered' by the Australian Society for Fish Biology, and a recovery plan has been prepared for it.

Natural history A shy fish found only in streams, swampy areas, and two lakes in coastal wallum (*Banksia* dominated heathlands), usually where there is dense vegetation; beds of the sedge *Eleocharis ochrostachys* seem preferred in some localities. Water always acidic (pH 5.4–5.7) and of very low conductivity (< 200 microsiemens per cm^2), often darkly stained with humic acids, over substrates of siliceous sand and plant debris. Prefers still to slow-flowing waters. Tolerates temperatures of 12–28°C and spawns from about October to May, when temperatures exceed 20°C. Females may breed at 30 mm, males at 27 mm. Spawning is protracted, with a few eggs laid daily, demersal, and scattered over the substrate or aquatic vegetation; aquarium-reared fish matured in 4–5 months. Young hatch in 3–4 days and feed on rotifers and protozoans; larger fish feed on copepods, cladocerans, caridinians and aquatic insects, especially chironomid midges, as well as diatoms, filamentous algae and a few terrestrial insects.

Utility Is an excellent display species in aquaria but requires care with water quality and prefers live foods.

Similar species Other pygmy perches and juveniles of other perch-like species.

Other names Common: none. Scientific: none.

Literature Arthington & Marshall, 1993; Arthington *et al.*, 1994; Kuiter & Allen, 1986; Sanger, 1978; Wager, 1992.

28
Family Cichlidae
Cichlids

A.H. ARTHINGTON AND P.L. CADWALLADER

This is a large family of fishes found naturally in warm fresh waters of Central and South America, Africa, Syria and Palestine, southern India and associated islands such as Sri Lanka, Cuba and Madagascar. Remarkable adaptive radiations have occurred within the family, particularly in the African Great Lakes, where hundreds of closely related species coexist. Cichlids are very popular as aquarium fish, and many attractive species have been distributed widely as part of the world trade in tropical fishes. Others cultivated as food fish now have a circumtropical distribution. Of the 150 or so cichlid species imported regularly into Australia through the aquarium trade, only three have established self-sustaining populations. The most widely distributed species is the Mozambique mouthbrooder, *Oreochromis mossambicus*.

Cichlids are distinguished by a deep, strongly compressed body, a large head, protrusible mouth, single nostril on each side of the head, a rounded or truncate caudal fin and a discontinuous lateral line.

Key to cichlids

1 Anal fin with 8–9 spines, 6–8 rays, 8–9 more or less prominent, dark vertical bands along body; black blotch on upper part of operculum, another at base of caudal fin. . . . . *Cichlasoma nigrofasciatum* p. 178; Fig. 28.2
Anal fin with only 3 spines, 9–12 rays. 2

2 Dorsal fin XV–XVI, 12–15; rays longer than spines, extending back to somewhat rounded tip well along caudal fin; dorsal fin base reaches base of caudal peduncle. Body dark olive green to light yellowish green, indistinct dark vertical banding, 2–6 large dark spots along sides of body between vertical bands. Eye deep red, fins greyish to pale green. . . . . *Tilapia mariae* p. 176; Fig. 28.1
Dorsal fin XV–XVII, 10–13; rays longer than spines, extending back to pointed tip, reaches to about $1/3$ the length of the caudal fin; dorsal fin base not reaching base of caudal peduncle. Dorsal surface dark grey to yellowish, body pale olive to silver grey, with 2–5 dark blotches on sides and sometimes several dark blotches towards the dorsal fin. Breeding males become dark olive to black, with white lower areas to head; black fins with red margins along dorsal and anal fins; pectoral fins become translucent red. . . . . *Oreochromis mossambicus* p. 179; Fig. 28.3

BLACK MANGROVE CICHLID
Tilapia mariae Boulenger

Description Small to medium-sized, body deep, strongly compressed and ovate. Head large, triangular; snout rounded; mouth small, terminal and protrusible, with gape not reaching back to eye; lips thick; jaws with 4–6 series of slender-shafted teeth. Single nostril on each side of head, eyes large. Single, long-based dorsal fin, XV–XVI, 12–15. Anal fin shorter-based, III, 10–12, rays long; large pectoral and pelvic fins (13 and I, 5). Caudal fin slightly rounded to truncate. Cycloid scales of moderate size cover the body and cheeks, 28–30 along lateral line, which is disjunct.
Sexual dimorphism Sexes can be distinguished

28.1 Black mangrove cichlid *Tilapia mariae*. (A. Günther)

by length of pelvic fins; in males extending back to 1st ray of anal fin; in females barely reaching 1st anal spine.
Colour Colour pattern changes with size, but there is no sexual dichromatism. Small fish olive green, with 8–9 dark, well-marked vertical bands on side of head and body. In large fish, body colour varies from dark olive green to light yellowish green, vertical banding much less distinct, 2–6 large dark spots along sides of body between vertical bands. Eye deep red; fins greyish to pale green.
Size May attain 300 mm; known to reach 235 mm and 325 g in Hazelwood pondage.
Introduction and distribution Is native to the lower reaches and coastal lagoons of rivers in West Africa, from the Ivory Coast to the Cameroons. Has become established in warmer waters of southern Florida.

Occurrence in south-eastern Australia is possibly due to deliberate releases by aquarists. Is known from the cooling pondage of the Hazelwood Power Station near Morwell, Victoria, Eel Hole Creek below the pondage, and in the La Trobe River, into which Eel Hole Creek drains via the Morwell River. Was first recorded at Hazelwood in 1978, and though it formed a large self-maintaining population, it has evidently not spread beyond heated areas of Eel Hole Creek and the La Trobe River near the power plant. Is considered unlikely to be able to survive in the low-temperature regimes of most Victorian waters. Also occurs north of Cairns in small estuaries and tidal creeks that interconnect during the wet season, and in the Barron River at Cairns.
Natural history Occurs in still and flowing waters, in rocky areas, bays with mud substrates, areas strewn with debris and beneath overhanging banks in the cooling pondage. Breeds at lengths of 100–150 mm. Eggs (600–3300 per female) are laid on leaves, submerged logs and debris at depths from 0.15 to 1.5 m; parent fish clean the egg deposition sites by mouth before spawning. Eggs olive green, adhesive, oval, 2.5 x 1.5 mm, guarded by one or both parents. They hatch after 1–3 days; yolk sac absorbed after about 10 days. Parental care continues until fry reach 25–30 mm. Adults and shoals of young fry are territorial but shoals of immature fish are nomadic. Diet includes diatoms, desmids, filamentous green algae and plant detritus.
Utility Valued as an aquarium species.
Similar species Convict cichlid, Mozambique mouthbrooder and other cichlids.
Other names Common: Niger cichlid. Scientific: none.
Literature Arthington, 1986; Cadwallader & Backhouse, 1983, Cadwallader *et al.*, 1980; Mather & Arthington, 1991; Trewavas, 1983.

CONVICT CICHLID *Cichlasoma nigrofasciatum* (Günther)

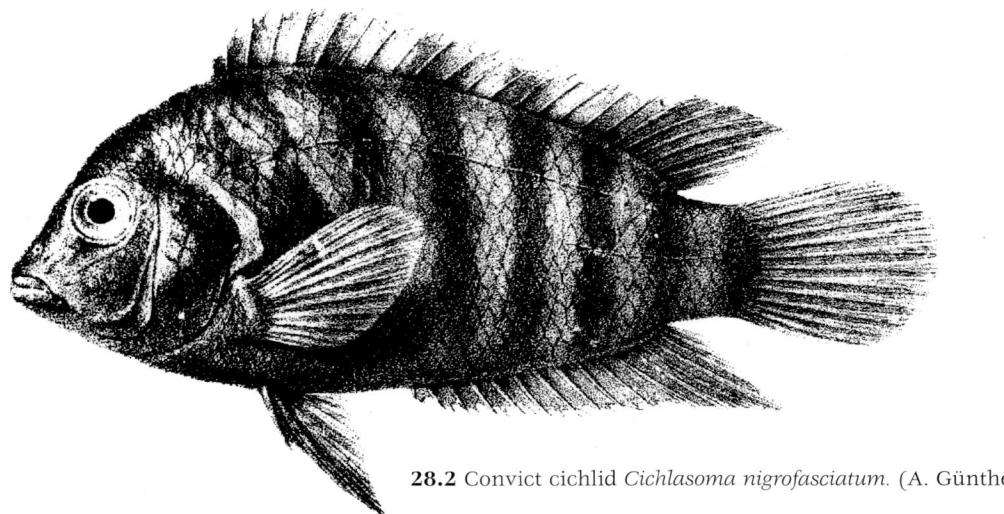

28.2 Convict cichlid *Cichlasoma nigrofasciatum*. (A. Günther)

Description A small, deep-bodied, strongly compressed and ovate fish. Head large, triangular, head profile steep. Mouth small, terminal and protrusible, with gape not reaching back to anterior border of eye; lips thick. Single nostril on each side of head; eyes moderate to large. A single, long-based dorsal fin, XVIII, 7–8. Anal fin IX, 6, long-based rays, posterior rays extended; large pectoral and pelvic fins (13–14, and I, V). Caudal fin slightly rounded to truncate. Ctenoid scales of moderate size cover the body and cheeks, 29–30 along lateral line which is disjunct.

Sexual dimorphism Sexes can be distinguished by deeper body and longer dorsal fin rays in males.

Colour Dorsal surface dark grey to bluish, sides grey often with violet sheen, belly pale grey; 8–9 dark vertical bands along body and nearly encircling body. More or less distinct dark blotch on upper part of operculum and at base of caudal fin. Eyes dark; fins pale greenish, with metallic sheen; dorsal and anal fins bordered with red. Vertical bands regress in males during breeding season, being replaced by gleaming metallic areas. Females generally duller, without band on dorsal fin but vertical bands become black during breeding season.

Size Reaches 150 mm; known up to 90 mm and 16.5 g in Hazelwood pondage.

Introduction and distribution Is native to Central America (Guatemala, El Salvador, Costa Rica and Panama) and has become established in ponds around Miami, Florida, in several springs in southern Nevada and in hot springs near Banff, Alberta. Is probably another aquarist's release. Is known only from the cooling pondage of the Hazelwood power station near Morwell and in Eel Hole Creek below the pondage, where it is not common. Has evidently not spread beyond Eel Hole Creek below the pondage and probably will be unable to survive in the low temperature regimes of most Victorian waters. Not found in the wild elsewhere in Australia. Was first recorded at Hazelwood in 1978.

Natural history Usually found in areas of Hazelwood pondage where cover is provided by rocks, vegetation or debris. Breeds at length of 80 mm, each female laying 130–400 adhesive eggs on cleared areas of substrate; eggs hatch after 3–4 days. Parents are territorial and aggressive, guarding the eggs and young fish and warding off intruders. Diet includes desmids, filamentous green algae and plant detritus.

Utility Known as an aquarium species but is aggressive and incompatible with other fish.

Similar species Black mangrove cichlid, Mozambique mouthbrooder and other cichlids.

Other names Common: zebra cichlid. Scientific: none.

Literature Arthington, 1986; Cadwallader & Backhouse, 1983; Mather & Arthington, 1991.

MOZAMBIQUE MOUTHBROODER *Oreochromis mossambicus* (Peters)

28.3 Mozambique mouthbrooder *Oreochromis mossambicus*. (A. Günther)

Description Medium-sized and deep bodied, compressed and ovate. Head large and triangular, head profile steep. Mouth small, terminal and protrusible, sometimes reaching back to anterior border of eye; lips thick. Eyes large. A single nostril on each side of head. A single, long-based dorsal fin (XV–XVII, 10–13); anal fin long-based (III, 9–12, usually 10–11), posterior rays extended; large pectoral and pelvic fins. Caudal fin truncate, fin tips often rounded. Scales in 3, occasionally 2, rows on cheeks, 30–32 in lateral line.

Sexual dimorphism Jaws of sexually mature males become enlarged, often causing the upper profile of the head to become concave.

Colour Dorsal surface dark grey to yellowish, body pale olive to silver grey, with 2–5 dark blotches on sides and sometimes several more dorsal blotches. Breeding males become very dark olive to black, with white lower areas to head, black fins with red margins along dorsal and anal fins; pectoral fins become translucent red.

Size May attain 360 mm and 2 kg.

Introduction and distribution Is native to brackish and freshwater estuaries of eastern Africa but now occurs widely in tropical and sub-tropical areas and in heated waters in Japan, USA and Europe. Evidence of problems with *O. mossambicus* in other countries has led to its declaration as 'noxious' in several

Australian states. Is established in impoundments and downstream creeks in Brisbane, Queensland; in creeks, drainage systems and artificial lakes throughout urban Townsville; and in small estuaries north of Cairns. Also occurs in the Gascoyne-Lyons River system near Carnarvon, Western Australia. The Cairns form is a hybrid of *O. mossambicus*, *O. niloticus*, *O. hornorum* and *O. aureus*; other populations are pure strains of *O. mossambicus*.

First recorded in south-eastern Queensland in 1977, Townsville in 1978, and Carnarvon, Western Australia, in 1981, where it has since spread in urban areas and along the Gascoyne-Lyons River system. There is concern that it could become established in other river systems, including the Murray-Darling; impacts on native species presently seem slight, but little research has been done.

Natural history Is found in well-vegetated, shallow areas of impoundments in Brisbane, shallow, vegetated drains, freshwater and tidal creeks in Townsville, melaleuca swamps, tidal creeks and isolated tidal pools near Cairns, and in relatively pristine riverine habitats in the arid-zone Gascoyne-Lyons River system characterised by intermittent flows. Members of this population are 'stunted', with sexual maturity reached at a mean standard length of 90 mm; usually breeds when over 150 mm. Male excavates and defends a basin-shaped pit where female lays (depending upon her size) 100–1700 pear-shaped, bright yellow eggs. Eggs and milt are sucked up by female and fry hatch in her mouth after 3–5 days; fry are released from the mouth 10–14 days after spawning;

FAMILY CICHLIDAE

Mozambique mouthbrooder *Oreochromis mossambicus*. (D. Bludhorn)

young fish stay near female, entering her mouth for safety and rest until about the 22nd day. Large *O. mossambicus* are essentially herbivorous, consuming plants, phytoplankton, filamentous algae and detritus; juveniles are microphagous omnivores.
Utility Is favoured in aquaculture in some countries, though it has largely been replaced by other more preferred species; remains important in capture fisheries in some countries. Is regarded as a pest in Australia.

Similar species Black mangrove cichlid, Niger cichlid and other cichlids.
Other names Common: official English name in southern Africa is Mozambique tilapia. Scientific: since 1984 has been referred variously to the genera *Chromis, Tilapia* and *Sarotherodon*; other specific names used include *niloticus, vorax*, and *natalensis*.
Literature Arthington, 1986; Cadwallader & Backhouse, 1983; Mather & Arthington, 1991; Trewavas, 1983.

29
Family Apogonidae
Cardinalfishes and mouthbrooders

D.A. POLLARD

Members of the family Apogonidae occur in most tropical and subtropical waters of the world and are most numerous in the Indo-Pacific region. Most are marine or estuarine, though a few enter fresh waters. Species of *Glossamia* live in fresh waters of northern Australia and Papua New Guinea. In many species the male carries the eggs and newly hatched young in its mouth, hence the common name 'mouthbrooder' that is sometimes used for them. The one species that occurs in the northern waters of south-eastern Australia is characterised by its huge mouth, two dorsal fins (the 1st with 6 spines) and large scales.

MOUTH ALMIGHTY *Glossamia aprion* (Richardson)

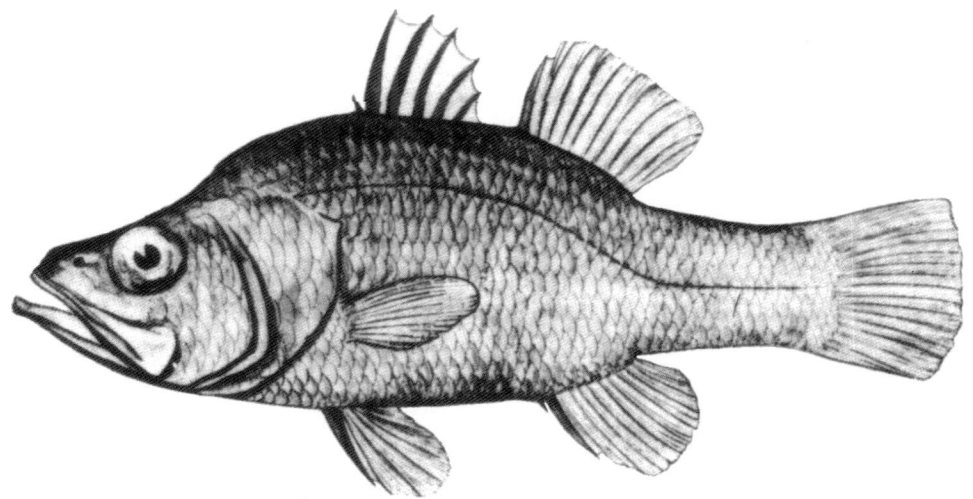

29.1 Mouth almighty *Glossamia aprion*. (F. Olsen)

Description A small, deep-bodied fish, somewhat laterally compressed, back sloping forwards to a long bluntly-pointed snout and a very large mouth that reaches back well below the large eyes; lower jaw protrudes slightly. There are two small dorsal fins, the 1st spinous (VI) and placed high on back and clearly separated from 2nd, soft-rayed dorsal (I, 10). Both dorsal fins are short-based, high and rounded; tail slightly forked to truncate, with rounded lobes. Pelvic fins located well forward, beneath pectoral fin bases (12-13 pectoral rays). Lateral line complete, with 25-31 scales.

Colour Creamy brown to reddish brown on back, with irregular darker-brown blotches on back and sides, belly paler. A dark bar passes forward from shoulder and down through eyes; outer half of 1st dorsal fin dusky.

Size Reputedly reaches a length of 180 mm and up to 600 g in weight in northern Australia and grows to at least 150 mm in more southern waters.

FAMILY APOGONIDAE

Mouth almighty *Glossamia aprion*. (N. Armstrong)

Distribution Is found in coastal drainages of northern New South Wales, Queensland, Northern Territory, northern Western Australia and southern rivers of Papua New Guinea.

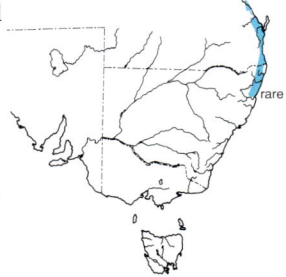

Conservation status Remains widely abundant throughout its range.

Natural history Is a strictly freshwater species, even though it belongs to a predominantly marine family. Inhabits streams, ponds and reservoirs, where it is usually found among dense vegetation. Spawns in summer at temperatures of about 22°C, with the eggs being about 3 mm in diameter, transparent and laid in a bundle enveloped by a thin membrane. The male broods the eggs in its mouth. Is carnivorous, and will readily take a variety of invertebrates and small fish. It will tolerate temperatures up to 38°C and pH levels of 4.5–8.

Utility A good eating fish despite its generally small size. Does not usually survive well in captivity, as it is prone to fungus infection after handling.

Similar species Small juveniles could be confused with pigmy perches, possibly chanda perches, and also the juveniles of some Australian 'perches', 'basses' and 'cods' (family Percichthyidae).

Other Names Common: Queensland mouthbrooder, Gill's cardinalfish, flabby, stinker. Scientific: *Apogon aprion*. Sometimes separated into two subspecies or even two species, the southern form (Clarence River in New South Wales to Fitzroy River in Queensland) being classified as *Glossamia aprion gillii*, or *Glossamia gillii*.

Literature Allen, 1989; Grant, 1991; Merrick & Schmida, 1984; Munro, 1967.

30
Family Percidae
Freshwater perches

R.M. McDOWALL

The family Percidae contains a small number of small to medium-sized fishes found naturally in the fresh waters of the cool-temperate Northern Hemisphere—in eastern North America (largely east of the Rockies), Europe, with the European perch, *Perca fluviatilis*, spreading across into Siberia; some North American percids are found in the temperate to subtropical southern states. No member of the family occurs naturally in Australia, but one species has been introduced successfully. The family is characterised by two dorsal fins, the long anterior dorsal supported by 13–17 stout spines; there are stout, firmly attached ctenoid scales.

REDFIN PERCH *Perca fluviatilis* Linnaeus

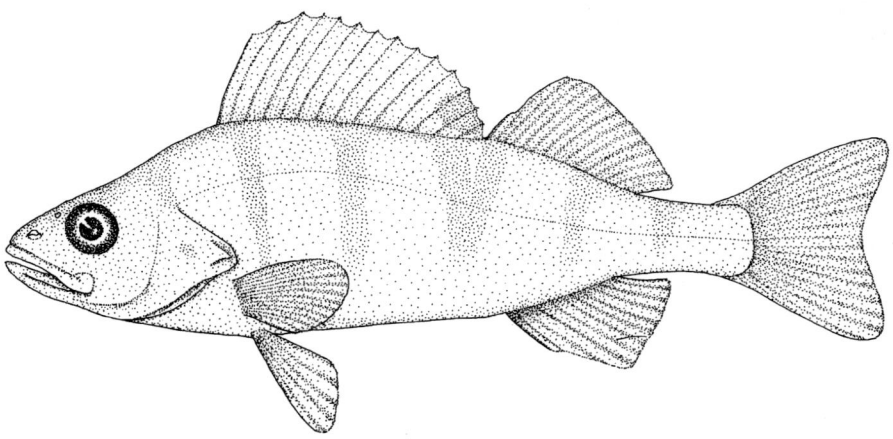

30.1 Redfin perch *Perca fluviatilis*. (R.M. McDowall)

Description A moderate-sized, deep-bodied and robust fish, with back humped behind head, especially in large fish. Two quite separate dorsal fins, the 1st high and rounded, middle spines longest (XIII–XVII), the 2nd a little smaller and more angular, anterior rays longest (I–II, 13–16); tail slightly forked. Anal fin similar to 2nd dorsal (I–II, 8–10); pectoral fins small and lateral (14 rays). Pelvic fins thoracic. Head large, gill covers ending in a broad, flat spine. Mouth large, reaching well below eyes. Scales large (58–68 along lateral line), strongly ctenoid, thick and hard, firmly implanted, and cover trunk, extending forward onto gill covers and cheeks; lateral line follows profile of back; vertebrae 39–41; gill rakers 22–25.

Colour A very handsome fish, olive greenish to grey on back, greenish to silvery on sides and white on belly. There are six or more vertical bands, broader across back and tapering on sides, prominent in small fish but becoming less so in larger ones. Dorsal and caudal fins greenish grey, 1st dorsal with a distinct black blotch at rear. Pelvic and anal fins, and lower margin of tail, bright reddish orange.

Size Commonly reaches 400–450 mm in length and 1–2 kg in weight; is known to have reached more than 10 kg in Australia, but this is unusually large.

Introduction and distribution Is an introduced species that was shipped from Europe early in Australian history—at least by 1862. Natural range includes the whole of western Europe and across through eastern Europe into subarctic Siberia. The American yellow perch, *P. flavescens*, is certainly very closely related and is considered by some to be the same species. New Zealand stocks of redfin originated in Tasmania.

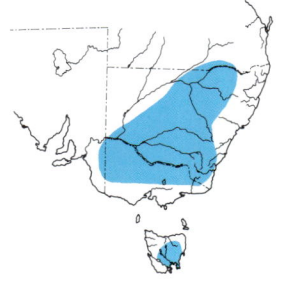

Is widespread in eastern Australia in cooler waters on both sides of the Great Dividing Range—central and southern New South Wales, throughout Victoria and in South Australia; is locally abundant in eastern and southern Tasmania.

Natural history Occurs mainly in still and slow-flowing waters, especially where aquatic vegetation abounds. Spawning occurs in spring; the females have only one ovary and no oviduct, and the yellowish eggs, 2–3 mm in diameter, are released through a temporary pore in the body wall. Thousands to several hundred thousand eggs are laid as long gelatinous ribbons dispersed among aquatic weeds, submerged logs, etc. This egg mass is unpalatable to other fish, and the eggs are thus protected from predation. The eggs develop in 1–3 weeks (depending on temperature), and the young form shoals for some time before becoming solitary. Redfin may take several years to mature, though fast-growing males may mature at the end of first year. Higher mortality of males at young age means that the populations of larger fish are mostly females. They feed on a wide range of aquatic organisms, including crustaceans, molluscs; larger redfin eat some small fish.

Fears have been expressed about redfin predation on native Australian species, such as pigmy perch, rainbowfishes and even such larger species as Murray cod, Macquarie perch and golden perch.

Utility Is regarded in Europe as a fine sporting fish and a good table fish (as is the related species in North America). Does not have such a wide and high general reputation in Australia, though

Redfin perch *Perca fluviatilis*. (R.H. Kuiter)

it is a desired sporting fish in Victoria and does provide good angling and is good eating. Populations tend to become very dense, resulting in small, stunted fish of little interest to anglers. In some localities, redfin have been blamed for the decline of previously good trout populations.

Is not well regarded as an aquarium fish, even though it is handsome, quite active and easily kept; this disfavour is due to its predatory habits, which make it unsuitable for keeping with other fish species.

Similar species Superficially resembles the larger native Australian perches, basses and cods; the young resemble pigmy perches and chanda perches.

Other names Common: European or English perch. Scientific: none.

Literature Anon., 1973; Collette *et al.*, 1977; Craig, 1987; Thorpe, 1977; Weatherley, 1963, 1977; Weatherley & Lake, 1967b.

31

Family Gadopsidae

Freshwater blackfishes

P.D. JACKSON, J.D. KOEHN, M. LINTERMANS AND A.C. SANGER

The family Gadopsidae is a small endemic family that occurs naturally on the mainland of south-eastern Australia and in northern Tasmania. A recent taxonomic review of the family has resulted in a second species, *Gadopsis bispinosus*, being described. Taxonomic review has also recognised 'northern' and 'southern' forms of *Gadopsis marmoratus*, with the 'southern' form growing much larger. The familial relationships remain unclear, though recent work suggests some affinity with the Percichthyidae and other basal percoids. Both described species are characterised by jugular pelvic fins reduced to one branched ray in each, also long and rather low dorsal and anal fins, tiny scales, and a lateral line following the profile of the back.

Key to freshwater blackfishes

6–13 stout spines in the dorsal fin . . . . ***Gadopsis marmoratus*** p. 186; Fig. 31.1
2 slender spines in the dorsal fin (occasionally 1 or 3) . . . . ***Gadopsis bispinosus*** p. 188; Fig. 31.2

RIVER BLACKFISH *Gadopsis marmoratus* (Richardson)

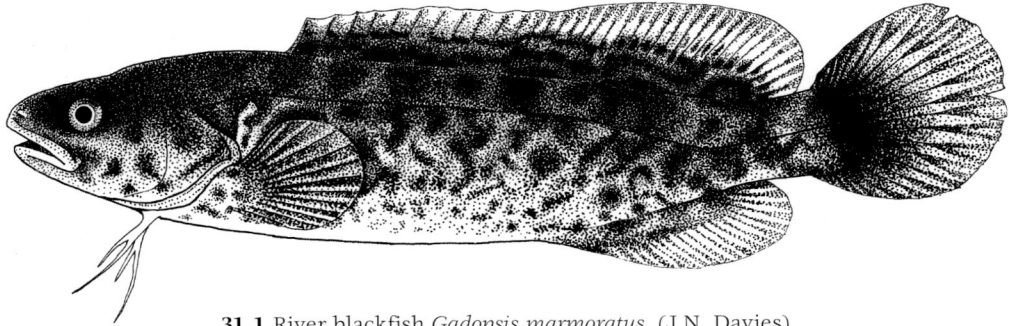

31.1 River blackfish *Gadopsis marmoratus*. (J.N. Davies)

Description Small to moderate in size, elongated and slender, moderately compressed; head, mouth and snout large, with lower jaw shorter than upper, both jaws with an outer large row of conical teeth and an inner narrow band of teeth. One long, low dorsal fin originating just behind head (VI–XIII, 22–31), anterior spinous portion lower than soft-rayed portion. Dorsal and anal fins enveloped in skin, anal fin III, 16–19. Pelvic fins beneath opercular openings (jugular), each consisting of a single divided ray; 15–19 pectoral rays. Tail rounded. Scales very small and cycloid; lateral line follows curvature of back. Vertebrae 40–50.

Colour Varies from pale olive green or yellowish brown to almost black on back and sides, marbled with numerous irregular darker blotches. Ventrally varies from pale yellow to bluish or purplish grey.

Size Varies from region to region; on the mainland reaches greatest size in southern (coastal) Victoria, up to 600 mm and 5.5 kg; however,

River blackfish *Gadopsis marmoratus*. (R.H. Kuiter)

rarely this big, with specimens up to 450 mm more common. Smaller inland of the Great Dividing Range in Victoria and New South Wales, growing to only 300–350 mm and about 250 g. Reaches at least 625 mm and 5 kg in Tasmania.

Distribution Occurs in many west-flowing streams in New South Wales, and north into southern Queensland as far as the Condamine River; in the north only at altitudes above 600–900 m. Further south is found much further west and as low as 150 m elevation. In Victoria occurs in tributaries of the Murray system and in south-flowing streams, but is missing from many of the smaller coastal streams, such as those in the Otway Ranges. Range once extended as far west as the Torrens and Onkaparinga Rivers in South Australia, and remnant populations still remain in streams draining into the Murray River from the Adelaide Hills, though total number is thought to be less than 500. A separate group of populations exists in drainage areas in the south-east corner of South Australia, including Ewens Ponds. This may be a taxonomically distinct group, though further work is needed to clarify the situation. Is native to northern Tasmanian rivers draining into Bass Strait, the Arthur River in the north-west and the Anson River in the north-east. Has been introduced into other rivers, such as the Huon and Derwent in the south; populations in the North and South Esk Rivers also possibly introduced. No records from Bass Strait islands. Early records suggest that in Victoria 'southern' variety blackfish were moved north of the Great Dividing Range in the early 1900s for stocking some farm dams.

Conservation status Range is apparently considerably reduced, particularly at extremities of former distribution. A decline in abundance is also evident in areas where habitat degradation has occurred, but remains common in many areas. Appears susceptible to increased sediment loads in rivers. Without parental care, eggs become smothered with a thin layer of silt and die. Increased sediment also kills many juvenile fish, and a big drop in numbers has been reported from a Victorian stream subjected to high levels of suspended sediment from a weir desilting

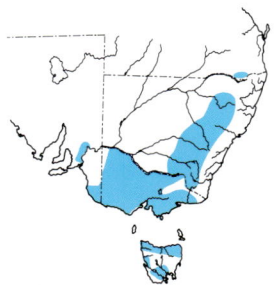

operation. Sediment may also reduce substrate variability; other threats include removal of wood debris and reductions in water temperatures during the spring and summer spawning seasons.
Natural history Frequents diverse stream types, from clear, fast-flowing mountain streams to more sluggish, lowland rivers; prefers streams with abundant cover, such as snags and boulders. Occurs in some Victorian reservoirs, though never in high numbers, and populations have been established in farm dams.

Life cycle is completed entirely in fresh water. Spawning occurs in spring or early summer, when water temperature reaches at least 16°C. Eggs are large (about 4 mm in diameter), demersal, adhesive and yellow to orange in colour. Fecundity is comparatively low, usually less than 500 eggs per female, although it shows a linear increase with body length. A female 320 mm long will have about 750 eggs. The fish appear to form pairs before spawning, and aggression appears to increase. Spawning takes place inside hollow logs, where water speed is low. This is the most important criterion for site selection, although spawning sites are also normally close to instream cover, such as wood debris and boulders. River blackfish spawn readily in PVC pipes and may spawn in farm dams, particularly if there are hollow logs or PVC pipes. The male guards the fertilised eggs and rarely leaves the spawning site. It actively cleans the eggs and is aggressive towards intruders. Time to hatching depends on water temperature, but is about 14 days at 15°C. Newly hatched larvae are about 6–8 mm long. After hatching, the large yolk sacs remain inside the ruptured egg cases, allowing the larvae to stay in a stationary upright position for about 19 more days. They do not lose their yolk sac for about 26 days after hatching, by which time they are about 15 mm long.

River blackfish are nocturnal and relatively sedentary, there is no apparent migration, and the normal home range is about 20 m. They prefer habitat with water speeds of less than 0.2 m/s and accumulations of wood debris. Depth preferences change with fish size, the large fish occupying deeper waters. Depth and velocity preferences are shared with the introduced brown trout, and coexistence appears to depend on the benthic and nocturnal habits of river blackfish, which contrast with the mid-water and diurnal habits of trout. Diets also overlap when the two species occur together, river blackfish being opportunistic carnivores and feeding on a variety of aquatic insects, crustaceans, terrestrial invertebrates and occasionally other fish.
Utility Can be caught readily on hooks baited with earthworms or grubs, but its small size in northern parts of range reduces its value there. In the south is a popular angling species, though fewer large specimens have been caught in recent times. Is possibly unable to withstand heavy angling pressure, partly owing to low fecundity. Reported to be good eating. Has been kept successfully in aquaria, becoming quite tame, but is pugnacious and predatory; where there is more than one in the tank, savage fighting often occurs.
Similar species Two-spined blackfish, young Murray cod and trout cod.
Other names Common: freshwater blackfish, slippery, slimy, greasy, tailor, marbled river cod. Scientific: none.
Literature Davies, 1989; Jackson, 1978a, b; Johnson, 1984; Koehn, 1986; Koehn & O'Connor, 1990a, b.

TWO-SPINED BLACKFISH
Gadopsis bispinosus Sanger

Description A small to medium-sized species, body narrow and slightly compressed, head narrow and slightly elongated. A single long dorsal fin inserted posterior to pectoral fins, spines slender and weakly calcified (I–III, 35–38). Anal fin long (III, 17–20). Pelvic fin reduced to a single branched ray, tail rounded, scales small and cycloid. Lateral line prominent, with a distinct dorsal curvature. Vertebrae 46–49.
Colour Variable, but typically consisting of pale yellowish brown basal colouration overlayed by two or three rows of brown blotches along the length of the body and extending onto the dorsal, caudal and anal fins. Ventral surface uniformly pale to the original of the anal fin. Outer edge of the dorsal, caudal and anal fins colourless (white in preserved specimens), often bordered by an intense dark stripe. Fin rays in dorsal, caudal and anal fins bright yellow in live specimens. Colour pattern more distinct in juveniles, becoming less well defined in older specimens.
Size Up to 300–320 mm and 250 g, but typically much smaller, rarely larger than 250 mm.
Distribution Distribution covers a band on the north of the Great Dividing Range running through

FRESHWATER BLACKFISHES

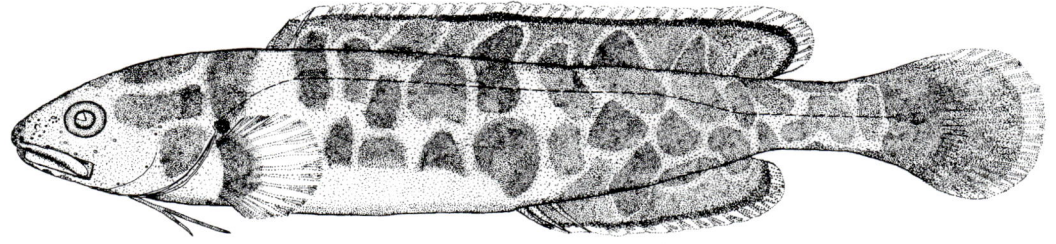

31.2 Two-spined blackfish *Gadopsis bispinosus*. (A.C. Sanger)

north-eastern Victoria into south-eastern New South Wales and the Australian Capital Territory. Is widespread and often abundant in north-eastern Victoria at altitudes of 200–700 m. In New South Wales and the Australian Capital Territory, occurs in the lower-altitude streams of the Murray River above Lake Hume, such as the Swampy Plains, Geehi and Tooma Rivers, also in a few streams of the upper Murrumbidgee drainage, including the Goodradigbee, Goobarrangandra, Cotter and Tumut Rivers.

Conservation status Very abundant in many streams across its range.

Natural history Shows a distinct preference for clear, cool streams with coarse gravel, cobble and boulder substrates, where it is often very abundant. Its preference for specific instream habitats has been clearly demonstrated and dramatic increases in abundance have occurred where boulders and woody debris have been added to otherwise uniform substrates. Such instream cover provides shelter from high water speeds. Is often found in sympatry with river blackfish in the lower-altitude areas of its range. Spawning occurs in spring and early summer, large females spawning earlier in the season than smaller ones. In the Australian Capital Territory, spawning is known

Two-spined blackfish *Gadopsis bispinosus*. (N. Armstrong)

to occur from November to mid-December, with a small number of eggs (120–320) laid by each female. Spawning site is unknown but is thought to be among cobble/boulders or other sheltered positions associated with instream cover. Will spawn in PVC pipes introduced to the instream habitat, where eggs are demersal and adhesive and are laid in a single egg mass. Eggs are large (about 3.5 mm diameter), yolky, and golden yellow and hatch in 15–17 days at 15°C. Eggs and larvae are guarded by the male until the larvae leave the spawning site about 3 weeks after hatching. Juveniles are often found in large numbers sheltering and feeding in accumulations of leaf litter and woody debris on the edge of quieter stretches of streams. The species is carnivorous, feeding mainly on benthic aquatic insect larvae, particularly mayflies, caddises, stoneflies and dipterans, though terrestrial insects, crustaceans, earthworms, other fish and fish eggs are also consumed. Little seasonal variability is evident, though terrestrial insects comprise a greater proportion of the diet in spring and summer and as fish size increases. The species is extremely sedentary, with a home range of about 15 m. Appears able to co-exist with introduced trout species despite some predation pressure.

Utility Is readily caught on small hooks baited with worms and occasionally caught on a fly; its small size does not make it a desirable angling species. Is a very attractive aquarium species and easily kept, provided cover is available and water temperature does not exceed 22–24°C, above which fungal infections are troublesome. Generally only a single adult can be kept per aquarium as fighting often occurs, although several juveniles will co-exist.

Similar species River blackfish, young Murray cod and trout cod.

Other names Common: as for river blackfish. Scientific: none.

Literature Koehn, 1986, 1987, 1990; Koehn & O'Connor, 1990; Lintermans & Rutzou, 1990a, b; Sanger, 1984, 1990; Waters *et al.*, 1994.

32
Family Mugilidae
Grey mullets

J.M THOMSON

The family Mugilidae is widespread in tropical and temperate seas and sometimes in fresh waters. One Australian species can be regarded as substantially freshwater in habit, and several others enter fresh water from time to time, though all reproduce at sea. Mullets are characterised by having 2 dorsal fins that are well separated, the 1st with four slender spines, an anal fin with three spines, scales present, no lateral line, and pelvic fins that are midabdominal. They are typically shoaling, silvery fishes that are herbivorous or detritus feeders and have a muscular, gizzard-like stomach that helps break down their food.

Key to mullets

1 Axillary processes at bases of pectoral fins; adipose eyelid strongly developed 2
   No axillary processes at bases of pectoral fins; adipose eyelid rudimentary or absent 3

2 Lateral scales 30–32; 15 pectoral rays; teeth present on tongue; 22 pyloric caeca . . . .
   . . . . *Valamugil georgii* p. 196; Fig. 32.5
   Lateral scales 38–42; 16 pectoral rays; no teeth on tongue; 2 pyloric caeca . . . .
   . . . . *Mugil cephalus* p. 193; Fig. 32.3

3 Lateral scales 54–59; 12 rays in anal fin . . . . *Aldrichetta forsteri* p. 195; Fig. 32.4
   Lateral scales 50 or fewer; 10 or fewer anal fin rays 4

4 Lateral scales 35–38; 10 anal fin rays . . . . *Liza argentea* p. 196; Fig. 32.6
   Lateral scales more than 40; 8–9 anal fin rays 5

5 Lateral scales 43–46; 16 pectoral rays; 9 anal rays; no teeth on tongue . . . .
   . . . . *Myxus elongatus* p. 193; Fig. 32.2
   Lateral scales 47–50; 15 pectoral rays; 8 anal rays; teeth present on tongue . . . .
   . . . . *Myxus petardi* p. 191; Fig. 32.1

FRESHWATER MULLET
Myxus petardi (Castelnau)

Description A small to moderate-sized fish, stout in build and deep-bellied. Head small and pointed, its length less than body depth; eye moderately large. First dorsal (IV) set well back, in hind ½ of body; 2nd dorsal I, 8; tail deeply forked; anal fin III, 9. Pectoral fin (15 rays) does not nearly reach level of 1st dorsal fin. Teeth fine and hair-like; 2 pyloric caeca; scales ctenoid, 47–50 along side; 24 vertebrae.

Colour Dark olive green above, silvery below, fins pale yellow, upper edge of operculum and eye golden to pinkish.
Size Has been recorded to more than 800 mm in length and 7.5 kg in weight, but is generally 400 mm or less.
Distribution Inhabits fresh waters of New South Wales, from

FAMILY MUGILIDAE

32.1 Freshwater mullet, *Myxus petardi* (F. Steindachner)

Georges River north into Queensland as far as the Burnett River. Adults have been taken in the sea after heavy floods; distribution of young unknown.
Conservation status Remains widespread and common.
Natural history Generally found where deep pools slow the flow of rivers, often occurring in small shoals. Reaches maturity at age 4 and around 300 mm long, females a little larger than males; fecundity very high, with larger females producing about 1–3 million eggs. Mature adults move downstream to estuaries and the sea in late summer and early autumn, with most spawning occurring during February. Feeds on microscopic plants and animals, and detritus.
Utility Can be caught on a fine hook using dough baits or even a piece of earthworm; is a good food fish but often develops an earthy flavour in semi-stagnant waters. Is said to be easily kept in captivity.
Similar species Sea mullet and at small sizes other mullets; possibly hardyheads when small.
Other names Common: pinkeye. Scientific: *Trachystoma petardi*.
Literature Humphrey, 1979; Lake, 1971; Merrick & Schmida, 1984; Thomson, 1954a.

Freshwater mullet *Myxus petardi*. (D. Rodgers)

GREY MULLETS

SAND MULLET *Myxus elongatus* Günther

32.2 Sand mullet *Myxus elongatus*. (E.R. Waite)

Description A small to moderate-sized fish with a slender body, head longer than body is deep. Origin of 1st dorsal fin (IV) nearer to tail base than to tip of snout; 2nd dorsal I, 8; anal III, 9; no axillary process in upper angle of pectoral fin (16 rays); 2 pyloric caeca. Scales cycloid, 43–46 along side; 24 vertebrae.

Colour Dark olive green on back, silvery on sides, shading to white below; pectoral and dorsal fins and tail dusky, anal fin white-edged. Eye yellow with iridescent hues. A dark spot at pectoral fin base.

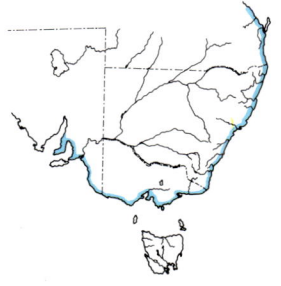

Size Is known to reach 260 mm.
Distribution Southern Australia from Fremantle (Western Australia) to Maryborough (Queensland); also Lord Howe Island.
Conservation status Quite common in most estuaries.
Natural history Spawning probably occurs near mouths of estuaries. Fish in their first year often enter fresh water but seldom thereafter; more commonly seen in brackish water.
Utility Taken in gill and hauling nets; usually marketed with other rare species of mullet.
Similar species Other mullets, and hardyheads when small.
Other names Common: tellegalene. Scientific: none.
Literature Thomson, 1954a, 1966.

SEA MULLET *Mugil cephalus* Linnaeus

Description A moderate-sized, robustly built fish with a broad, flat head that is longer than the body is deep; eye moderately large with an adipose eyelid. Dorsal fins (IV, I, 8) similar to those of freshwater mullet but 1st dorsal not set as far back, origin nearer snout tip than base of tail; anal fin III, 8. Pectoral fin (16 rays) with a long, pointed axillary process in its upper angle (lacking in freshwater mullet); 2 pyloric caeca. Scales feebly ctenoid, 38–42 along side; 24 vertebrae.

Colour In the sea olive green dorsally, with bright silvery sides, shading to off-white below. After a few weeks in fresh water the olive green back becomes suffused with deep blue or dirty brown and the sides become dull. When freshly caught, 6–7 faint brown stripes run down body. Fins dusky except pelvics, which are pale yellow. A golden ring around eye; a dark blue to violet blotch at pectoral fin base.

Size Is known to grow to 750 mm and may exceed 5 kg weight; often reaches 500 mm.

Distribution Is found all around the coastline of the Australian mainland but only in northern Tasmania; is not common north of Townsville, Queensland. Is worldwide in the tropics and warm temperate zones, anywhere between about 40°N and 40°S, but is less common in the tropics than in cooler latitudes.

Conservation status Continues to abound in

193

FAMILY MUGILIDAE

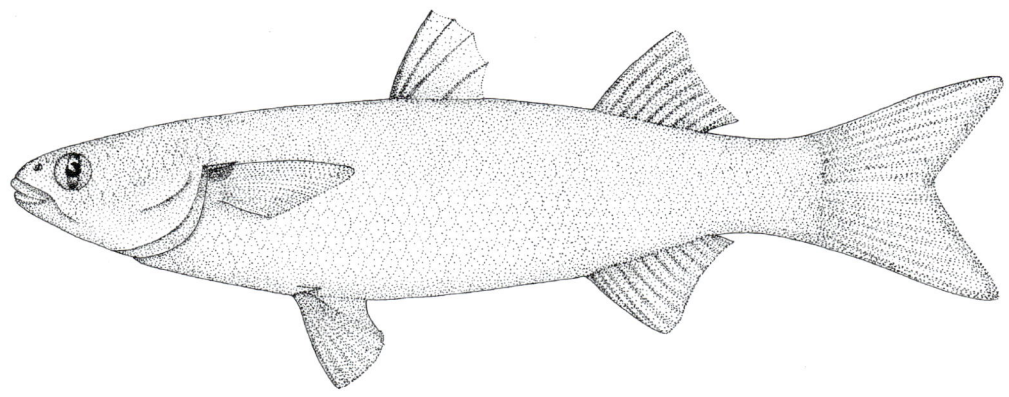

32.3 Sea mullet *Mugil cephalus*. (R.M. McDowall)

Australian coastal seas and rivers.
Natural history Occurs in the lower and estuarine reaches of larger river systems in great roving shoals. Reaches maturity at age 3, when about 320 mm long. Spawns in the sea, probably in winter. Eggs are small, about 0.75 mm in diameter, and may number many hundred thousand to several million; they hatch in a few days, producing larvae 2–3 mm long. Little is known about spawning behaviour or site, but the mature fish must go to sea to spawn; ovaries are resorbed in fish confined to fresh water. However, small juveniles may be found in fresh water, where they may feed and grow for several years, though a freshwater phase is not obligatory, and not all spend time there. Feeds on algae, detritus, and small organisms ingested with these materials.

Utility Is taken commercially in nets and is a prime market fish, being commonly served in fish-and-chip shops. However, is notorious for not taking a bait, so is less important to anglers, though is more susceptible in fresh water and may be taken on a fine hook baited with dough or a small piece of earthworm.
Similar species Freshwater mullet at all sizes; also other mullets that sometimes enter fresh water at small sizes, though not as adults.
Other names Common: bully mullet, mangrove mullet, hardgut mullet, paddy mullet, river mullet. Scientific: *Mugil dobula*.
Literature Chubb *et al.*, 1981; Grant and Spain, 1975; Merrick and Schmida, 1984; Thomson, 1953, 1954a, b, 1955, 1957, 1963, 1966; Wells, 1984.

Sea mullet *Mugil cephalus*. (R.H. Kuiter)

YELLOWEYED MULLET *Aldrichetta forsteri* (Cuvier and Valenciennes)

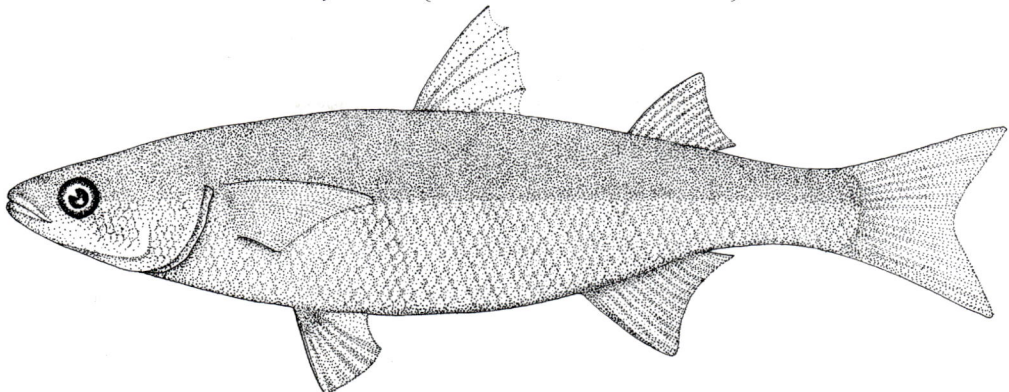

32.4 Yelloweyed mullet *Aldrichetta forsteri*. (R.M. McDowall)

Description A small to moderate-sized fish of rather slender build; head pointed, rather longer than body is deep, and with a rather large eye. First dorsal fin (IV) with origin nearer base of tail than tip of snout; 2nd dorsal I, 9; anal fin III, 12; no axillary process at upper angle of pectoral fin (16 rays); 2 pyloric caeca. Scales cycloid except on cheeks, where they become ctenoid, 54–58 along side; 24 vertebrae.

Colour Olive brown above, silvery to yellowish white in sides and belly, fins with brownish margins; eyes bright yellow. No axillary spot on pectoral fins.

Size Attains at least 500 mm, but rarely more than about 300 mm.

Distribution Occurs in southern Australia, from Geraldton in Western Australia to the Hunter River in New South Wales; also in New Zealand. Is mainly marine and estuarine, but does enter fresh water, though not as far upstream as sea mullet.

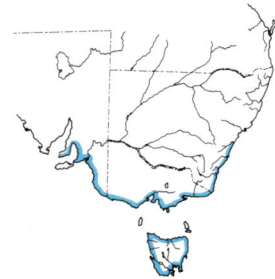

Conservation status Remains abundant throughout range.

Natural history Is a schooling fish that moves into estuaries and lower reaches of rivers on the

Yelloweyed mullet *Aldrichetta forsteri*. (R.H. Kuiter)

rising tide and may remain there for several tidal cycles; also frequents brackish low-elevation lakes. Spawns at sea, eggs about 3 mm in diameter and numbering several hundred thousand; little is known about reproduction but probably occurs in late summer and autumn, when schools of mature fish assemble in coastal waters. Feeds on detritus, algae and small animals.

Utility Is the main mullet species marketed in southern Australian states, from Western Australia to New South Wales and Tasmania, where it is taken in beach seines and gill nets. Will take a bait and is a valued quarry for children around wharves and estuaries.

Similar species Other mullets, and hardyheads when small.

Other names Common: sea mullet (Victoria, South Australia, Tasmania). Scientific: *Agonostomus forsteri*.

Literature Chubb *et al.*, 1981; Gorman, 1962; Harris, 1968; Manikiam, 1963; Thomson, 1954a, b, 1956, 1957a, 1966.

FANTAIL MULLET *Valamugil georgii* (Ogilby)

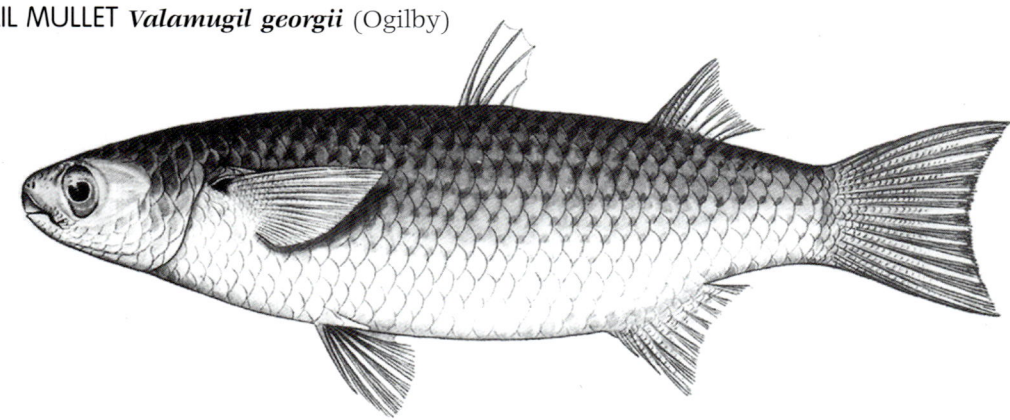

32.5 Fantail mullet *Valamugil georgii*. (A.R. McCulloch)

Description A small species, but deep-bodied, head not as long as body is deep. First dorsal (IV) nearer to base of tail than to tip of snout; 2nd dorsal III, 8; a large axillary process in upper angle of pectoral fin (15 rays); 22 pyloric caeca. Scales cycloid on back and flanks but ctenoid on belly, 30–32 along side; 24 vertebrae.

Colour Greenish brown on back, sides silvery shading to off-white below; fins pale, but 2nd dorsal, anal and caudal fins with dusky borders, membrane of 2nd dorsal speckled with brown; a golden segment over upper $1/3$ of eye and a smaller segment below; a deep purple axillary spot on base of pectoral fin.

Size Grows to 260 mm.

Distribution Known from Port Hacking (New South Wales) to Burnett River (Queensland), estuarine and coastal, the young fish entering fresh water.

Conservation status Common in estuaries within range.

Natural history Little known of habits, but moves in schools in coastal seas and estuaries.

Utility Because it is a small species, only limited quantities reach fish markets, usually mixed with other, rarer mullet species.

Similar species Other mullets, and hardyheads when small.

Other names Common: silver mullet. Scientific: none.

Literature Thomson, 1954a, 1966.

FLAT-TAIL MULLET
Liza argentea (Quoy and Gaimard)

Description A small to moderate-sized fish, rather slender, but with a deep and rather compressed caudal peduncle. Head about as long as body is deep in larger individuals, but longer in small individuals. Origin of 1st dorsal (IV) nearer to tip of snout than to base of tail; 2nd dorsal fin I, 9; anal III, 10. No axillary process at upper angle

GREY MULLETS

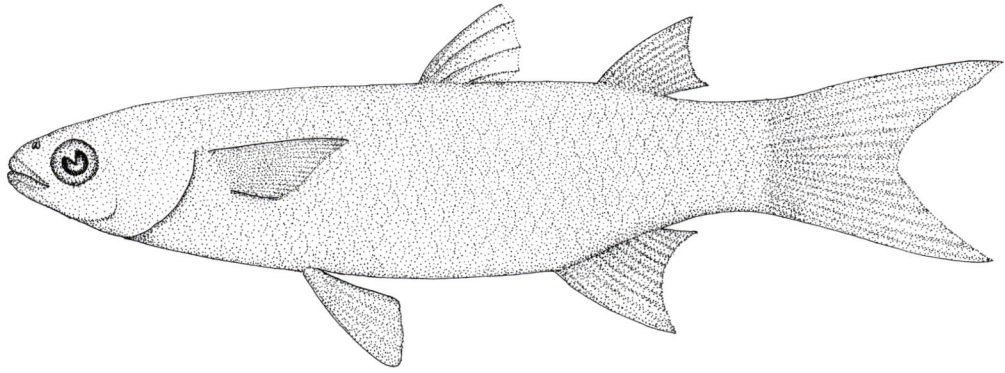

32.6 Flat-tail mullet *Liza argentea*. (R.M. McDowall)

of pectoral fin (16 rays); 2 pyloric caeca. Scales feebly ctenoid, 35–38 along side; 24 vertebrae.
Colour Light brown above, silvery below, dorsal fin and tail dusky, anal and pectorals less so, a patch of bright gold at upper, posterior corner of operculum; eyes purple with golden flecks.
Size Reaches 300 mm.
Distribution Southern and eastern Australia, from Fremantle (Western Australia) to Cooktown (Queensland).

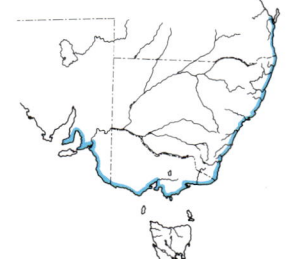

Conservation status Abundant in some estuaries, especially in New South Wales and southern Queensland.
Natural history Young fish enter fresh water but seldom, if ever, after age of 12 months; schools common in estuaries and along sea beaches.
Utility Is caught in gill nets and hauling nets and is marketed in all states. Will take a bait, though not readily; worms and molluscs are effective bait.
Similar species Other mullets, and hardyheads when small.
Other names Common: tiger mullet. Scientific: none.
Literature Thomson, 1954a, 1966.

Flat-tail mullet *Liza argentea*. (R.H. Kuiter)

33

Family Bovichtidae

Congolli

A.P. ANDREWS

This family consists mainly of Antarctic and subantarctic marine fishes and is only sparingly represented in the Australian region. One species, however, has the unusual ability to enter and live in fresh water. Members of the family have characteristically elongated tubular bodies and broad flattened heads, with the eyes set close together, almost on top. Two dorsal fins are present, the 1st short and separated from the 2nd, which is long and low and extends almost to the tail. The pelvic fins are well forward, about level with the pectorals. The scales are moderate in size, and a lateral line is present.

CONGOLLI *Pseudaphritis urvillii* (Cuvier and Valenciennes)

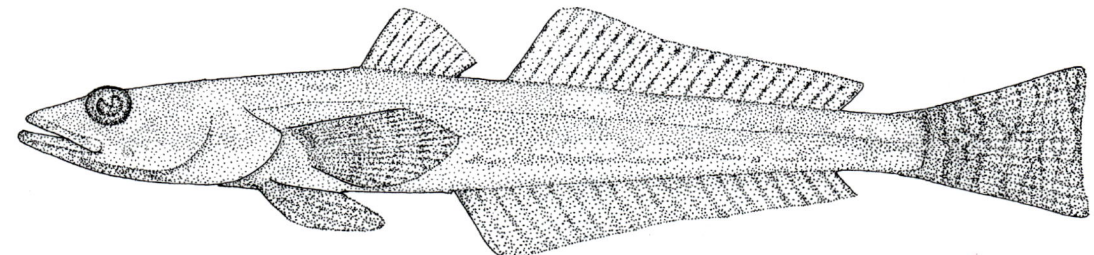

33.1 Congolli *Pseudaphritis urvillii*. (R.M. McDowall)

Description A small to medium-sized fish with a slender, almost cylindrical body, slightly compressed towards tail; head conical in shape and flattened on top, with smallish eyes set close together above corners of wide mouth. Lower jaw longer than upper, and both jaws have bands of small teeth. First dorsal fin (VII–VIII) origin well behind head, 2nd dorsal long and low (19–22 rays). Anal fin origin below last rays of 1st dorsal and resembles 2nd dorsal in shape (II, 21–22); rays in both 2nd dorsal and anal fin low and almost equal in length. Pelvic fins (I, 5) originate a little ahead of pectorals (18 rays). Head and body covered with moderate-sized ctenoid scales, 59–65 along lateral line.

Colour Colour variable, depending on substrate colour, mostly reddish or greenish brown on back, irregularly marked on sides, depth of colour variable; undersurface a uniform yellowish white. Dorsal, pectoral and tail fins striated with dark lines.

Size Can attain at least 340 mm, but mostly to about 100–150 mm.

Distribution

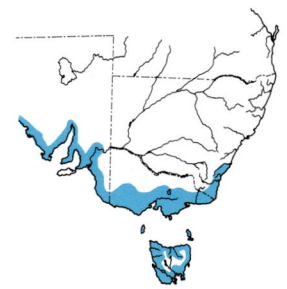

Although primarily a marine species, is common in fresh water, particularly near mouths of slow-moving streams emptying into estuaries. Has been recorded up to 120 km inland and is known to move considerable distances up the Murray River and in rivers of the Derwent drainage basin in Tasmania. Is widespread in coastal New South Wales south of Bega, Victoria, South Australia and all coastal areas of Tasmania.

Congolli *Pseudaphritis urvillii*. (R.H. Kuiter)

Conservation status Remains widespread and abundant through its range.

Natural history Favoured habitat is on the beds of slow-flowing streams in leaf litter, where it often remains partly buried, among rocks and under sunken logs and overhanging banks. Juveniles are most abundant in the lower reaches of rivers during spring and summer, while larger specimens are generally more common further upstream. There may be some habitat differences between the sexes. Knowledge of the spawning and migration patterns is limited. In Victoria, adult fish are believed to migrate from the upper reaches down to the estuaries to spawn during autumn and winter. Food consists of insect larvae, worms, small crustaceans and fish. Some plant material is also consumed.

Utility Although excellent eating, is usually too small to be utilised as table fish. Is usually caught only by recreational anglers and is of no commercial importance.

Makes an excellent, although not very active, aquarium fish, and wild fish can be transferred directly to either salt- or freshwater aquaria without signs of distress. Acclimatises rapidly, and captive specimens have been known to live for 4–5 years. However, if the aquarium is to be shared with other species, care must be taken to ensure that congolli are small and comparable in size owing to the carnivorous nature of the congolli. Spawning in captivity has not been recorded.

Similar species Has distinctive shape and is seldom confused with other freshwater fish. However, in the sea and estuaries could be confused with juvenile marine flathead (Family Platycephalidae).

Other names Common: tupong, flathead, sandy, marble fish and sand trout. Scientific: referred to as *Pseudaphritis bursinus* in some early literature.

Literature Hale, 1920; Hortle, 1979; Hortle & White, 1980; Koehn & O'Connor, 1990; Lake & Fulton, 1983; Last *et al.*, 1983; Merrick & Schmida, 1984.

34
Family Gobiidae, subfamilies Eleotridinae and Butinae
Gudgeons

H.K. LARSON AND D.F. HOESE

The family Eleotridae, as has been usually understood, is now considered to consist of three subfamilies within the family Gobiidae. Two of these subfamilies occur in Australia (Butinae and Eleotridinae) and contain the greatest number of freshwater species there, about 40–50. Most occur in fresh or brackish water, but some occur in the sea, around mangroves or in grass flats. Most Australian species are small, reaching 30–100 mm, but the sleeper (*Oxyeleotris lineolatus*) reaches 500 mm and a weight of several kilograms in northern Queensland. Because of their small size, many of the species are hard to identify; for instance, it has been found that what was known as the 'western carp gudgeon' is actually three separate species. Similarly, early workers have overlooked the dwarf flathead gudgeon.

Little is known about the natural history of Australian gudgeons, although they live well in aquaria, and some work has been done on their breeding biology. All species attach their eggs to the bottom or to vegetation, and the larvae hatch in a few days. All species feed on small insects, crustaceans and sometimes small fishes. Species of *Hypseleotris* and *Gobiomorphus* prey on mosquito larvae. Although early fish workers pointed this out, the eastern gambusia (*Gambusia holbrooki*, see p. 116) was unfortunately and unnecessarily introduced into Australia.

The purple-spotted gudgeons (*Mogurnda adspersa* and *Mogurnda mogurnda*) are common aquarium fish in both Australia and Europe; because gudgeons are small, often colourful and adapt well to aquaria, amateur aquarists are increasingly keeping them.

Key to gudgeons

1. Top of head with bony ridges above and behind eyes; snout long and flat, 3–4 times eye diameter . . . . **Butis butis** p. 207; Fig. 34.8
   Top of head without bony ridges; snout short, rounded or flattened, 1–2 times eye diameter . . . 2

2. Head and body strongly compressed, head much narrower than deep; pectoral fin base narrow and reduced; upper pectoral ray well below upper attachment of gill membrane 8
   Head rounded or depressed, width about equal to or greater than depth; pectoral fin base broad, upper pectoral ray behind or slightly below upper attachment of gill membrane 3

3. Head strongly depressed, width about 1.5 times depth; mouth large in adults, reaching to or beyond middle of eye; sides of head without scales; several transverse rows of sensory papillae below eye 4
   Head rounded or truncate in side view, width 0.8–1.2 times depth; mouth small, ending below or before pupil; sides of head scaled; 2–3 longitudinal rows of sensory papillae below eye 5

4. Gill openings extensive, reaching forward to below eye; sides of belly usually with 4–5 thin, transverse bands; bar at base of pectoral fin absent or faint; body uniformly coloured or with

a few irregular blotches; pectoral rays usually 18–19; total rakers on 1st arch 14–20; reaches 110 mm . . . . **Philypnodon grandiceps** p. 202; Fig. 34.1
Gill openings narrow, reaching to below posterior preopercular margin; sides of belly without thin transverse bands; bar at base of pectoral fin usually distinct, body mottled or with faint broad, irregular bands; pectoral rays 15–16; total gill rakers on 1st arch 11–12; reaches only 50 mm . . . . **Philypnodon sp.** p. 203; Fig. 34.3

5   Interorbital narrow, about equal to eye diameter; no transverse groove on chin; 2nd dorsal I, 8–9; preopercular margin with 3–5 pores; body usually with 1 or more distinct longitudinal stripes. 6
    Interorbital wide, about twice eye diameter; a shallow transverse groove on chin; 2nd dorsal I, 10–14; preopercular margin without pores; body with a series of distinct spots 7

6   Sides of body with a single, broad midlateral stripe; mouth ends before eye; 1st dorsal VI; 2nd dorsal and anal I, 9; pectoral rays 18–19; lateral scales 36–40. . . . . **Gobiomorphus coxii** p. 204; Fig. 34.4
    Sides of body with 5–7 thin, longitudinal lines; mouth reaches to below anterior part of eye; 1st dorsal VII; 2nd dorsal and anal I, 8; pectoral rays 14–16; lateral scales 29–33 . . . . **Gobiomorphus australis** p. 206; Fig. 34.5

7   Scales along side 30–36; scales from anal origin upward and forward to base of 1st dorsal 9–13 . . . . **Mogurnda adspersa** p. 209; Fig. 34.9
    Scales along side 37–46; scales from anal origin to dorsal base 14–18 . . . . **Mogurnda mogurnda** p. 210; Fig. 34.11

8   No scales before 1st dorsal fin or on top of head, before pelvic fins or on midline of belly; 1st dorsal very low; 3rd to 5th pelvic rays about equal in length (4th a little longer), giving fin an obtusely pointed margin; inland drainages of southern Queensland, New South Wales and South Australia. . . . . **Hypseleotris sp. 5** p. 218; Fig. 34.18
    Predorsal, top of head, prepelvic and midline of belly scaled; 1st dorsal fin low to high; 3rd and 5th pelvic fin rays distinctly shorter than 4th, 5th distinctly shorter than 3rd, giving fin an acutely pointed margin; inland and coastal 9

9   Sides with irregular cross-hatching forming Xs; 2nd dorsal with a series of black and white spots posteriorly; faint spot on caudal peduncle before tail, below midline; preoperculum with 2–4 pores; 2nd dorsal usually I, 9; anal usually I, 10; 14–16 gill rakers on 1st arch; coastal drainages of Victoria, New South Wales, Queensland, Northern Territory and north-western Western Australia . . . . **Hypseleotris compressa** p. 212; Fig. 34.12
    Sides plain or with faint vertical bars; no white and black spots on 2nd dorsal; spot on caudal peduncle present or absent; preoperculum without pores; 2nd dorsal usually I, 10–13; anal usually I, 11–14; gill rakers 7–12 10

10 A series of transverse papillary rows above and below eyes; adults with basal half of dorsal and anal fins darker (orange in life) than outer half; a faint vertical bar on caudal peduncle just before tail, below midline; no black or iridescent blotches on belly; coastal rivers, central Queensland to central New South Wales, and inland drainages of Queensland, New South Wales and South Australia . . . . **Hypseleotris klunzingeri** p. 215; Fig. 34.15
    Sometimes a single transverse papillary row above eye, otherwise no transverse rows, but longitudinal rows above and below eyes; adults with basal $2/3$ of dorsal and anal fins darker than outer $1/3$, or with two narrow orange or grey longitudinal stripes; no vertical bar on caudal peduncle below midline; midline of belly with large iridescent and black patches below skin 11

11 Females with a black urinogenital papilla; a very thin black line on midline of breast before pelvic fins; iridescent marks on belly faint; bar at base of pectoral fin covers almost entire base; predorsal scales 8–14; coastal streams of southern New South Wales and Queensland . . . . **Hypseleotris galii** p. 214; Fig. 34.14
    Females with a pale urinogenital papilla; no thin black line on midline of breast; iridescent marks on belly prominent; bar at base of pectoral fin covers upper half of base only; predorsal scales typically 13–18; inland drainages of South Australia, New South Wales, southern Queensland, also coastal drainages of southern and central Queensland . . . . **Hypseleotris sp. 4** p. 217; Fig. 34.17

FAMILY GOBIIDAE

FLATHEAD GUDGEON *Philypnodon grandiceps* (Krefft)

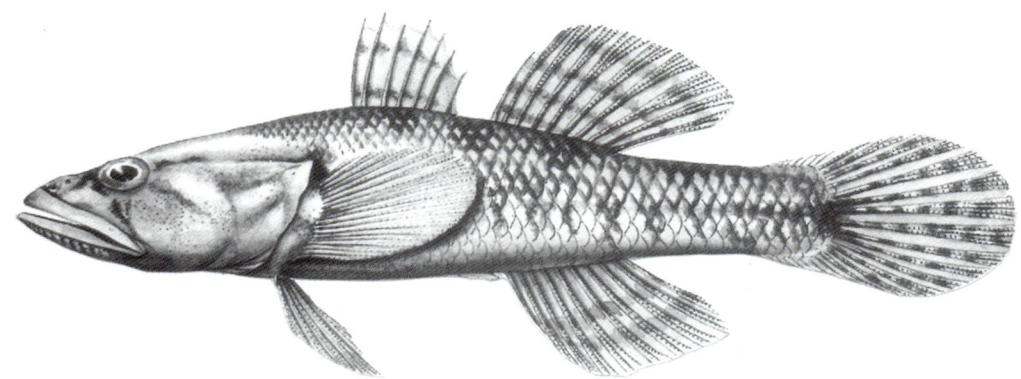

34.1 Flathead gudgeon *Philypnodon grandiceps*. (A.R. McCulloch)

Description A small species, body slender, head broad and flat; mouth large, reaching to below middle to hind of eye; gill openings broad, extending forward on lower surface of head to below eyes. Tip of tongue has a slight notch. No spines on head or body. First dorsal fin (VI–VIII, usually VII) origin well behind level of pelvic fin bases; 2nd dorsal (I, 8–10, usually I, 9). Tail broadly rounded, shorter than head; 15 segmented caudal rays. Pectorals with 16–20 rays (usually 18–19). Body covered with medium-sized ciliated scales (33–44 along side). Top of head scaled forward to near eyes except in northern populations, in which scales end above gill covers. Gill rakers 14–20 (usually 16–18); vertebrae 29. Rows of papillae on head as in Fig. 34.2.

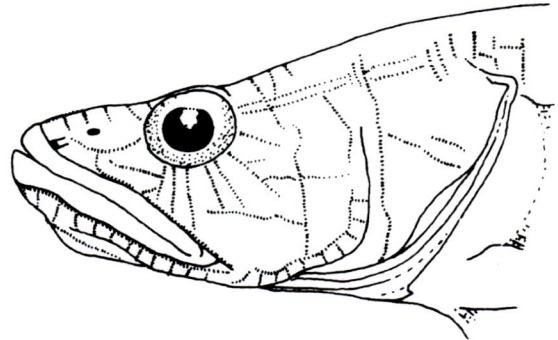

34.2 Rows of papillae on head of flathead gudgeon *Philypnodon grandiceps*. (H.K. Larson)

Sexual dimorphism Large male in breeding condition has a larger mouth, a more bulbous head, a broader interorbital and larger pelvic fins than the female. In the male the mouth generally reaches to below pupil, while in the female it ends under front of eye (the two sexes have been treated as separate species by some workers). Urinogenital papilla similar in the two sexes. Breeding females develop several flaps around opening of elongate papilla, while males have only small bumps around opening. Breeding males often become very dark.

Colour Varies considerably; head and body may be grey, brown, reddish brown, yellowish or black. A series of faint, dark blotches on back behind head and below dorsal fins. Ventral surfaces lighter and may be yellowish. Usually 4–5 thin dark lines on side of belly. Often a series of blotches forming an irregular and broken midlateral stripe. A stripe from behind eye extends back over gill covers, and a second stripe below it extends onto preoperculum. A thin, darker bar from below eye to corner of mouth. First dorsal fin with 2 grey stripes, 2nd dorsal with 3 stripes; fin between stripes clear to yellowish. A thin median stripe on anal fin, a broad stripe at tip. Tail with 5–7 thin bands formed by black spots. Pectoral and pelvic fins yellowish to grey; a faint bar at pectoral fin base. All bands and stripes on fins may be very faint. Usually a transverse bar on back just before dorsal fin, and a dark spot at tail base, which characterises this species.

Size Reaches 115 mm, but normally grows to only about 80 mm.

Distribution Occurs in coastal drainages from the MacKenzie River, in Queensland, to South Australia; ranges northwards inland in the Murray River and tributaries, west of the Murrumbidgee and throughout the Lachlan; also north coast of Tasmania.

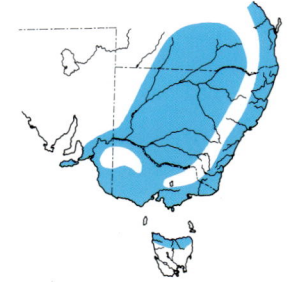

Conservation status Abundant through most of range, though rare in Tasmania.
Natural history Prefers quiet waters, particularly lakes and dams, usually on weedy or mud bottoms; frequently occurs in estuaries. Breeds in spring and summer in inland drainages, apparently over a longer period (through to winter) in northern coastal rivers. Feeds on small fishes, crustaceans and insects. Little work done on natural history (though captive breeding has been successful).
Utility Birds and larger fishes prey on this gudgeon; lives well in aquaria but preys on smaller fishes, including its own kind.
Similar species Easily confused with dwarf flathead gudgeon; young may be confused with Cox's gudgeon and carp gudgeons.
Other names Common: big-headed gudgeon, bullhead, collundera, Yarra gudgeon. Scientific: *Philypnodon angustifrons*, *P. nudiceps*.
Literature Koehn & O'Connor, 1990; Last *et al.*, 1983; Lake, 1978; Merrick & Schmida, 1984.

DWARF FLATHEAD GUDGEON *Philypnodon* sp. (undescribed)

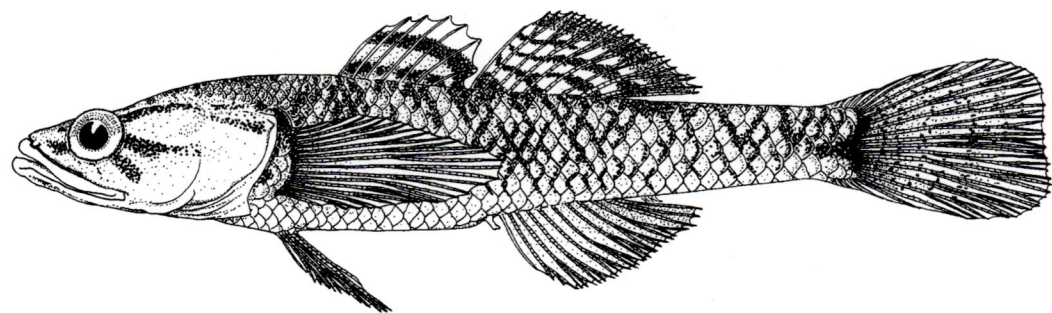

34.3 Dwarf flathead gudgeon *Philypnodon* sp. (H.K. Larson)

Description Similar in general appearance to flatheaded gudgeon, but smaller and differing in having narrower gill openings and fewer gill rakers, and in colouration. Gill openings broad, extending forward to below gill covers or preoperculum, well behind eyes. Development of scales on top of head varies considerably: in northern New South Wales females have scales forward to above posterior end of preoperculum; in southern areas both sexes have scales on top of head forward to just behind eyes; specimens from Bathurst have no scales on top of head in either sex.

Other features similar to *P. grandiceps*. First dorsal VI–VII; 2nd dorsal I, 8–9; anal I, 7–9; pectorals 15–16; lateral scales 32–36; gill rakers 11–12.

Sexual dimorphism Adult males have a much larger mouth, a more bulbous head, a broader interorbital and larger pelvic fins than do females. In males the mouth reaches beyond rear of eye, while in females it ends below middle of eye. During breeding, males are often darker than females.

Colour Head and body brown to black, body covered with large, irregular blotches, often resembling broad saddles; a black vertical bar at tail base. Chin usually grey, the rest of lower surface of head and body white to grey, lighter than sides. Two oblique bars extend from behind eyes, across gill covers. Lips generally black. First dorsal with 2 black stripes, area between stripes and tip of fin whitish or orange. Second dorsal with 3–4 oblique black lines sloping towards body posteriorly. Tail with 5–6 wavy lines; basal two-thirds of fin orange or grey, and margin grey or blue. Anal fin dusky to clear. Pectoral and pelvic fins clear to whitish, a black vertical bar on pectoral fin base.

Size Reaches 50 mm but rarely grows to more than 40 mm.

Distribution Occurs in coastal streams of southern Queensland and New South Wales, Victoria and South Australia; also known from a few localities in the Murray River in South Australia and New South Wales, also from Bathurst, New South Wales. Found from brackish waters in estuaries to altitudes of a few hundred metres.

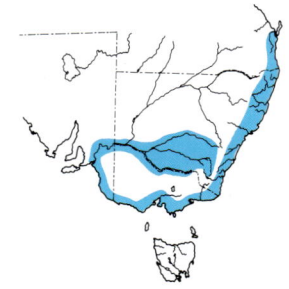

Conservation status Common in coastal northern areas of range.

Dwarf flathead gudgeon *Philypnodon* sp. (R.M. McDowall)

Natural history Tends to prefer relatively calm waters and lives over mud or rocks or in weedy areas. Nothing is known of breeding and life history. Regularly occurs with flathead gudgeon (*P. grandiceps*).
Utility Lives well in aquaria and feeds on live or dead foods; its bright colouration makes it an attractive aquarium fish.
Similar species Easily confused with flathead gudgeon (was long not recognised as distinct); also confused with young Cox's gudgeon and striped gudgeons.
Other names Common: none. Scientific: none, though has been confused with *P. grandiceps*, and some literature references under this name may be dwarf flathead gudgeon.
Literature Allen, 1989.

COX'S GUDGEON *Gobiomorphus coxii* (Krefft)

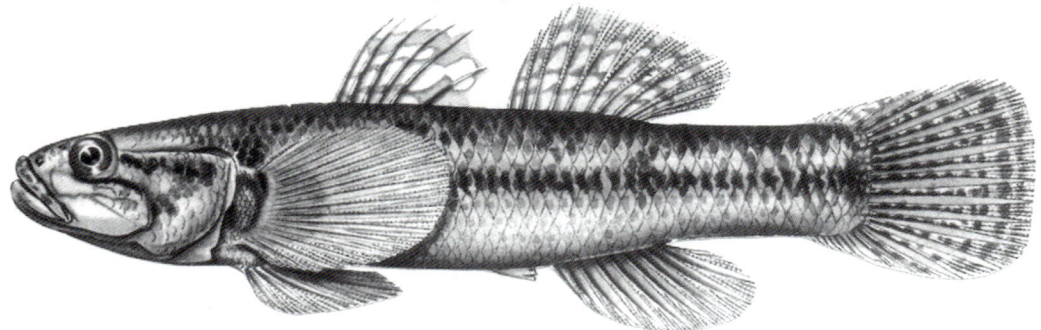

34.4 Cox's gudgeon *Gobiomorphus coxii*. (A.R. McCulloch)

Description A small species, moderately slender, with a rounded head; snout as broad as long, mouth oblique, reaching to about anterior eye margin. First dorsal spines (VI) all shorter than rays of 2nd dorsal (I, 9). Anal fin (I, 9) below 2nd dorsal. Pectorals with 18–19 rays; 15 segmented caudal rays. Body covered with large ciliated scales (36–40 along sides). Top of head has small scales reaching to above back of eyes; scales present on gill covers but not on cheeks. Gill rakers 9–12; vertebrae 26–27.
Sexual dimorphism In breeding condition, males are brighter than females, although females are larger and, when gravid, have swollen bellies, giving them a robust appearance. Urinogenital papilla slender and pointed in

males, broad and truncate in females, with numerous small flaps surrounding the opening.
Colour Varies from light to dark brown, purple, or olive green, belly generally a pale tan, blue or gold. Sides often scattered with yellow. A single black or brown stripe along midside. In juveniles this may be broken into a series of elongate blotches posteriorly. Chin and lower surface of head dusky to black; 1–2 faint, dark stripes from behind eye across gill cover; a distinct black spot at upper pectoral fin base, with a spot at bottom of fin base in large individuals. First dorsal fin with 2 dark grey or black stripes separated by a clear, yellow or orange stripe; edge of fin clear to yellow. Second dorsal with 3 dark stripes separated by yellow or orange. Pectoral, pelvic and anal fins clear, yellow or dusky grey. Tail yellowish brown with 5–6 irregular bands formed by grey to black spots.
Size Commonly reaches 150 mm and is known to reach 190 mm.
Distribution Occurs in coastal rivers and streams from the Richmond River in New South Wales to Wilsons Promontory in Victoria.

Conservation status Abundant in rivers of southern New South Wales.
Natural history Normally prefers flowing waters and often occurs in rapids. Is rarely found close to the ocean, and ranges inland into mountain areas to an altitude of at least 700 m. Juveniles generally found lower in the rivers; it is thought that the young are washed downstream and later migrate upriver. Has been seen out of water, climbing up waterfalls and dam walls. Female lays eggs on rock; after they have been fertilised, the male guards the nest and fans the eggs. Eggs develop in a few days. Feeds on small aquatic insects, including mosquito larvae.
Utility Forms a food source for birds and larger fishes and is thought important in controlling mosquitos. Lives well in aquaria.
Similar species Can be confused with striped gudgeon, the juveniles of the two species being difficult to differentiate; juveniles may also be confused with small empirefish.
Other names Common: Mulgoa gudgeon. Scientific: none.
Literature Koehn & O'Connor, 1990; Lake, 1978; Merrick & Schmida, 1984; Cadwallader and Backhouse, 1983.

Cox's gudgeon *Gobiomorphus coxii*. (N. Armstrong)

STRIPED GUDGEON *Gobiomorphus australis* (Krefft)

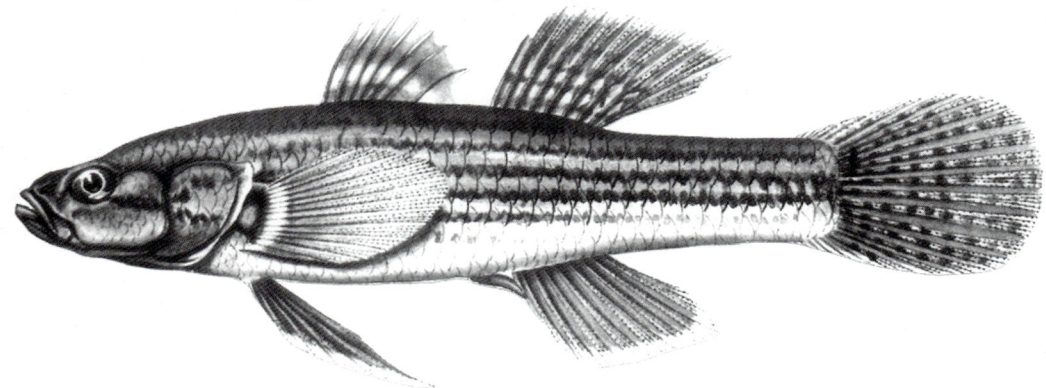

34.5 Striped gudgeon *Gobiomorphus australis*. (A.R. McCulloch)

Description Small, stout and robust; head rounded, the snout broader than long; mouth oblique, reaching just below anterior of eye. First dorsal (VII, rarely VI or VIII) with flexible spines, all slightly shorter than rays of 2nd dorsal (I, 8). Anal fin (I, 8) below 2nd dorsal; pectorals with 14–16 rays; 15 segmented caudal rays. Body covered with large ciliated scales (30–34 along sides); top of head with large scales reaching forward to eyes, with several smaller scales between eyes; none on snout. Sides of head completely scaled, but those on cheeks small and not easily seen. Gill rakers 10–12; vertebrae 28. Rows of pores and papillae on head as in Fig. 34.6.

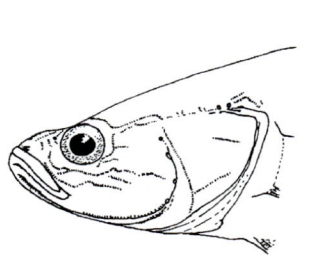 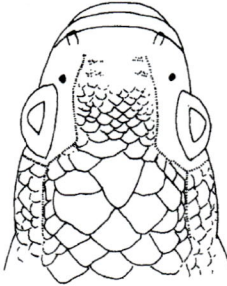

34.6 Rows of pores and papillae on head of striped gudgeon *Gobiomorphus australis*. (H.K. Larson)

Sexual dimorphism Breeding male much brighter in colour than female, its anal fin orange with a lilac border; that of female golden yellow. Urinogenital papilla lanceolate and nearly twice as long as broad in male; oblong, truncated and barely longer than broad in female. Posterior rays of 2nd dorsal and anal fins, and also 4th ray of pelvic fin, longer in males.

Colour Background colour varies from a rich chocolate-brown to dark grey dorsally, merging to cream or grey ventrally. Usually 5–7 prominent dark stripes along sides, running from behind pectoral fin to tail base. Stripes darker on midside; in life the stripes can be lightened and darkened rapidly. More colourful in breeding season, a greenish gold background laterally, with yellow in paler areas, and small violet or purplish spots over much of body. There is a prominent stripe from eye to pectoral base that is generally faint on gill covers; there may be a very faint stripe above and below this stripe. A dark stripe from front of eye to tip of snout; a dark spot at upper pectoral base, a thin oblique line ventrally, and a distinct white bar across fin base. Fins with chestnut spots, 1st dorsal with 2 rows, 2nd with 4–6 rows, tail with 6–10 vertical rows forming wavy vertical bands. Anal, pectoral and pelvic fins yellowish to grey and may have faint grey or purple bands.

Size Occasionally reaches 175 mm, but more commonly seen up to about 120 mm.

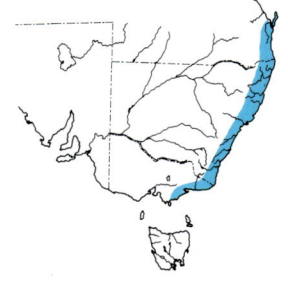

Distribution Occurs in coastal streams from Maryborough (Queensland), south through New South Wales to about Wilsons Promontory in eastern Victoria.

Conservation status Common in some streams during summer.

Natural history Appears to prefer muddy waterholes and sluggish creeks, rather then swifter, clearer waters. Generally most abundant at lower elevations, close to the coast. Juveniles common in estuaries, and it is thought that the young are carried downstream, migrating back upstream later in life. Spawns in late summer and autumn at temperatures of around 21°C. After spawning,

Striped gudgeon *Gobiomorphus australis*. (N. Armstrong)

becomes rare until the following spring. Adults feed on mosquitofish and aquatic insects.

Utility Is an important source of food for larger fish such as bass in coastal streams. Can be kept readily in aquaria, but attempts to breed it in captivity have been unsuccessful.

Similar species Most easily confused with Cox's gudgeon, particularly when small. Young can also be confused with purple-spotted gudgeon and empirefish.

Other names Common: none. Scientific: has generally been called *Mogurnda australis*, but studies reveal closer affinities with Cox's gudgeon (*Gobiomorphus coxii*); similarities to species of *Mogurnda* are only superficial.

Literature Bayly *et al.*, 1975; Koehn & O'Connor, 1990; Lake, 1978; Merrick & Schmida, 1984; Whitley, 1961.

CRIMSON-TIPPED GUDGEON
Butis butis (Buchanan)

Description A small species, head depressed, snout long and rather flattened. The fairly large mouth (extending under or behind eye) and a bony knob at snout tip give this fish a distinctive appearance. Body covered with scales, largely ciliated, although those on belly, breast and pectoral fin bases cycloid (29–30 scales along side). Fine ciliated scales cover head except for a pattern of exposed bony ridges, the most distinct being a V-shaped ridge on top of snout. Small axillary scales at bases of body scales. Dorsal fins VI, I, 8; anal I, 8; pectorals 18–21; gill rakers 12; segmented caudal rays 15. Rows of papillae on head as in Fig. 34.7.

Sexual dimorphism Sexes very similar; genital papilla of female short and fat, with a few small, fine flaps at tip; that of male slender and flat, with fine scalloped flaps at tip.

Colour Generally blackish to blackish brown

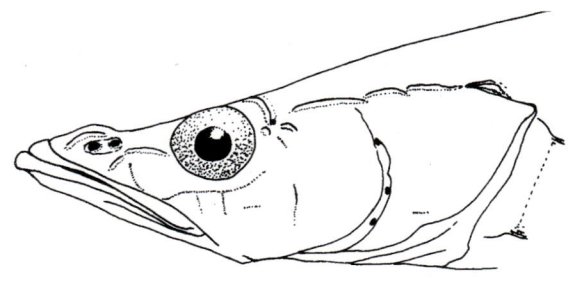

34.7 Rows of papillae on head of crimson-tipped gudgeon *Butis butis*. (H.K. Larson)

FAMILY GOBIIDAE

34.8 Crimson-tipped gudgeon *Butis butis*. (P. Clarke)

above, often a lighter brown below head and on belly. Sides with black lines following scale rows, and a light-brown spot in the centre of each scale. Pattern on sides may be obscured by darker transverse blotches. On head 3–5 oblique bars radiate from eyes, the most distinct running back across cheeks, another from front of eyes to middle of upper jaw. The most distinctive body marking is a dense black (sometimes double) spot in centre of pectoral base, this spot bordered above and below by bright orange-red blotches. Dorsal fins blackish basally and white above (1st dorsal may have an orange to crimson outer margin). Anal fin similar to 1st dorsal, but may also have a basal orange-red stripe. Tail blackish brown with an oblique white area that may occupy dorsal margin only, or up to almost $1/3$ of fin. Lower margin of tail may be red. Pelvics blackish, with white margins.
Size May reach 145 mm, with about 95 mm an average size.
Distribution Occurs from Broome north to Darwin, around to the Queensland coast and northern New South Wales, the southernmost record being Yamba; also occurs throughout the South-East Asian archipelago to New Guinea and northern Australia.
Conservation status Reasonably abundant.

Crimson-tipped gudgeon *Butis butis*. (R.H. Kuiter)

Natural history Not strictly a freshwater-dwelling fish; is nearly always found in brackish water and coastal mangroves, but may extend up coastal rivers. A tropical to warm–temperate species that seems to prefer slower-moving waters.
Utility Not much kept in aquaria but lives quite well if fed on live small fish and crustaceans. Exhibits interesting prey-stalking behaviour, and grips sticks and leaves with its pelvic fins (even upside down).
Similar species Unlikely to be confused with other species in area.
Other names Common: bony-snouted gudgeon, nosy parker, crimson-tipped flathead gudgeon. Scientific: none.
Literature Allen, 1991.

SOUTHERN PURPLE-SPOTTED GUDGEON *Mogurnda adspersa* (Castelnau)

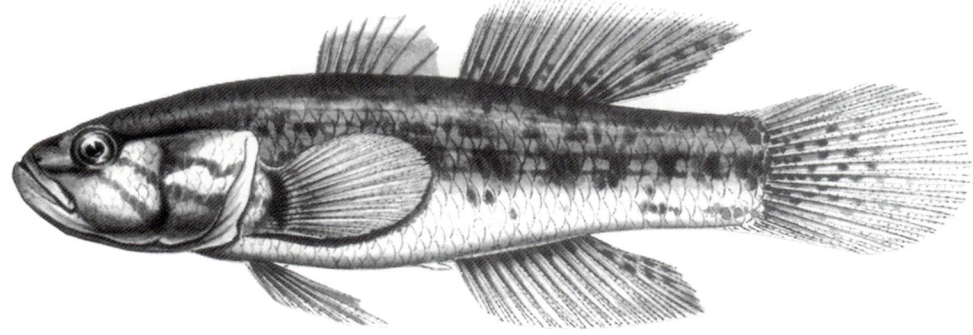

34.9 Purple-spotted gudgeon *Mogurnda adspersa*. (A.R. McCulloch)

Description A small, robust species with a rounded head and slightly compressed body; mouth small and slightly oblique, reaching to below front of eyes. First dorsal fin (VI–IX, usually VII–VIII) rounded, and lower than 2nd dorsal (I, 11–13). Anal fin I, 11–13 (occasionally 14); pectorals 14–16; fin ray counts vary considerably from locality to locality, but usually consistent at one locality. Tail rounded, 15 segmented caudal rays. Body covered with medium-sized ciliated scales (30–36 along side). Gill rakers 10–12; vertebrae 31. Rows of papillae on head as in Fig. 34.10.

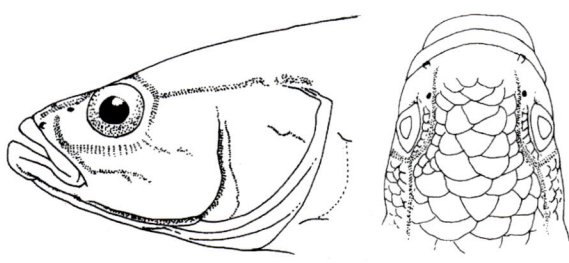

34.10 Rows of papillae on head of purple-spotted gudgeon *Mogurnda adspersa*. (H.K. Larson)

Sexual dimorphism Males have a pronounced bulge in the head profile above eyes, and urinogenital papilla tapers to a point and is brightly coloured. Females lack bulge in head, and papilla of breeding female is broader, with a rugose edge and is pale in colour.

Colour Background colour fades from dark chocolate dorsally to pale fawn or buff ventrally; 9–12 large, black to grey patches on sides, sometimes extending dorsally as bars; where dorsal extensions are absent, lighter patches replace them. Around these patches are numerous white and brick-red spots that become more prominent at onset of breeding. All fins have a background yellowish colour, and all except pectorals and pelvics become darker towards extremities. Dorsal fins, tail and pelvic fins have numerous brick-red spots that are more numerous at bases; these brighten during breeding. Males with 3 pairs of brown and white stripes running diagonally across cheeks and gill covers, while the female generally has 2 pairs of paler stripes.

Size Reaches 120 mm, but commonly around 70 mm; males mature at 45 mm, females at 49 mm.

Distribution Until recently, occurred patchily in the Murray, Murrumbidgee and Lachlan River systems and tributaries of the Darling in inland New

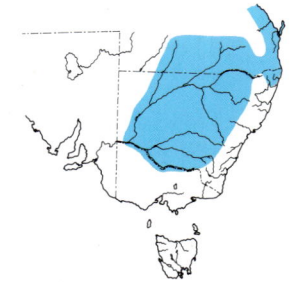

FAMILY GOBIIDAE

Purple-spotted gudgeon *Mogurnda adspersa*. (R.H. Kuiter)

South Wales. Is now considered extinct in Victoria but is found locally in coastal streams from northern New South Wales to northern Queensland. Its exact distribution in the Darling River system remains to be determined.

Conservation status Is rarely caught in large numbers; it has been suggested that this species has declined in number as a result of high densities of eastern gambusia (*Gambusia holbrooki*), which often frequents similar habitats. Is listed as 'restricted' by the Australian Society for Fish Biology.

Natural history Generally found in slow-flowing waters among weed, where suitable hard substrates are available for spawning. Breeds between December and February, when temperatures are 20–34°C and food is abundant. Flooding is not essential for stimulating spawning. Elaborate breeding display occurs between the sexes. Females, which apparently may spawn repeatedly, produce between 280 and 1300 ova at each spawning. Eggs are transparent and elliptical, measuring 1.1–1.3 mm by 2.0–3.8 mm; the eggs have an adhesive disc at the pointed end. Oil globules are small and numerous with the yolk. Eggs are deposited in a cluster on solid objects and are fanned and cared for by the male.

Hatching takes from 3–8 days at temperatures of 29–20°C. Newly hatched larvae are about 4 mm long and later possess a number of longitudinal pigment bands. Yolk is fully absorbed in about 6 days, when feeding on zooplankton commences. Juveniles have opercular stripes when 12–20 mm long. Adults feed on worms, mosquitofish, dragonfly larvae, chironomid and mosquito larvae.

Utility An attractive fish frequently kept in aquaria. Can readily be induced to breed in captivity. Redfin perch (*Perca fluviatilis*) are known to prey upon this species.

Similar species Frequently confused with northern trout gudgeon; juveniles can be confused with flathead gudgeon, Cox's gudgeon and striped gudgeon.

Other names Common: purple-striped gudgeon, chequered gudgeon, trout gudgeon, koerin and kurrin. Scientific: some authors have referred this species to *Mogurnda striata*; however, the original description of this species is inadequate to properly determine its identity.

Literature Blewett, 1929; Gale, 1914; Hansen, 1988; Harris, 1987b; Koehn & O'Connor, 1990; Lake, 1978; Merrick & Schmida, 1984; Hamlyn-Harris, 1931; Lake, 1971, 1978.

NORTHERN TROUT GUDGEON
Mogurnda mogurnda (Richardson)

Description Very similar to purple-spotted gudgeon, differing mainly in scale counts. In profile, a slight bulge in head in region of nostrils. Oblique mouth slightly larger than in purple-spotted gudgeon; jaws generally reach to below middle of eye. Outer row of teeth markedly larger than other jaw teeth. Snout scaleless, scales with fewer ctenii than in *M. adspersa*. Fin ray counts vary with locality, dorsals VII–IX, I, 10–13 (rarely

34.11 Northern trout gudgeon *Mogurnda mogurnda*. (M. Weber)

14); anal I, 10–13; pectorals 15–16; 15 segmented caudal rays; lateral scales 37–45; gill rakers 9–12; vertebrae 31.
Sexual dimorphism Sexes very similar; urinogenital papilla pointed in males, broad with a fringed margin in females.
Size Known to reach 175 mm, but generally grows to about 100 mm.
Distribution Occurs in streams in north-western Australia to the Gulf of Carpentaria and the west coast of Cape York; also present in some localities in the Central Australian drainage; has also been found in the Herbert River in eastern Queensland.

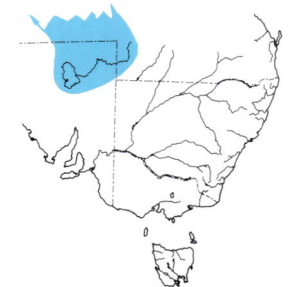

Conservation status Most common in the north-west and gulf.
Natural history Is known to survive in dams and billabongs, in muddy waters; also occurs in bores and rivers. Information purporting to describe the breeding of this species by several Australian and overseas aquarists is probably based on southern purple-spotted gudgeon; their breeding biology is probably quite similar.
Utility An attractive fish with considerable value as an aquarium species.
Similar species Southern purple-spotted gudgeon.
Other names Common: purple spotted gudgeon, trout gudgeon, Australian spotted gudgeon, chequered gudgeon. Scientific: none.
Literature Bishop *et al.*, 1978; Lake, 1978; Merrick & Schmida, 1984; Pollard, 1974.

Northern trout gudgeon *Mogurnda mogurnda*. (R.H. Kuiter)

FAMILY GOBIIDAE

EMPIREFISH *Hypseleotris compressa* (Krefft)

34.12 Empirefish *Hypseleotris compressa*. (A.R. McCulloch)

Description A small species, its head and body very compressed, body slender and elongate but becoming deeper with growth. Mouth small and slightly oblique, reaching to almost anterior eye margin; gill openings broad, ending ventrally below posterior opercular margin. Tongue with a slight indentation at tip. First dorsal fin (VI) origin just behind the level of the pelvic fin bases; origin of 2nd dorsal (I, 8–9) above vent; anal fin I, 9–10; pectorals with 14–15 rays. Tail truncate to slightly rounded (15 segmented rays). Body covered with ciliated scales (26–29 along side, 15–17 predorsal scales); top of head with cycloid scales forward to above middle of eyes; cheeks and gill covers with small cycloid scales that are difficult to discern. No scales on snout. Two to four pores along posterior preopercular margin; large fish often have two pores above each eye (Fig. 34.13); vertebrae 25; gill rakers 13–13.

Sexual dimorphism Males grow larger than females, and large males develop a slight hump on top of head, although the hump is not as well

Empirefish *Hypseleotris compressa*. (G. Schmida)

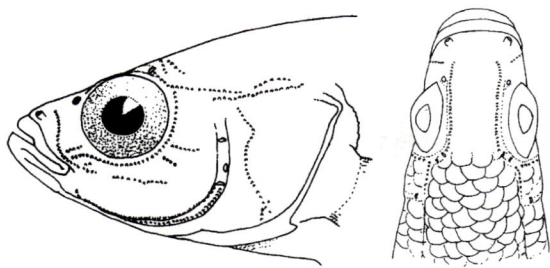

34.13 Rows of pores and papillae on head of empirefish *Hypseleotris compressa*. (H.K. Larson)

developed as in other species of *Hypseleotris*. Urinogenital papilla of males broad and smooth-edged, while in females it is broad and fringed with small lobes at tip. Breeding males have more brightly coloured vertical fins than females.

Colour Varies considerably, and individual fish can rapidly lighten or darken colours. Head and body vary from fawn to light golden brown, with belly lighter; can reach almost chocolate brown. A dense black or brown spot at top of pectoral base, and may form a vertical bar. Along midside a nebulous dusky stripe that varies from dark to almost absent. Scales along sides have margins outlined in dark brown, especially among the midside stripe. Upon preservation, a reticulated pattern of dark-margined scales appears over the whole body, forming about 10 short bars along sides. This pattern may be visible in live fish. At tail base a small black spot which may not be obvious. On head a nebulous dusky stripe runs around snout above upper lip, through eye and across side of head to edge of gill cover. On gill covers, this stripe may be underlain by an iridescent green to gold blotch.

Unpaired fins not usually brightly coloured in females or juveniles, so the following applies chiefly to breeding males. First dorsal with four bands of colour: a basal brownish stripe, then one of bright orange-red, a dense black stripe and a narrow, pearly white stripe along margin. The amount of black and red varies between individuals. Anal fin markings similar. Second dorsal similar but for a dense black spot at posterior end (remains conspicuous regardless of breeding condition). Also a series of bright white spots along base of 2nd dorsal, and several more around posterior black spot on fin. Tail dusky, especially towards base, with light spots in irregular transverse rows. Pelvics and pectorals clear (sometimes dusky).

All colours in males in breeding condition intensify considerably, and develop bright orange colour under head, belly and on lower sides. This, and intensified golden brown body colour, gives the impression that the fish is glowing from inside.

Size Reaches 100 mm, but matures at about 40–50 mm.

Distribution Occurs in low-elevation coastal streams in New South Wales, Queensland, Northern Territory and north-western Western Australia; also recorded from southern New Guinea. Southern limit on east coast is Towamba River, New South Wales.

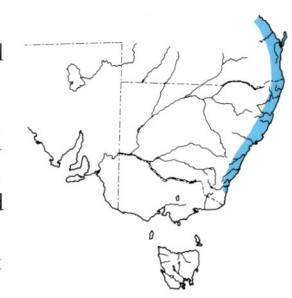

Conservation status Relatively common throughout range.

Natural history Normally found in the lower reaches of coastal rivers and streams; schools of juveniles are often found in estuaries. Spawning normally occurs during warmer months, the eggs being deposited on rock, weed or sand and guarded by male; up to 3000 eggs about 0.3 mm long are laid and hatch in 10–13 hours. In aquaria, females often spawn several times a season, with spawning occurring every 2–7 days. One female may spawn with several males.

Feeds on aquatic invertebrates, particularly cladocerans, and larvae of insects such as mosquitos; also grazes algae and detritus, generally feeding in mid-water.

Utility Attractive colouration makes this species a highly desirable aquarium fish; is readily kept and will breed in aquaria. Has considerable potential for the control of mosquito and midge larvae.

Similar species Distinctive colour of adults readily differentiates empirefish from other gudgeons. Juveniles are particularly difficult to separate from firetailed gudgeon in New South Wales and southern Queensland; may also be confused with western carp gudgeon and Midgley's gudgeon in these areas and with juveniles of *Gobiomorphus* spp.

Other names Common: carp gudgeon, empire gudgeon. Scientific: *Carassiops compressus*.

Literature Allen, 1989; Anderson *et al.*, 1971; Auty, 1978; Hoese & Allen, 1983; Lake, 1978; Merrick & Schmida, 1984.

FIRETAILED GUDGEON *Hypseleotris galii* (Ogilby)

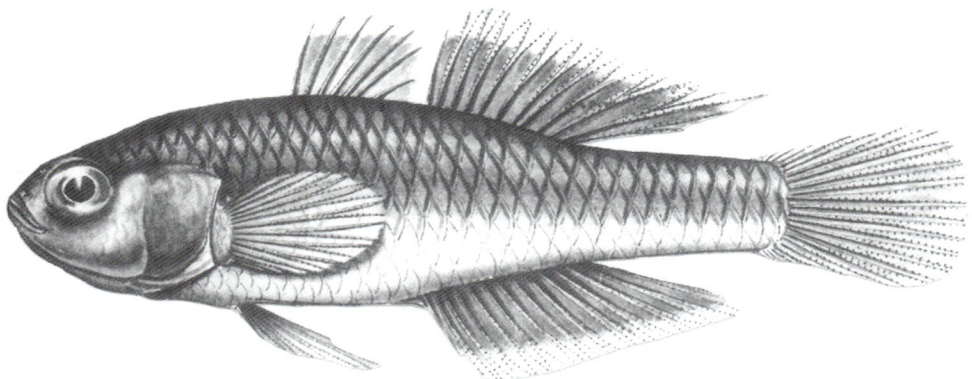

34.14 Firetailed gudgeon *Hypseleotris galii*. (A.R. McCulloch)

Description A tiny species. Head and body very compressed, body moderately slender, tapering towards tail. Mouth oblique, small, reaching to below anterior margin of eye. Gill openings broad, ending below posterior preopercular margin; tip of tongue truncate. First dorsal fin (VI–VIII, usually VII) with origin just behind level of pelvic fin bases, low, with a rounded margin in females and a rectangular margin in males. Second dorsal (I, 10–12) origin above vent, higher than 1st dorsal, anterior rays elevated, in males the posterior rays elongated. Anal fin I, 11–13; pectorals 14–15. Tail truncate to slightly rounded. Body covered with ciliated scales (30–32 along side, 8–12 predorsal scales); top of head scaled forward to above posterior half of eye. Cheeks and gill covers with small cycloid scales, none on snout. No head pores. Vertebrae 29–30; gill rakers 9–12.

Sexual dimorphism Breeding males develop a slight hump on top of head behind eyes; also have longer posterior rays in dorsal and anal fins and a differently shaped dorsal fin. Urinogenital papilla of male brown; that of female black. Breeding males often become black, and the

Firetailed gudgeon *Hypseleotris galii*. (R.H. Kuiter)

reddish-orange fin colouration intensifies; males larger than females.

Colour Varies considerably with size, sex, locality and season. Head and body pale grey to bronze or almost black. A black bar on body above pectoral fin base and generally covering upper $2/3$ of base, becoming narrower and paler ventrally. A faint stripe from spot above pectoral base to tail base (fades in preservative). Scales edged in black. Belly generally lighter than body. Urinogenital papilla black or brown. Fins vary from clear to dusky; basal $2/3$ of dorsal and anal fins darker than margins; in large fish the bases dusky and the tips reddish orange. Tail varies from clear to reddish orange.

Size Reaches 55 mm (males) and 40 mm (females); a tiny species.

Distribution Occurs in coastal streams from Fraser Island, Queensland, to Eden in southern New South Wales; scattered populations in northern Queensland (Tully River). Is rarely found in the lower reaches and rarely with empirefish, but sometimes occurs with western carp gudgeon in northern New South Wales and Queensland. Introduced into Bolgu Island, Torres Strait (for mosquito control).

Conservation status Remains common.

Natural history Restricted to fresh water and typically found in association with vegetation. Spawns in warm months—October to January—the eggs being deposited under stones, leaves or shells. Female lays 5–8 batches of 200–400 eggs at intervals of a week or more. Male guards eggs until they hatch in 4–5 days. Reaches maturity at age one.

Normally feeds on small aquatic invertebrates, particularly cladocerans and ostracods.

Utility Lives well in aquaria, and the brightly coloured males of this species are attractive in captivity.

Similar species Females are readily separated from other species of *Hypseleotris* by black urinogenital papilla; males easily confused with other species, particularly Midgley's carp gudgeon and Lake's carp gudgeon.

Other names Common: Gale's gudgeon. Scientific: *Carassiops galii*.

Literature Anderson *et al.*, 1971; Grant, 1978; Lake, 1967, 1971, 1978; McKay, 1973a, b; Merrick & Schmida, 1984;

WESTERN CARP GUDGEON *Hypseleotris klunzingeri* (Ogilby)

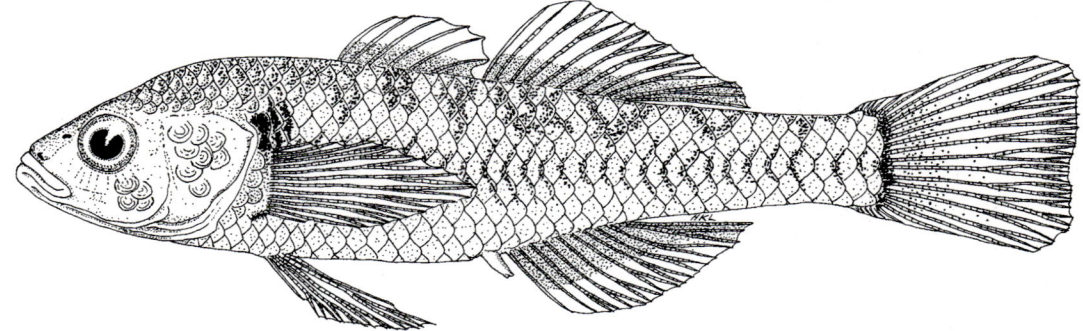

34.15 Western carp gudgeon *Hypseleotris klunzingeri*. (H.K. Larson)

Description A tiny fish with compressed head and body, tapering towards tail; fish from areas in South Australia and New South Wales often quite elongated. Mouth small, oblique, reaching below anterior margin of eye; eye about equal to interorbital space. Tongue truncate. Gill openings extend to below posterior margin of preoperculum. First dorsal (VI–VII) low, rounded and begins behind pectoral base; 2nd dorsal (I, 9–11) originates above vent; rays of 2nd dorsal and anal (I, 9–11) longer towards front of fin although in males the last few rays are elongated. Pelvic fins may reach vent, 4th ray long and pointed. Pectorals with 13–15 (usually 14) rays. Tail truncate or slightly rounded (15 segmented rays). Ciliated scales cover body (27–32 along sides. Gill covers and preopercles with small, cycloid scales; no pores on head; rows of papillae on head as in Fig. 34.16; vertebrae 26–29; gill rakers 9–12.

Sexual dimorphism Males tend to be larger and

FAMILY GOBIIDAE

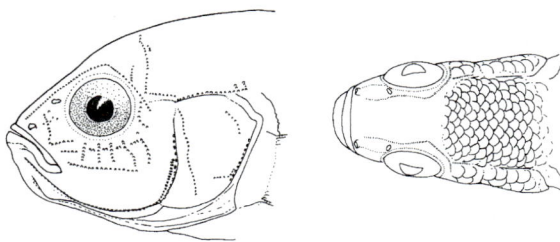

34.16 Rows of papillae on head of western carp gudgeon *Hypseleotris klunzingeri*. (H.K. Larson)

more slender than females, and during breeding develop a hump on head just behind the eyes; this hump is not as well developed as in firetailed gudgeon. In males the 1st dorsal fin has a more rectangular margin and is more brightly coloured than the more rounded fin of female. Urinogenital papilla of male a flat, conical flap with a concave tip; that of female a shorter, more bulbous lobe with fine projections and fringes at tip.
Colour Both sexes pearly to yellowish grey, darker above; most conspicuous marking an elongate, dark brown to black spot, bordered anteriorly by pearly white, on upper pectoral base. Along midside a row of thin, short, slightly curved brown bars, one at each scale base. Tail base with a short blackish bar that may become more conspicuous on preservation. Below the midside row of bars there may be a little dusky pigmentation but no other markings. Abdomen light yellowish with no black lines or iridescent spots on belly midline. Scales on upper sides and back outlined in brownish black. On preservation pattern on sides and back becomes more apparent. The brown scale margins are intensified in groups of scales to form 8–10 irregular bars across back and nape. Upper sides may have more blotchy bars at staggered intervals between bars across back, forming a vague chequered pattern. Behind pectoral fin these upper side blotches may merge with the row of short, midside brown bars, but blotches irregular and not forming distinct vertical bars. Head grey with a blue-mauve blotch on gill cover. Eyes silvery gold. First dorsal fin has a narrow, dusky basal stripe, with rest of fin red, bordered above with a bright white stripe, leaving edge of fin clear. Second dorsal has a similar, basal dusky stripe, and a white-bordered red band occupying the lower half of fin, leaving upper half clear. Anal fin similar, but basal stripe wider and more yellowish, white-bordered band duskier, and clear outer half of fin may be faintly pinkish. Tail has basal half reddish, the remainder clear. Pelvics and pectorals clear. Fin colouration most distinct in breeding males.
Size Reaches 45 mm.

Western carp gudgeon *Hypseleotris klunzingeri*. (N. Armstrong)

Distribution Occurs throughout the inland drainages of South Australia, New South Wales and Queensland and in coastal drainages from the Hunter River in New South Wales to the Fitzroy in central Queensland. Populations in the Wimmera and Wannon Rivers, in western Victoria, are introduced.

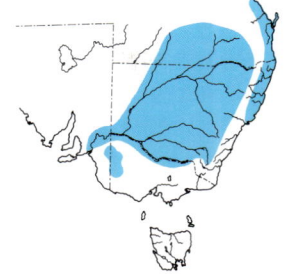

Conservation status Continues to be widespread and abundant.
Natural history Normally lives around vegetation; information has been published on breeding the 'western carp gudgeon', but the species studied may not be this one, as two further inland *Hypseleotris* species were not then known. Feeds on small aquatic invertebrates such as copepods and insect larvae; also eats weeds.
Utility Has considerable potential as an aquarium fish.
Similar species Distribution overlaps that of other species of *Hypseleotris* in south-eastern Australia, and these are easily confused with each other.
Other names Common: none. Scientific: *Carassiops klunzingeri*.
Literature Anderson *et al.*, 1971; Cadwallader and Backhouse, 1983; Koehn & O'Connor, 1990; Lake, 1967, 1971.

MIDGLEY'S CARP GUDGEON *Hypseleotris* sp. 4 (undescribed)

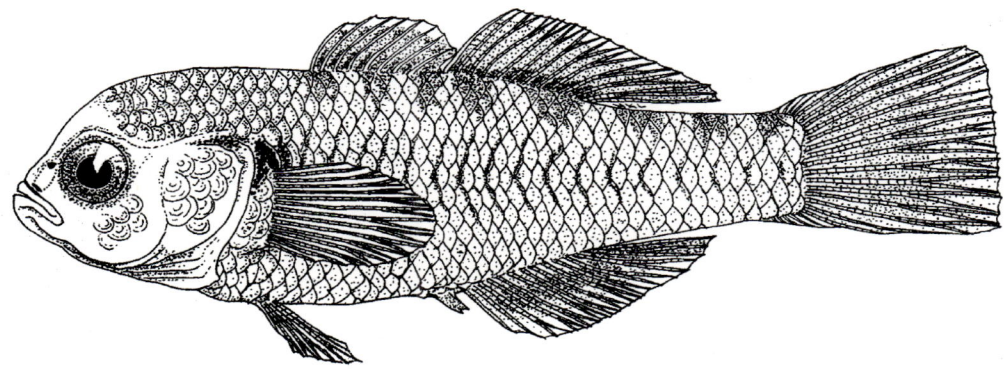

34.17 Midgley's carp gudgeon *Hypseleotris* sp. 4 (H.K. Larson)

Description A tiny fish with a very compressed head and body. First dorsal fin (VI–VIII) originates behind level of pelvic fin bases, lower than 2nd dorsal and has a rounded margin. Second dorsal (I, 11–13) originates above vent, anterior rays slightly elevated, posterior rays short in females and elongate in males. Anal fin I, 11–13. Tail truncate to slightly rounded (15 segmented caudal rays). Pectorals with 14–16 rays. Fourth pelvic ray longest, 3rd ray longer than 5th, giving fin an acutely pointed margin. Pelvic fins reach halfway or more of distance to vent. Body covered with ciliated scales (31–36 along side, usually 31–34, 13–20 predorsal scales). Top of head covered with small cycloid scales to above posterior margin of eyes. Sides of head, pectoral bases, breast and midline of belly covered with small cycloid scales; no scales on snout. No head pores; 29–30 vertebrae; 11–13 gill rakers.

Sexual dimorphism In males the 1st dorsal is only slightly rounded and long-based, reaching to base of 2nd dorsal; the fin in females with a rounded margin and widely separated from 2nd dorsal. Mature males develop a distinct hump extending from about front of eyes back to 1st dorsal fin; also have the posterior dorsal and anal rays elongated. Males typically grow slightly larger than females. As in all species of *Hypseleotris*, the urinogenital papilla margin is smooth in males and fringed in females.

Colour Varies considerably with size, sex, locality and season. Head and body vary from pale grey to brown; a distinct black spot on upper third of pectoral base, scales edged in black. Often charcoal-grey vertical bands on sides; no black spots or bars on caudal peduncle. Belly distinctly silvery in life, and has black along or to sides of midline. First dorsal has a basal dusky stripe, above which is a wider reddish orange stripe. Just above midline of fin another dusky stripe bordered above by a white marginal stripe. Second dorsal with a similar orange stripe, but dusky bands above and below nearly as wide as orange band. A thin submarginal orange stripe present, becoming pearly white posteriorly. Anal fin very like 2nd dorsal, but bands of colour less distinct. Tail fin with basal half yellowish or dusky. Pelvics and pectorals clear. Fin colours most marked in breeding males and indistinct in females and young males. This colour description is based on samples from one area, and fish from others may display slightly different colour patterns.

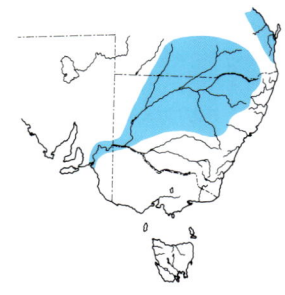

Size Reaches 60 mm.
Distribution Occurs in inland drainages of South Australia, New South Wales and southern Queensland, also in coastal drainages of southern and central Queensland.
Conservation status Common across range.
Natural history Often found around vegetation, but also schools in caves or other sheltered areas; feeds on aquatic crustaceans, particularly cladocerans and copepods.
Utility Has considerable potential as an aquarium fish.
Similar species Can be confused with other species of *Hypseleotris* from south-eastern Australia; frequently confused with western carp gudgeon.
Other names Common: none. Scientific: none.
Literature Allen, 1989.

LAKE'S CARP GUDGEON *Hypseleotris* sp. 5 (undescribed)

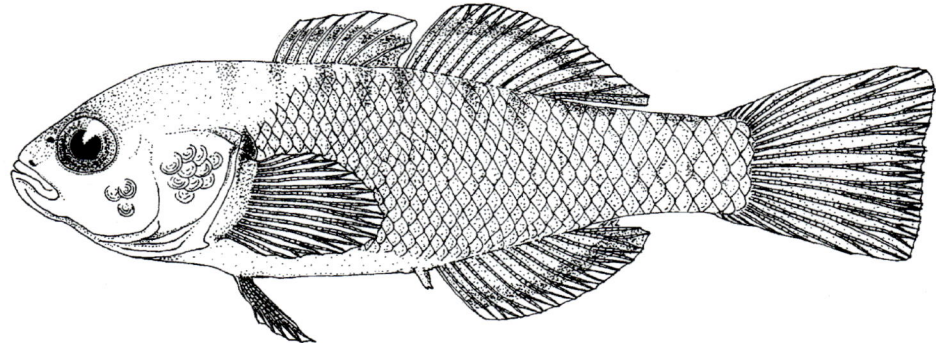

34.18 Lake's carp gudgeon *Hypseleotris* sp. 5 (H.K. Larson)

Description Very similar to Midgley's carp gudgeon, differing only in a few features, most notably the lack of scales on the nape, breast and belly; also sometimes a naked strip under 1st dorsal fin (31–36 lateral scales). First dorsal (VII–IX, usually VIII) very low in both sexes, that of male basically rectangular in shape and reaching base of 2nd dorsal. In females the fin has a rounded margin and is much shorter-based, a wide gap between the two dorsals. Second dorsal I, 10–12; anal I, 10–13. Anterior rays of 2nd dorsal and anal fins highest, posterior rays elongated slightly in males. Pelvic fins short, reaching about $1/2$ to $3/4$ distance to vent. Fourth pelvic ray only slightly longer than other rays, 3rd and 5th rays about equal in length, giving fin an obtusely pointed margin; 15–17 pectoral rays; 15 segmented caudal rays; vertebrae 30–31; gill rakers 7–10. Rows of papillae on head as in Fig. 34.19.

Sexual dimorphism Males tend to be slightly deeper-bodied and stockier than females. Breeding males become very dark and develop a hump on nape from anterior eye margins to 1st dorsal fin. Also slight differences between sexes in fin shape.

Colour Little known of colour in life. Head and

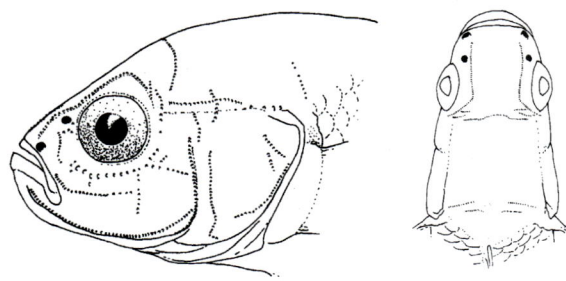

34.19 Rows of papillae on head of Lake's carp gudgeon *Hypseleotris* sp. 5 (H.K. Larson)

Lake's carp gudgeon *Hypseleotris* sp. 5 (N. Armstrong)

body colour almost the same as in Midgley's carp gudgeon, although individuals vary. Fin colouration distinct from that of Midgley's carp gudgeon, with generally a single distinct dusky band running across upper $1/3$ of dorsal and anal fins; the white marginal band rather wider. In some males basal $2/3$ of all fins may be quite dark, with only a white marginal band.

Size Known to only 40 mm.

Distribution Occurs throughout inland drainages of South Australia, New South Wales and southern Queensland.

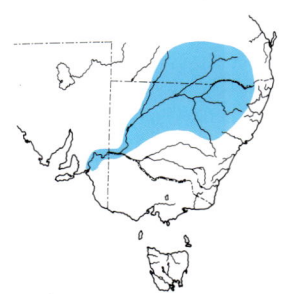

Conservation status Common across range.

Natural history Normally found around vegetation; breeding unknown but ripe males are found between October and December. Feeds on a variety of aquatic invertebrates, such as copepods, mosquito larvae and other aquatic insect larvae and eggs.

Utility Has considerable potential as an aquarium fish.

Similar species Frequently found with and easily confused with western carp gudgeon and Midgley's carp gudgeon.

Other names Common: none. Scientific: none.

Literature Allen, 1989.

35

Family Gobiidae, subfamilies Gobiinae and Gobiinellinae

Gobies

H.K. LARSON AND D.F. HOESE

True gobies (as in the subfamilies Gobiinae and Gobiinellinae) are generally small fishes, some only 15–30 mm long as adults. Their small size often makes identification difficult. They are easily separated from other fish families in having two dorsal fins, no lateral line and the pelvic fins usually connected to form a cup-shaped disc; they resemble and are closely related to the gudgeons, which always lack the connected pelvics. The family Gobiidae contains some 1800 species, about 400 in Australia, though most are marine, with about six tropical Australian species restricted to fresh water. A few southern Australian species may breed in fresh water, but most are essentially marine species that tolerate fresh water. We include here only species regularly found in fresh water, although all of them also occur in brackish water and the sea.

Gobies attach their eggs to the bottom or to objects on the bottom; the eggs hatch in a few days into planktonic larvae. The adults are found from fresh water to sea depths of 200 m, living on mud, sand, in seagrass beds, or on rocky or coral reefs. Many species live in burrows, and some bury in the sand or mud. A few are pelagic as adults. Most prey on small invertebrates, but some feed on microscopic algae. They are generally very hardy, are easily kept in aquaria, and some can be kept in fresh water. Many species are sexually dimorphic, especially during breeding.

Several taxonomic problems have not yet been resolved for the freshwater gobies, and further studies are needed to clarify them.

Key to gobies

1   Body largely without scales, sometimes a few on caudal peduncle; 1st dorsal VI–VIII . . . . ***Tasmanogobius lordi*** p. 225 ; Fig. 35.7
    Body completely scaled, 1st dorsal VI 2

2   Head without scales, none before 1st dorsal fin; anal rays I, 8–10; lateral scales 30–35 3
    Head scaled on sides and top, predorsal scaled; anal rays I, 6–7; lateral scales 27–30 4

3   Head distinctly depressed; no prominent stripes on head; anal rays I, 8; faint brown
    saddle-shaped marks on back . . . . ***Afurcagobius tamarensis*** p. 227; Fig. 35.12
    Head compressed, with 1–2 dark longitudinal stripes; anal rays I, 10; no saddles
    on back . . . . ***Amoya bifrenatus*** p. 226 ; Fig. 35.9

4   Head strongly compressed, especially in males, mouth oblique and may be greatly
    enlarged in males, snout not round and never overhanging mouth; anal rays I,
    6; 17 segmented caudal rays; body with irregular dark transverse bars . . . .
    . . . . ***Redigobius macrostoma*** p. 223; Fig. 35.5
    Head rounded, mouth horizontal and may be slightly enlarged in males; anal rays I,
    7–9; 16 segmented caudal rays; back and sides with series of small dark saddles and spots 5

5 Dorsal rays I, 8–9; anal rays I, 8–9; predorsal scales 8–12 (usually 10–11); transverse scale count 8–10; distributed from Western Australia to western Victoria .... ***Pseudogobius olorum*** p. 221; Fig. 35.1

Dorsal rays I, 7–8; anal rays I, 7 (rarely I, 8); predorsal scales 6–8 (up to 10 in Victoria); transverse scale count 8–9; distributed from western Victoria to southern Queensland .... ***Pseudogobius* sp.9** p. 222; Fig. 35.3

SWAN RIVER GOBY *Pseudogobius olorum* (Sauvage)

35.1 Swan River goby *Pseudogobius olorum*. (A.R. McCulloch)

Description A small, rather cylindrical goby with only a few short lateral line canals on head around eyes, a small mouth that tends to be overhung by the rounded, plump snout, scales extending forward over nape up to eyes, and rather short fins. Intestine and stomach coiled in a characteristic corkscrew manner about each other. Dorsal rays VI; I, 8 (sometimes 7 or 9) and anal I, 8 (sometimes 7 or 9); these fins relatively low. Pectoral rays 15–16 (sometimes 17). Only 16 segmented caudal fin rays. Pelvic fins disc-shaped, with frenum connecting the soft spines across front; fins do not reach anus. Caudal fin an asymmetric oval, with upper to central rays longest. Body scales rather large, ctenoid; cycloid scales cover nape, opercles and a few may be present on cheeks. Lateral scale count 27–30, predorsal scales 8–12 (average 10), transverse scale count 8–10 (average 9). Rows of papillae on head as in Fig. 35.2.

Sexual dimorphism Females have a short bulbous genital papilla and males have a slender, pointed papilla. Mature males may have slightly larger mouths than females; jaw extends to below mid-eye or further back (in females jaw may reach to below anterior half of eye to mid-eye). Mature females usually have rounded bellies, and those females ready to spawn look quite swollen.

Colour Head and body light brown to sandy above, dull to pearly white below, scales on body and name with edges outlined in brown; live fish appear to have gold speckles over sides of body. Five brown, roughly rectangular blotches across mid-dorsal line (one on nape, one below each dorsal fin and two on caudal peduncle). Five elongate dark brown blotches (formed by 2 confluent spots) along midside of trunk, with a 6th denser spot centrally on tail base. Above and below this 6th spot may be 2 further similar spots, forming a rough Y or V. Sides between the rows of elongate blotches and back may have further variable brown spots and blotches. Head marked irregularly with brown, usually short lines or bars radiating from around eye. A broad dusky stripe follows upper lip. Two elongate spots at top of pectoral fin. First dorsal with 2–3 broad dusky bars and a dense black spot posteriorly, near this may be a small bright blue spot (black spot may be small or absent). Second dorsal fin and tail barred with rows of

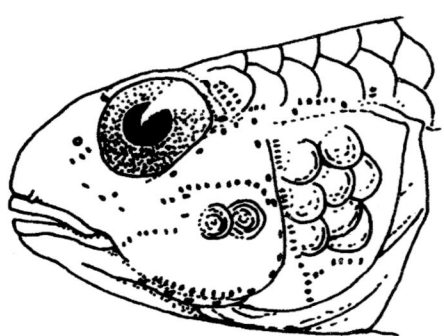

35.2 Rows of papillae on head of Swan River goby *Pseudogobius olorum*. (H.K. Larson)

FAMILY GOBIIDAE

Swan River goby *Pseudogobius olorum*. (R.H. Kuiter)

dark spots. Pectoral, pelvic and anal fins translucent to dusky; anal fin may have a blue margin; dusky pelvic fins may have a white to sandy coloured edge.

Size Can grow to over 55 mm, but average adults are about 45 mm.

Distribution Found from western Victoria westward along the coast to Western Australia. Common in protected estuaries and coastal lakes.

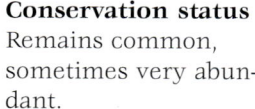

Conservation status Remains common, sometimes very abundant.

Natural history Spawns in upper reaches of estuaries (in less than 30% salinity), usually where aquatic vegetation is thick.

Utility Is preyed upon by birds such as cormorants and herons, which is probably why it prefers areas of thick vegetation, which offers protection. It is a hardy and adaptable fish and can be kept successfully in aquaria (although it is not very colourful or lively).

Similar species Can be easily confused with the slightly smaller *Pseudogobius* sp. 9, with which it has been confused in the literature.

Other names Common: Galway's goby, blue spot goby, southern goby. Scientific: *Lizagobius olorum, Ellogobius olorum, Mugilogobius galwayi, Lizagobius galwayi, Gobius olorum*; misidentified as *Parvigobius immeritus* and *Stigmatogobius javanicus*.

Literature Gill and Potter, 1993; Humphries and Potter, 1993.

BLUE SPOT GOBY
Pseudogobius sp. 9

Description A small, rather cylindrical goby with only a few short lateral line canals on the head around the eyes (Fig. 35.4), a small mouth that tends to be overhung by the rounded, plump snout, scales extending forward over the nape to close behind eyes, and rather short fins. The intestine and stomach are coiled in a characteristic corkscrew manner about each other. Dorsal rays VI; I, 7 (sometimes 8) and anal I, 7 (sometimes 8); these fins relatively low. Pectoral rays 15–17. Only 16 segmented caudal fin rays. Pelvic fins disc-shaped, with frenum connecting the soft spines across the front; fins do not reach the anus. Caudal fin oval to rectangular, with upper or central rays longest. Body scales rather large,

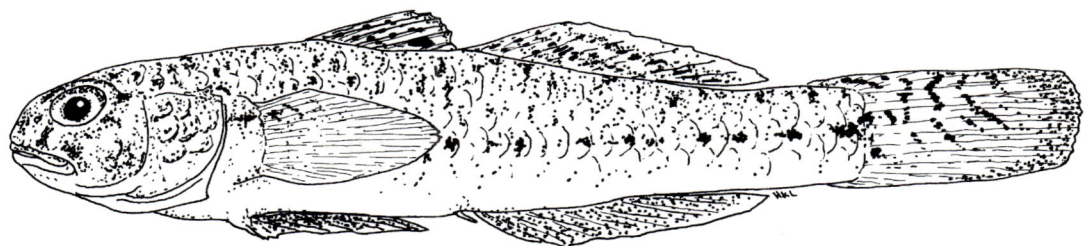

35.3 Blue spot goby *Pseudogobius* sp. 9 (H.K. Larson)

ctenoid; cycloid scales cover nape (sides of nape often ctenoid), opercles and a few may be present on the cheeks. Lateral scale count 25–27, predorsal scales 7–10 (average 8), transverse scale count 8–9 (average 8). Specimens from Victoria tend to have more predorsal scales than do those from New South Wales (up to 10).

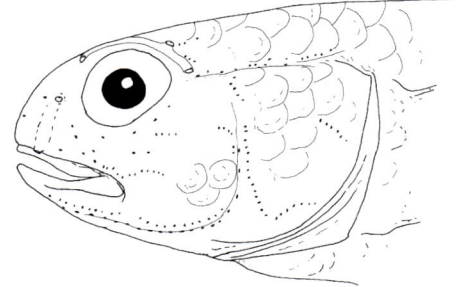

35.4 Rows of papillae on head of blue spot goby. (H.K. Larson)

Sexual dimorphism Females have a short bulbous genital papilla and males have a slender, pointed papilla. Mature males may have slightly larger mouths than females; jaw extends to below mid-eye or further back (in females jaw may reach to below anterior half of eye to mid-eye). Mature females usually have rounded bellies, and those females ready to spawn look quite swollen.
Colour Very similar in colour to *Pseudogobius olorum* although looks 'duller'; mature males may have heavier blotches and more defined bands on the dorsal fins.
Size Average adults are about 35 mm (can grow to about 45 mm).

Distribution From Moreton Bay, Queensland, south along the coast to western Victoria and Tasmania. Common in protected estuaries and coastal lakes and can be very abundant over muddy to sandy substrates or where aquatic vegetation is thick.
Conservation status Remains common in known range.
Natural history Probably spawns in upper reaches of estuaries where salinity is lower than in the marine environment.
Utility Is preyed upon by birds such as cormorants and herons and may be found in areas of thick vegetation, which offers protection. A hardy and adaptable fish; can be kept successfully in aquaria, though is not very colourful or lively.
Similar species Has been confused with the slightly larger *Pseudogobius olorum*.
Other names Common: Swan River goby. Scientific: none.
Literature None, though confused in literature with *Pseudogobius olorum*.

LARGEMOUTH GOBY
Redigobius macrostoma (Günther)

Description A small goby with a very compressed body and head; large males distinctive in having a very large mouth that extends back beyond eye and may be more than half head length. Eyes close to snout tip; snout may be gently rounded or rather pointed. First dorsal (VI) pointed anteriorly; 2nd spine may be elongated in males. Second dorsal (I, 7) and anal (I, 6) fins short-based. Pelvic fins possess a frenum and form an elongate disc that reaches vent. Pectoral rays 16–18. Body covered with ctenoid scales, scales also present on head to behind eyes; a large patch of scales on gill covers; 25–30 midlateral scales; gill rakers 6. Rows of papillae on head as in Fig. 35.6.
Sexual dimorphism Adult males tend to be larger than females and, when mature, the mouth extends back behind the eye; in females the mouth reaches to about middle of eye. Snout usually more pointed in females than males; in large

FAMILY GOBIIDAE

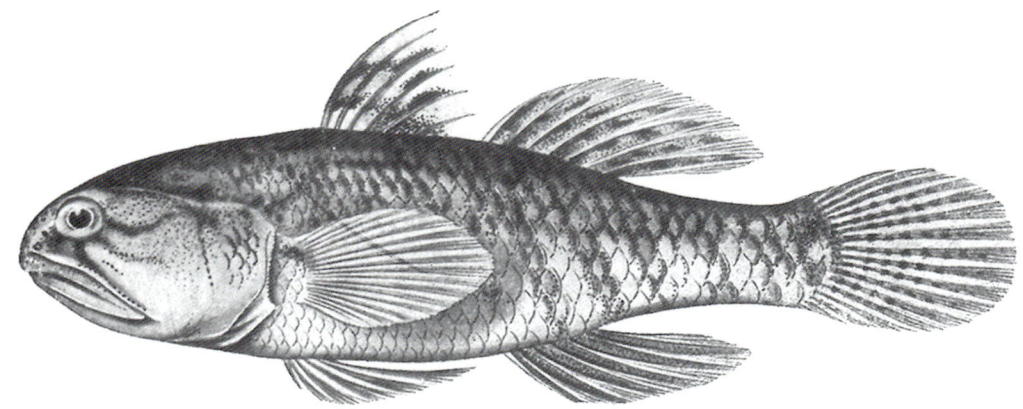

35.5 Largemouth goby *Redigobius macrostoma*. (A.R. McCulloch)

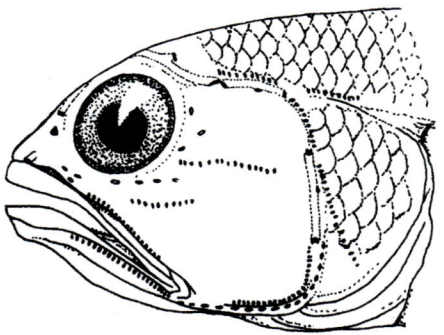

35.6 Rows of pores and papillae on head of largemouth goby *Redigobius macrostoma*. (H.K. Larson)

males the rounded snout may overhang the upper lip a little. Colours often brighter during the breeding season.

Size Reaches only about 50 mm, occasionally more.

Distribution Occurs in estuaries and coastal rivers from southern Queensland to Victoria (Glenelg River), and apparently also north-eastern Tasmania.

Largemouth goby *Redigobius macrostoma*. (R.H. Kuiter)

Conservation status Remains common.
Natural history Occurs mostly in muddy, weedy or sea-grass estuaries, especially in muddy areas, sometimes in rocky river mouths; may be found in purely freshwater streams.
Utility This abundant fish is probably a food for birds in estuaries; is hardy and lives well in aquaria, accepting dried foods.
Similar species Juveniles could be confused with young of estuarine or marine gobies—*Pandaka* or *Bathygobius*; highly compressed form of adults quite distinctive.
Other names Common: none. Scientific: none.
Literature Kuiter, 1993.

TASMANIAN GOBY *Tasmanogobius lordi* Scott

35.7 Tasmanian goby *Tasmanogobius lordi.* (E.O.G. Scott)

Description Very tiny and slender, scaleless but for a few embedded scales on caudal peduncle and a small patch behind pectoral fin. Head small with bluntly pointed snout; mouth extends obliquely below front of eye. Anterior nostril large, tubular. Gill openings extend to just past pectoral base, or to level of posterior preopercular margin. First dorsal (VI–VIII, rarely VI) low and slightly pointed; 2nd dorsal (I, 14–15) and anal (I, 13–15) fins fairly long. Tail rounded, about as long as head. Several transverse rows of papillae across caudal peduncle with a few embedded scales between them; gill rakers 8. Rows of papillae on head as in Fig. 35.8.

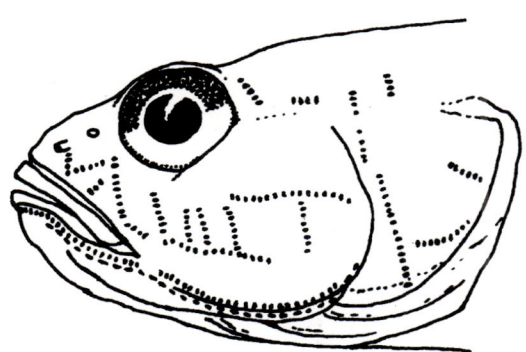

35.8 Rows of papillae on head of Tasmanian goby *Tasmanogobius lordi.* (H.K. Larson)

Sexual dimorphism Genital papilla of female large and swollen when breeding occurs, the male's a triangular flap.
Colour Light brown with 4–6 vertically elongate black spots along midside of body; thin black bar from below eye to end of jaws.
Size Is known to reach only about 35 mm.
Distribution Known only from coastal Tasmania.

Conservation status Uncommon, but little known.
Natural history Occurs in streams, at the mouths of rivers, and has been collected in sandy, estuarine tide pools; buries itself in the sand.
Utility An interesting goby about which little is known; not much kept in aquaria.
Similar species Easily confused, in estuaries, with *Tasmanogobius lasti*, which has more scales on the body; lack of body scales is distinctive among other freshwater gobies.
Other names Common: none. Scientific: none.
Literature Hoese, 1991.

FAMILY GOBIIDAE

BRIDLED GOBY *Amoya bifrenatus* (Kner)

35.9 Bridled goby *Amoya bifrenatus*. (F. Kner)

Description Smallish and elongate, with a slightly compressed head, snout broadly rounded in lateral view, mouth oblique, moderate, reaching below front of small eyes. Cheeks somewhat bulbous. First dorsal (VI) acutely pointed and slightly higher than 2nd; 2nd dorsal (I, 10) and anal (I, 10) fins long, with their posterior rays elongate. Tail elongate, becoming lanceolate in large males. Pectoral fins (17–18 rays) long and pointed. Pelvic fins fused into a broad, cup-shaped disc. Body covered with large, ctenoid scales, those below 1st dorsal fin, on belly and before pelvic fins cycloid. No scales on head; 35–39 along side. Jaws with rows of fine teeth, the outer series the largest; gill rakers 9–11. Pores and papillae on head as in Fig. 35.10.

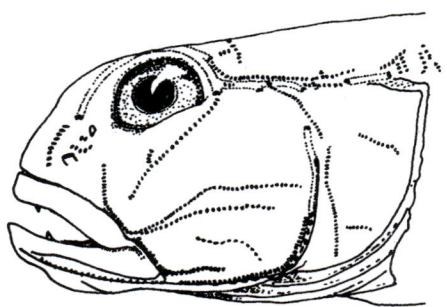

35.10 Rows of pores and papillae on head of bridled goby *Amoya bifrenatus*. (H.K. Larson)

Sexual dimorphism Mature males generally have darker fins than females, the lower surface of the head black in males, which reach a larger

Bridled goby *Amoya bifrenatus*. (N. Armstrong)

size than females and have more elongated rays in the posterior dorsal fin and tail. Urinogenital papilla short and broad in females, slender and pointed in males.
Colour Head and body pale grey to light brown, often a short black bar just above end of mouth which may connect with a thin line from edge of eye. Upper lip often black. A dark brown stripe runs from lower eye margin, backwards and down across cheek and gill cover and pectoral fin base to lower part of body, where it breaks up into a series of irregular dark spots above anal fin. A second stripe extends from behind eye to end of head, where it meets a large, dark brown blotch above pectoral fin base, continuing back as a faint brown stripe under 2nd dorsal fin. A series of small iridescent bluish green spots above and below upper brown stripe. Often similar iridescent spots scattered on top of head. Fins vary from pale grey to black, a black longitudinal stripe on lower third of dorsal fin and a series of 3–5 curved black bars on upper half of tail; upper margin of tail white.
Size Reaches about 150 mm, but rarely seen larger than 100 mm.
Distribution Is found throughout coastal areas of southern Queensland, New South Wales, Victoria and Tasmania, also South Australia and southern Western Australia.

Conservation status Common in estuaries and the sea but rare in fresh water.
Natural history Although occurring in fresh water, prefers mud and sand bottoms in brackish and marine environments, where it seeks shelter in burrows. Feeds on a variety of small invertebrates and fishes, also taking diatoms and algae. Spawns in spring and early summer, the adhesive eggs attached to objects on bottom, possibly in the burrows, Lives for 3–4 years.
Utility Does well in aquaria, but will survive for only a few months on dried food and is best fed on small invertebrates. Is preyed upon by birds, particularly cormorants.
Similar species Is unlike other gobies in fresh water but may be confused with the marine *Amoya frenatus*, which has spots on top of head and irregular, vertical bands on body.
Other names Common: none. Scientific: *Arenigobius bifrenatus*.
Literature Kuiter, 1993.

TAMAR RIVER GOBY
Afurcagobius tamarensis (Johnston)

Description Small, with a flattened head and snout, mouth oblique reaching to below eye in females and to well beyond in large males, in which the cheeks are very bulbous. First dorsal fin (VI) slightly lower than 2nd (I, 8) which is short-based and rounded, the rays becoming progressively longer, posterior rays reaching to middle of caudal peduncle in females and small males but to base of tail in large males. Anal fin I, 8. Tail rounded, about as long as head. Pectoral fins (17–18) broad with rounded margins. Pelvic fins fused into large, cup-shaped disc. Body covered with medium-sized ctenoid scales (30–35 along side), cycloid on belly, none on head or on nape in front of 1st dorsal. Gill rakers 9–11. Rows of papillae on head as in Fig. 35.11.
Sexual dimorphism Strongly developed, differences most marked in large males during breeding. Mature males have a much larger mouth and more bulbous cheeks than females, and the posterior dorsal and anal rays are more elongate in males. A black spot near end of 1st dorsal is faint

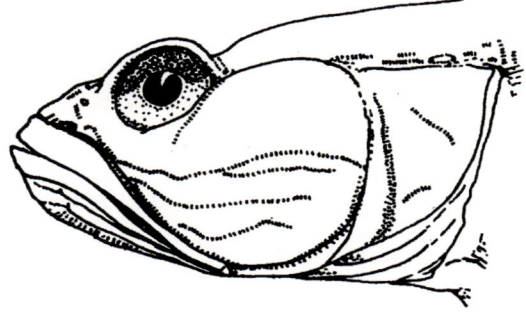

35.11 Rows of papillae on head of Tamar River goby *Afurcagobius tamarensis*.

or absent in males but prominent in females, while males generally have darker pelvic and anal fins. Males reach a larger size than females.
Colour Head and body pale grey to brown, sides of head covered with irregular brown markings, giving a mottled appearance; often a faint brown bar from front of eye extending forward and down to middle of upper lip. A series of brown blotches, about size of eye, along middle of sides. Along back below dorsal fins are 3–5 dark brown saddles. Fins clear, pale grey or dusky. A dark

FAMILY GOBIIDAE

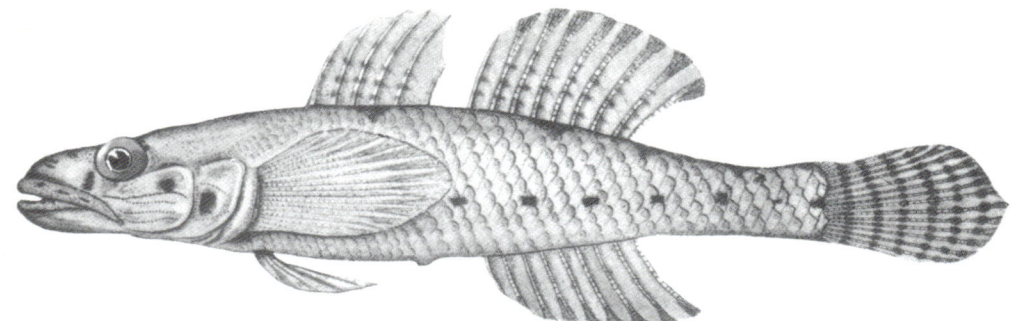

35.12 Tamar River goby *Afurcagobius tamarensis*. (E.O.G. Scott)

brown stripe occurs on lower part of 1st dorsal, which darkens posteriorly forming a dark black spot, this generally absent in males. On 2nd dorsal 2–3 thin, brown longitudinal stripes. Tail with several thin, brown, wavy vertical bands. Pelvic and anal fins become black in males. Colour pattern can vary locally.
Size Reaches about 110 mm.
Distribution Occurs in estuaries of New South Wales, Victoria, Tasmania and South Australia; replaced by a related species *A. suppositus*, in Western Australia.
Conservation status Remains common.
Natural history Is little studied, but prefers silt and mud bottoms in quiet waters of estuaries and coastal lakes and occurs often in the lower parts of rivers, in fresh water. Apparently breeds in spring.
Utility Is preyed upon by birds.
Similar species Juveniles may be confused with Swan River goby and largemouth goby.
Other names Common: none. Scientific: *Arenigobius tamarensis*, *Favonigobius tamarensis*.
Literature Gill, 1993; Kuiter, 1993.

Tamar River goby, *Afurcagobius tamarensis* (R.M. McDowall)

References

Allen, G.R. 1989, *Freshwater fishes of Australia*, T.F.H., Neptune City.

Allen, G.R. & Burgess, W.E. 1990, A review of the glassfishes (*Chandidae*) of Australia and New Guinea, *Records of the Western Australian Museum Supplement* 34: 139–206.

Allen, G.R. & Cross, N.J. 1982, *Rainbow fishes of Australia and Papua-New Guinea*, Angus & Robertson, Melbourne.

Allen, G.R. & Ivantsoff, W. 1982, *Pseudomugil mellis*, le honey blue-eye, une nouvelle espèce du blue-eye (Melanotaeniidae) d'Australie orientale, *Revue Française Aquariologie* 9(3): 83–86.

Allen, K.R. 1951, The Horokiwi Stream—a study of a trout population, *New Zealand Marine Department Fisheries Bulletin* 10: 1–238.

Allen, S. 1984, Occurrence of juvenile weatherfish *Misgurnus anguillicaudatus* (Pisces: Cobitidae) in the Yarra River, *Victorian Naturalist* 101(5): 240–242.

Allport, M. 1870, no title, *Papers and Proceedings of the Royal Society of Tasmania* 1896: 6.

Anderson, H.K., & Whitley, G.P. 1925, The story of the freshwater eel, *Australian Museum Magazine* 2: 266–270.

Anderson, J.R., Lake, J.S. & McKay, N.J. 1971, Notes on reproductive behaviour and ontogeny in two species of *Hypseleotris* (= *Carassiops*) (Gobiidae: Teleostei), *Australian Journal of Marine and Freshwater Research* 22(2): 139–145.

Anderson, J.R., Morison, A.K. & Ray, D.J. 1992a, Age and growth of Murray cod, *Maccullochella peelii* (Perciformes: Percichthyidae), in the lower Murray-Darling Basin, Australia, from thin-sectioned otoliths, *Australian Journal of Marine and Freshwater Research* 43(5): 983–1014.

Anderson, J.R., Morison, A.K. & Ray, D.J. 1992b, Validation of the use of thin-sectioned otoliths for determining the age and growth of golden perch, *Macquaria ambigua* (Perciformes: Percichthyidae), in the lower Murray-Darling Basin, Australia. *Australian Journal of Marine and Freshwater Research* 43(5): 1013.

Anon. 1970, Prepare for some brook trout fishing, *The Fisherman (NSW)* 3(9): 1–2.

Anon. 1973, The redfin episode at Lake Canobolas, *The Fisherman (NSW)* 4(7): 3–4, 22–23.

Armstrong, N. 1993, Re-discovering *Galaxias fuscus*, *Fishes of Sahul* 7(4): 328–329.

Arthington, A.H. 1986, Introduced cichlid fish in Australian waters, pp. 239–248 in Dekker, P. de & Williams, W.D. (eds), *Limnology in Australia*. Junk, Dordrecht.

Arthington, A.H. 1989, Diet of *Gambusia affinis holbrooki*, *Xiphophorus helleri*, *X. maculatus* and *Poecilia reticulata* (Pisces: Poeciliidae) in streams of south-eastern Queensland, *Asian Fisheries Science* 2: 193–212.

Arthington, A.H., Chandica, A. & Marshall, C.J. 1994, Geographic information system and distributions for the honey blue-eye, *Pseudomugil mellis*, and other freshwater fishes in south-eastern Queensland, final report to the Australian Nature Conservation Agency, vol. 2, 50 pp.

Arthington A.H. & Lloyd, L.N. 1989, Introduced poeciliids in Australia and New Zealand, pp. 333–347 in Meffe, G.F. & Snelson, F.F. (eds), *Evolution and Ecology of Live-bearing fishes*, Prentice-Hall, New York.

Arthington, A.H., McKay R.J. & Milton, D.H. 1986, The ecology and management of exotic and endemic freshwater fishes in Queensland, pp. 224–225 in Hundloe, T. (ed.) *Fisheries management—theory and practice in Queensland*, Griffith University/University of Queensland, Brisbane.

Arthington, A.H. & Marshall, C.J. 1993, Distribution, ecology and conservation of the honey blue-eye, *Pseudomugil mellis*, in south-eastern Queensland, final report to the Australian Nature Conservation Agency, vol. 1, 158 pp.

Auty, E.H. 1978, Reproductive behaviour and early development of the empire fish, *Hypseleotris compressus* (Eleotridae), *Australian Journal of Marine and Freshwater Research* 29(5): 585–597.

Bacher, G.J. & O'Brien, T.A. 1989, Salinity tolerance of the eggs and larvae of the Australian grayling, *Australian Journal of Marine and Freshwater Research* 40: 227–230.

Backhouse, G.N. 1983, The dwarf galaxias *Galaxiella pusilla* (Mack, 1936) in Victoria, *Fishes of Sahul* 1(1): 5–8.

Backhouse, G.N. & Vanner, R.W. 1978, Observations on the biology of the dwarf galaxiid *Galaxiella pusilla* (Mack) (Pisces: Galaxiidae), *Victorian Naturalist* 95: 128–132.

Baker, W.J. 1933, *Goldfish in Australia*, Graham, Sydney.

Banarescu, P. & Coad, B.W. 1991, Cyprinids of Eurasia, pp. 127–155 in Winfield I.J. & Nelson J.S. (eds), *Cyprinid fishes: systematics, biology and exploitation*, Chapman & Hall, London.

Barnham, C.A. 1977, Ten year effort to acclimatise quinnat salmon pays off, *Australian Fisheries* 36(9): 14–15, 19.

Battaglene, S.C., Beevers, P.J. & Talbot, R.B. 1989, The artificial propagation of Australian bass (*Macquaria novemaculeata*), *New South Wales Agriculture and Fisheries Bulletin* 3: 1–11.

Bayly, I.A.E., Ebsworth, E.P. & Wan, H.F. 1975, Studies on the lakes of Fraser Island, Queensland, *Australian Journal of Marine and Freshwater Research* 26(1): 1–13.

Beck, R.G. 1985, Field observations upon the dwarf galaxiid *Galaxiella pusilla* (Mack) (Pisces: Galaxiidae) in the south-east of South Australia, Australia, *South Australian Naturalist* 60(1): 12–22.

REFERENCES

Bell, J.D., Berra, T.M., Jackson, P.D., Last, P.R. & Sloane, R.D. 1980, Recent records of the Australian grayling *Prototroctes maraena* Günther (Pisces: Prototroctidae) with notes on its distribution, *Australian Zoologist* 20(3): 419–431.

Berra, T.M. 1974, The trout cod, *Maccullochella macquariensis*, a rare freshwater fish of eastern Australia, *Biological Conservation* 6(1): 53–56.

Berra, T.M. 1982, Life history of the Australian grayling, *Prototroctes maraena* (Salmoniformes: Prototroctidae) in the Tambo River, Victoria, *Copeia* 1982(4): 795–805.

Berra, T.M. 1981, *An atlas of distribution of the freshwater fish families of the world*, University of Nebraska, Lincoln, 197 pp.

Berra, T.M. & Cadwallader, P.L. 1983, Age and growth of the Australian grayling *Prototroctes maraena* Günther (Salmoniformes: Prototroctidae) in the Tambo River, Victoria, *Australian Journal of Marine and Freshwater Research* 34: 451–460.

Berra, T.M. & Weatherley, A.H. 1972, A systematic study of the Australian freshwater serranid fish genus *Maccullochella*, *Copeia* 1971(1): 53–64.

Bertin, L. 1956, *Eels—a biological study*, Cleaver-Hume, London.

Beumer, J.P. 1979a, Reproductive cycles of two Australian freshwater fishes: the spangled perch, *Therapon unicolor* Günther, 1859, and the east Queensland rainbowfish, *Nematocentris splendida* Peters, 1866, *Journal of Fish Biology* 15: 111–134.

Beumer, J.P. 1979b, Temperature and salinity tolerance of the spangled perch *Therapon unicolor* Günther, 1859, and the east Queensland rainbowfish *Nematocentris splendida* Peters, 1866, *Proceedings of the Royal Society of Queensland* 90: 85–91.

Beumer, J.P. 1979c, Feeding and movement of *Anguilla australis* and *A. reinhardti* in Macleods Morass, Victoria, Australia, *Journal of Fish Biology* 14(6): 573–592.

Beumer, J. & Sloane, R. 1990, Distribution and abundance of glass-eels *Anguilla* spp. in east Australian waters, *International Revue gesamten Hydrobiologie* 75(6): 721–736.

Bird, D.J. & Potter, I.C. 1981, Proximate body composition of the larval metamorphosing and downstream migrant stages in the life cycle of the Southern Hemisphere lamprey *Geotria australis*, *Environmental Biology of Fishes* 6(3/4): 285–297.

Bishop, K.A. & Bell, J.D. 1978a, Aspects of the biology of the Australian grayling *Prototroctes maraena* Günther (Pisces: Prototroctidae), *Australian Journal of Marine and Freshwater Research* 29(6): 743–761.

Bishop, K.A. & Bell, J.D. 1978b, Observations on the fish fauna below Tallowa Dam (Shoalhaven River, New South Wales) during river flow stoppages, *Australian Journal of Marine and Freshwater Research* 29(4): 543–549.

Blackburn, M. 1950, The Tasmanian whitebait *Lovettia seali* (Johnston) and the whitebait fishery, *Australian Journal of Marine and Freshwater Research* 1(2): 155-198.

Blewett, C.L. 1929, Habits of some Australian freshwater fishes, *South Australian Naturalist* 10(2): 21–29.

Brown, P. 1992, Occurrence of a neosilurid catfish (*Neosilurus* sp.) in the Paroo River, Murray-Darling basin, *Proceedings of the Linnean Society of New South Wales* 113(4): 341–343.

Brumley, A.R. 1991, Cyprinids of Australia, pp. 265–283 in Winfield, I.J. & Nelson, J.S. (eds), *Cyprinid fishes: systematics, biology and exploitation*, Chapman & Hall, London.

Brumley, A.R., Morison, A.K. & Anderson, J.R. 1987, Revision of the conservation status of several species of warmwater native fish after surveys of selected sites in northern Victoria (1982–1984), *Arthur Rylah Institute for Environmental Research Technical Report* 33: 1–93.

Burchmore, J., Farragher, R. & Thorncraft, G. 1990, Occurrence of the introduced oriental weather loach (*Misgurnus anguillicaudatus*) in the Wingecarribee River, New South Wales, *Bureau of Rural Resources Proceedings* 8: 38–46.

Butcher, A.D. 1947, Quinnat salmon in Victorian inland waters, *Victoria Fish and Game Department Fisheries Pamphlet* 4: 1–28.

Cadwallader, P.L. & Backhouse, G.N. 1983, *A guide to the freshwater fish of Victoria*, Government Printer, Melbourne.

Cadwallader, P.L., Backhouse, G.N. & Fallu, R. 1980, Occurrence of exotic fish in the cooling pondage of a power station in temperate south-eastern Australia, *Australian Journal of Marine and Freshwater Research* 32: 51–546.

Cadwallader, P.L., Backhouse, G.N., Gooley, G.J. & Turner, J.A. 1979, New techniques for breeding and raising Murray cod, *Australian Fisheries* 38(9): 9–16.

Cadwallader, P.L. & Eden, A.K. 1979, Observations on the food of Macquarie perch, *Macquaria australasica* (Pisces: Percichthyidae), in Victoria, *Australian Journal of Marine and Freshwater Research* 30(2): 401–409.

Cadwallader, P.L., Eden, A.K. & Hook, R.A. 1980, The role of streamside vegetation as a food source for *Galaxias olidus* Günther (Pisces: Galaxiidae), *Australian Journal of Marine and Freshwater Research* 31: 257–262.

Cadwallader, P.L. & Rogan, P.L. 1977, The Macquarie perch, *Macquaria australasica* (Pisces: Percichthyidae), of Lake Eildon, *Australian Journal of Ecology* 2(4): 409–418.

Castle, P.H.J. 1963, Anguillid leptocephali in the south-west Pacific, *Zoology Publications from Victoria University of Wellington* 33: 1–14.

Caughley, A. 1992, Some aquarium notes on *Galaxias truttaceus*, *Fishes of Sahul* 7(3): 319.

Chessman, B.C. & Williams, W.D. 1975, Salinity tolerance and osmoregulatory ability of *Galaxias maculatus* (Jenyns) (Pisces, Salmoniformes, Galaxiidae), *Freshwater Biology* 5(2): 135–140.

Chubb, C.F., Potter, I.C., Grant, C.J., Lenanton, R.C.J. & Wallace, J. 1981, Age structure, growth rates and movements of sea mullet, *Mugil cephalus* L., and yellow-eye mullet, *Aldrichetta forsteri* (Valenciennes), in the Swan-Avon River system, Western Australia, *Australian Journal of Marine and Freshwater Research* 32: 605–628.

Coates, D. 1987, Observations on the biology of tarpon, *Megalops cyprinoides* (Brousonnet) (Pisces: Megalopidae), in the Sepik River, northern Papua New Guinea, *Australian Journal of Marine and Freshwater Research* 38(4): 529–535.

Collette, B.B., Ali, M.A., Hokanson, K.E.F., Nagiec, M., Smirnov, S.A., Thorpe, J.E., Weatherley, A.H. & Willemson, J. 1977, Biology of the percids, *Journal of the Fisheries Research Board of Canada* 34: 1890–1899.

Craig, J.F. 1987, *The biology of perch and related fish*, Croom Helm, Beckenham.

Cribb, T.H. 1988, Two new digenetic trematodes from Australian freshwater fishes with notes on previously described species, *Journal of Natural History* 22: 27–43.

Crowley, L.E.L.M. 1990, Biogeography of the endemic freshwater fish *Craterocephalus* (Family Atherinidae), *Memoirs of the Queensland Museum* 28(1): 89–98.

Crowley, L.E.L.M. & Ivantsoff, W. 1988, A new species of Australian *Craterocephalus* (Pisces: Atherinidae) and the redescription of four other species, *Records of the Western Australian Museum* 14(2): 151–169.

Crowley, L.E.L.M. & Ivantsoff, W. 1989, An historical overview of the genus *Craterocephalus* with special reference to the hardyheads of Dalhousie Springs, pp. 113–118 in Zeidler W. & Ponder W.F. (eds), *Natural history of Dalhousie Springs*, South Australian Museum, Adelaide.

Crowley, L.E.L.M. & Ivantsoff, W. 1990a, A review of species previously known as *Craterocephalus eyresii* (Pisces: Atherinidae), *Proceedings of the Linnean Society of New South Wales* 112(2): 87–103.

Crowley, L.E.L.M. & Ivantsoff, W. 1990b, A second hardyhead, *Craterocephalus gloveri* (Pisces: Atherinidae), from Dalhousie Springs, central Australia, *Ichthyological Exploration of Freshwaters* 1: 113–122.

Crowley, L.E.L.M. & Ivantsoff, W. 1992, Redefinition of the freshwater fish genus *Craterocephalus* (Teleostei: Atherinidae) of Australia and New Guinea with an analysis of three species, *Ichthyological Exploration of Freshwaters* 3: 273–287.

Crowley, L.E.L.M., Ivantsoff, W. & Allen, G.R. 1986, Taxonomic position of two crimson-spotted rainbowfish, *Melanotaenia duboulayi* and *Melanotaenia fluviatilis* (Pisces: Melanotaeniidae), from eastern Australia with special reference to their early life stages, *Australian Journal of Marine and Freshwater Research* 37: 385–398.

Dakin, J.W. & Kesteven, G.L. 1938, The Murray cod (*Maccullochella macquariensis* (Cuv. and Val.)), *Bulletin of New South Wales State Fisheries* 1: 1–18.

Davies, P.E. 1989, Relationships between habitat characteristics and population abundance for brown trout, *Salmo trutta* L., and blackfish, *Gadopsis marmoratus* Rich., in Tasmanian streams, *Australian Journal of Marine and Freshwater Research* 40: 341–359.

Davies, P.E. & Sloane, R.D. 1986, Validation of aging and length back-calculation in rainbow trout, *Salmo gairdneri* Rich., from Dee Lagoon, Tasmania, *Australian Journal of Marine and Freshwater Research* 37: 289–295.

Davies, P.E., Sloane, R.D. & Andrew, J. 1989, The effects of hydrological change and the cessation of stocking on a stream population of *Salmo trutta* L., *Australian Journal of Marine and Freshwater Research* 39: 337–354.

Davis, T.L.O. 1977a, Age determination and growth of the freshwater catfish, *Tandanus tandanus* Mitchell, in the Gwydir River, Australia, *Australian Journal of Marine and Freshwater Research* 28(2): 119–137.

Davis, T.L.O. 1977b, Food habits of the freshwater catfish, *Tandanus tandanus* Mitchell, in the Gwydir River, Australia, and effects associated with impoundment of this river by the Copleton Dam, *Australian Journal of Marine and Freshwater Research* 28(4): 455–465.

Davis, T.L.O. 1977c, Reproductive biology of the freshwater catfish, *Tandanus tandanus* Mitchell, in the Gwydir River, Australia. I. Structure of the gonads, *Australian Journal of Marine and Freshwater Research* 28(2): 139–158.

Davis, T.L.O. 1977d, Reproductive biology of the freshwater catfish, *Tandanus tandanus* Mitchell, in the Gwydir River, Australia. II. Gonadal cycle and fecundity, *Australian Journal of Marine and Freshwater Research* 28(2): 159–169.

Douglas, J.W., Gooley, G.J. & Ingram, B.A. 1994, *Trout cod,* Maccullochella macquariensis *(Cuvier) (Pisces: Percichthyidae): resource handbook and research and recovery plan*, Department of Conservation and Natural Resources, Melbourne.

Dufty, S. 1986, *Genetic and morphological divergence between populations of Macquarie perch (*Macquaria australasica*) east and west of the Great Dividing Range*, unpublished BSc Hons thesis, University of New South Wales, Sydney.

Ege, V. 1939, A revision of the genus *Anguilla* Shaw. A systematic, phylogenetic and geographical study, *Dana Report* 3(16): 1–256.

Elliott, J.M. 1994, *Quantitative ecology and the brown trout*, Oxford, London.

Finlay, H.J. 1972, Report on the examination of the scales of quinnat salmon (*Oncorhynchus tshawytscha* (Walbaum)) for the determination of age and growth-rate, *New Zealand Marine Department Fisheries Technical Report* 66: 1–27.

Fletcher, A.R. 1979, *Effects of* Salmo trutta *on* Galaxias olidus *and macroinvertebrates in stream communities*, unpublished MSc thesis, Monash University, Clayton.

Fortes, R.D. 1980, Tarpon as predator to control Java tilapia young in brackish water ponds, *Fisheries Research Journal of the Philippines* 5(2): 22–35.

Francois, D.D. 1965, Atlantic salmon for New South Wales, *Australian Natural History* 15(2): 61–64.

Frankenberg, R.S. 1968, Two new species of galaxiid

REFERENCES

fishes from the Lake Pedder region of southern Tasmania, *Australian Zoologist* 14(3): 168–174.

Frost, W.E. & Brown, M.E. 1967, *The trout*, Collins, London.

Fulton, W. 1978a, A description of a new species of *Galaxias* (Pisces: Galaxiidae) from Tasmania, *Australian Journal of Marine and Freshwater Research* 29(1): 109–116.

Fulton, W. 1978b, A new species of *Galaxias* (Pisces: Galaxiidae) from the Swan River, Tasmania, *Records of the Queen Victoria Museum, Launceston* 63: 1–8.

Fulton, W. 1982, Observations on the ecology of four species of the genus *Paragalaxias* (Pisces: Galaxiidae) from Tasmania, *Australian Journal of Marine and Freshwater Research* 33(6): 999–1016.

Fulton, W. 1984, Whitebait whiteout, *Tasmanian Inland Fisheries Commission Newsletter* 13(3): 2.

Fulton, W. 1990, Tasmanian freshwater fishes, *Fauna of Tasmania Handbook* 7: 1–80.

Fulton, W. & Pavuk, N. 1988, The Tasmanian whitebait fishery: summary of present knowledge and outline of future management plans, *Tasmanian Inland Fisheries Commission Occasional Report* 88-01: 1–27.

Gaffney, R.F., Hamr, P. & Davies, P.E. 1992, *The Pedder galaxias recovery plan: management phase*, Department of Parks, Wildlife and Heritage, Hobart.

Gale, A. 1914, Notes on the breeding habits of the purple striped gudgeon, *Krefftius adpersa*, *Australian Zoologist* 1(1): 2526.

Gehrke, P.C. and Harris, J.H. 1994, The role of fish in cyanobaterial blooms in Australia, *Australian Journal of Marine and Freshwater Research* 45(5): 905-916.

Gerry, D. 1994, *Biochemical systematics of coastal New South Wales populations of the common tandan catfish,* (Tandanus tandanus: Plotosidae), unpublished BSc Hons thesis, University of New England, Armidale.

Gill, G.S. 1993, Description of new genus of goby from southern Australia, including osteological comparisons with related genera, *Records of the Western Australian Museum* 16(2): 175–210.

Gill, G.S. & Potter, I.C. 1993, Spatial segregation amongst goby species within an Australian estuary, with a comparison of the diets and salinity tolerances of the two most abundant species, *Marine Biology* 117: 515–526.

Glover, C.J.M. 1989, Fishes, pp. 89–109 in Zeidler, W. & Ponder, W.F. (eds), *Natural history of Dalhousie Springs*, South Australian Museum, Adelaide.

Glover, C.J.M. 1990, Fishes, pp. 189–195 in Tyler, M.J., Twidale, C.R., Davies, M. & Wells, C.B. (eds), *Natural history of the north-east deserts*, Royal Society of South Australia, Adelaide.

Gooley, G.J. 1986, Culture methods for Macquarie perch at Dartmouth, Victoria, *Australian Fisheries*, September 1986: 18–20.

Gooley, G.J. 1992, Validation of the use of otoliths to determine the age and growth of Murray cod, *Maccullochella peeli* (Mitchell) (Percichthyidae) in Lake Charlegrark, western Victoria, *Australian Journal of Marine and Freshwater Research* 43(5): 1091–1102.

Gooley, G.J. & McDonald, G.L. 1988, Preliminary study on the hormone-induced spawning of Macquarie perch, *Macquaria australasica* (Cuvier) (Percichthyidae), from Lake Dartmouth, Victoria, *Arthur Rylah Institute for Environmental Research Technical Report* 80: 1–13.

Gorman, T.B.S. 1962, Yellow-eyed mullet, *Aldrichetta forsteri* (Cuvier and Valenciennes) in Lake Ellesmere, *New Zealand Marine Department Fisheries Technical Report* 7: 1–20.

Grant, C.J. & Spain, A.V. 1975, Reproduction, growth and size allometry of *Mugil cephalus* Linnaeus (Pisces: Mugilidae) from north Queensland inshore waters, *Australian Journal of Zoology* 23: 181–201.

Grant, E.M. 1991, *Fishes of Australia*, Grant, Brisbane.

Green, K. 1979, Observations on rock climbing by the fish *Galaxias brevipinnis*, *Victorian Naturalist* 96: 230–231.

Hale, H.M. 1920, The Australian congolly, *Aquatic Life* 5(3): 25–26.

Hamlyn-Harris, R. 1931, A further contribution to the breeding habits of *Mogurnda (Mogurnda) adspersus* Castelnau, the trout gudgeon, *Australian Zoologist* 7(1): 55–58.

Hamr, P. 1982, The Pedder galaxias, *Australian Natural History* 23(12): 904.

Hamr, P. 1991, Conservation of *Galaxias pedderensis*, *Australian Society for Fish Biology Newsletter* 21(2): 35.

Hamr, P. 1992. Conservation of *Galaxias pedderensis*: report to the endangered species unit, ANPWS, *Tasmanian Inland Fisheries Commission Occasional Report* 92-01: 1–77.

Hansen, B. 1988, The male purple-spotted gudgeon—*Mogurnda adspersa*. *Fishes of Sahul* 5(1): 200–202.

Hansen, B. 1992, A forgotten jewel—*Rhadinocentrus ornatus*. *Fishes of Sahul* 7(3): 313–318.

Harris, J.A. 1968, The yellow-eye mullet: age structure, growth rate and spawning cycle of a population of yellow-eye mullet *Aldrichetta forsteri* (Cuv. and Val.) from the Coorong Lagoon, South Australia, *Transactions of the Royal Society of South Australia* 92: 37–50.

Harris, J.H. 1985a, Age of Australian bass, *Macquaria novemaculeata* (Perciformes: Percichthyidae), in the Sydney Basin, *Australian Journal of Marine and Freshwater Research* 36: 235–246.

Harris, J.H. 1985b, Diet of the Australian bass, *Macquaria novemaculeata* (Perciformes: Percichthyidae), in the Sydney Basin, *Australian Journal of Marine and Freshwater Research* 36: 219–234.

Harris, J.H. 1986, Reproduction of the Australian bass, *Macquaria novemaculeata* (Perciformes: Percichthyidae) in the Sydney Basin, *Australian Journal of Marine and Freshwater Research* 37: 209–235.

Harris, J.H. 1987a, Growth of Australian bass, *Macquaria novemaculeata* (Perciformes: Percichthyidae), in the Sydney Basin, *Australian*

Journal of Marine and Freshwater Research 38: 351–361.

Harris, J.H. 1987b, *Proceedings of the conference on threatened fishes, Melbourne, 15–16 August, 1987, by the Australian Society for Fish Biology*, New South Wales Department of Agriculture, Sydney.

Harris, J.H. 1988, Demography of Australian bass *Macquaria novemaculeata* (Perciformes: Percichthyidae) in the Sydney Basin, *Australian Journal of Marine and Freshwater Research* 39: 355–369.

Harris, J.H. & Dixon, P.I. 1986, Hybridization between trout cod and Murray cod, *Isozyme Bulletin* 19: 39.

Harris, J.H. & Pearn, J. 1987, Bullrout stings, pp. 155–159 in Covachevich J., Davies, P. & Pearn, J. (eds), *Toxic plants and animals: a guide for Australia*, Queensland Museum, Brisbane.

Healey, M.C. 1991, Life history of chinook salmon (*Oncorhynchus tshawytscha*), pp. 311–393 in Groot, C. & Margolis, L. (eds), *Pacific salmon life histories.*, University of British Columbia, Vancouver.

Hoese, D.F. 1991, A revision of the temperate Australian gobiid (Gobioidei) fish genus, *Tasmanogobius*, with a comment on the genus *Kimberleyeleotris*, *Memoirs of the Museum of Victoria* 52(2): 361–376.

Hoese, D.F. & Allen, G.R. 1983, A review of the gudgeon genus *Hypseleotris* (Pisces: Eleotridae) of Western Australia, with descriptions of three new species, *Records of the Western Australian Museum* 10(3): 243–261.

Hortle, M.E. 1979, *The ecology of the sandy* Pseudaphritis urvillii *in south-east Tasmania*, unpublished BSc Hons thesis, University of Tasmania, Hobart.

Hortle, M.E. & White, R.W.G. 1980, Diet of *Pseudaphritis urvillii* (Cuvier and Valenciennes) (Pisces: Bovichthyidae) from south-eastern Australia, *Australian Journal of Marine and Freshwater Research* 31: 533–539.

Howe, E. 1987, Breeding behaviour, surface egg morphology and embryonic development of four Australian species of the genus *Pseudomugil* (Pisces: Melanotaeniidae), *Australian Journal of Marine and Freshwater Research* 38: 885–895.

Howes, G.J. 1991, Systematics and biogeography: an overview, pp. 1–33 in Winfield, I.J. & Nelson, J.S. (eds), *Cyprinid fishes: systematics, biology and exploitation*, Chapman & Hall, London.

Hughes, R.L. & Potter, I.C. 1968, Studies on gametogenesis and fecundity in the lampreys *Mordacia praecox* and *M. mordax* (Petromyzontidae), *Australian Journal of Zoology* 17: 447–464.

Hume, D.J., Fletcher, A.R. & Morison, A.K. 1983a, Final report, *Fisheries and Wildlife Department, Ministry for Conservation, Victoria, Carp Programme Report* 10: 1–213.

Hume, D.J., Fletcher, A.R. & Morison, A.K. 1983b, Interspecific hybridisation between carp (*Cyprinus carpio* L.) and goldfish (*Carassius auratus* L.) from Victorian waters, *Australian Journal of Marine and Freshwater Research* 34: 915–919.

Humphries, P.L. 1985, *Aspects of the biology of the dwarf galaxiid,* Galaxiella pusilla *(Mack) (Salmoniformes: Galaxiidae)*, unpublished BSc Hons thesis, Monash University, Clayton.

Humphries, P. 1986, Observations of the ecology of *Galaxiella pusilla* (Mack) (Salmoniformes: Galaxiidae) in Diamond Creek, Victoria, *Proceedings of the Royal Society of Victoria* 98(3): 133–137.

Humphries, P. 1989, Variation in the life history of diadromous and landlocked populations of the spotted galaxias, *Galaxias truttaceus* Valenciennes, in Tasmania, *Australian Journal of Marine and Freshwater Research* 40: 501–518.

Humphries, P. 1990, Morphological variation in diadromous and landlocked populations of the spotted galaxias, *Galaxias truttaceus*, in Tasmania, south-eastern Australia, *Environmental Biology of Fishes* 27(1): 97–105.

Humphries, P. & Potter, I.C. 1993, Relationship between the habitat and diet of three species of atherinids and three species of gobies in a temperate Australian estuary, *Marine Biology* 116: 193–204.

Humphrey, C.L. 1979, *Growth and reproduction of the freshwater or pink-eye mullet,* Trachystoma petardi *(Castelnau) in the Macleay River, NSW*, unpublished BSc Hons thesis, University of New England, Armidale, 120 pp.

Ingram, B.A. & Rimmer, M.A. 1992, Induced breeding and larval rearing of the endangered Australian freshwater fish trout cod, *Maccullochella macquariensis* (Cuvier) (Percichthyidae), *Aquaculture and Fisheries Management* 24: 7–17.

Ivantsoff, W. 1987, Description of three new species and one subspecies of freshwater hardyhead (Pisces: Atherinidae: *Craterocephalus*) from Australia, *Records of the Western Australian Museum* 13(2): 171–188.

Ivantsoff, W. 1994, Atherinidae, pp. 375–381 in Gomon, M.F., Glover, C.J.M. & Kuiter, R.H. (eds), *The fishes of Australia's south coast*, State Print, Adelaide.

Ivantsoff, W., Crowley, L.E.L.M. & Allen, G.R. 1987, Decription of three new species and one subspecies of freshwater hardyhead (Pisces: Atherinidae: *Craterocephalus*) from Australia, *Records of the Western Australian Museum* 13(2): 171–188.

Ivantsoff, W., Crowley, L.E.L.M., Howe, E.I. & Semple, G. 1988, Development of methodologies for the culture of fish species of the Alligator Rivers Region, and the supply of fish to the Office of the Supervising Scientist, *Technical Memorandum 22*, Australian Government Publishing Service, Canberra.

Ivantsoff, W., & Glover, C.J.M. 1974, *Craterocephalus dalhousiensis* n. sp., a sexually dimorphic freshwater teleost (Atherinidae) from South Australia, *Australian Zoologist* 18(2): 88–98.

Jackson, P.D. 1975, *Bionomics of brown trout (*Salmo trutta *Linnaeus, 1758) in a Victorian stream, with notes on interactions with native fishes*, unpub-

REFERENCES

lished PhD dissertation, Monash University, Melbourne, 269 pp.

Jackson, P.D. 1976, A note on the food of the Australian grayling, *Prototroctes maraena* Günther (Galaxioidei: Prototroctidae), *Australian Journal of Marine and Freshwater Research* 27(3): 525–528.

Jackson, P.D. 1978a, Benthic invertebrate fauna and feeding relationships of brown trout, *Salmo trutta* Linnaeus, and river blackfish, *Gadopsis marmoratus* Richardson, in the Aberfeldy River, Victoria, *Australian Journal of Marine and Freshwater Research* 29: 725–742.

Jackson, P.D. 1978b, Spawning and early development of the river blackfish, *Gadopsis marmoratus* Richardson (Gadopsiformes: Gadopsidae), in the McKenzie River, Victoria, *Australian Journal of Marine and Freshwater Research* 29(3): 293–298.

Jackson, P.D. 1980, Movement and home range of brown trout, *Salmo trutta* Linnaeus, in the Aberfeldy River, Victoria, *Australian Journal of Marine and Freshwater Research* 31: 837–845.

Jackson, P.D. 1981, Trout introduced into south-eastern Australia: their interaction with native fishes, *Victorian Naturalist* 98(1): 18–24.

Jackson, P.D., & Davies, J.N. 1982, Occurrence of the Tasmanian mudfish *Galaxias cleaveri* Scott on Wilsons Promontory—first record from mainland Australia, *Proceedings of the Royal Society of Victoria* 94(1–2): 49–52.

Jackson, P.D. & Koehn, J.D. 1988, A review of biological information, distribution and status of the Australian grayling *Prototroctes maraena* Günther in Victoria, *Arthur Rylah Institute for Environmental Research Technical Report* 52: 1–20.

Jackson, P.D. & Williams, W.D. 1980, Effects of brown trout, *Salmo trutta* L., on the distribution of some native fishes in three areas of southern Victoria, *Australian Journal of Marine and Freshwater Research* 31(1): 61–67.

Jhingran, V.G. 1975, *Fishes and fisheries of India*, Hindustan Publishing, Delhi.

Johnson, G.D. 1984, Percoidei: development and relationships, pp. 464–498 in Moser, G.H. & others (eds), *Ontogeny and systematics of fishes*, American Society of Ichthyologists and Herpetologists Special Publication Vol. 1.

Johnston, P.G., Potter, I.C. & Robinson, E.S. 1988, Electrophoretic analysis of populations of southern hemisphere lampreys *Geotria australis* and *Mordacia mordax*, *Genetica* 74: 113–117.

Jones, J.W. 1959, *The salmon*, Collins, London.

Kailola, P.J. 1983, *Arius graeffei* and *Arius armiger*, valid names for two common species of Australo-Papuan fork-tailed catfishes (Pisces: Ariidae), *Transactions of the Royal Society of South Australia* 107: 187–196.

Kailola, P.J., Williams, M.J., Stewart, P.C., Reichelt, R.E., McNee, A. & Grieve, C. 1993, *Australian fisheries resources*, Bureau of Resources Science and Fisheries Research and Development Corporation, Canberra.

Kaliyamurthy, N., Rao, G.R. & Rao, A.V.P. 1977, Ecological considerations concerning the seed of cultivable fishes of the Pulicat Lake, *Indian Journal of Fisheries* 24(1–2): 223–227.

Keenan, C.P., Watts, R. and Musyl, M.K. 1994, Unrecognised speciation in Australian native freshwater fishes with special reference to the *Tandanus tandanus* species complex, *Australian Society for Fish Biology Annual Conference Abstracts*, Canberra, ACT.

Kershaw, J.A. 1911, Migration of eels in Victoria, *Victorian Naturalist* 27: 196–201.

Koehn, J.D. 1986, Approaches to determining flow and habitat requirements for native freshwater fish in Victoria, pp. 95–115 in Campbell, I.C. (ed.), *Stream protection: the management of rivers for instream uses*, Water Studies Centre and Chisholm Institute of Technology, Melbourne.

Koehn, J.D. 1987, Artificial habitat increases abundance of two-spined blackfish (*Gadopsis bispinosus*) in Ovens River, Victoria, *Arthur Rylah Institute for Environmental Research Technical Report* 56: 1–20.

Koehn, J.D. 1990, Distribution and conservation status of the two-spined blackfish (*Galaxias bispinosus*) outside Victoria, *Victorian Naturalist* 106: 26–27.

Koehn, J.D. & O'Connor, W.G. 1990a, *Biological information for management of native freshwater fish in Victoria*, Department of Conservation and Environment, Melbourne.

Koehn, J.D. & O'Connor, W.G. 1990b, Distribution of freshwater fish in the Otway Region, south-western Victoria, *Proceedings of the Royal Society of Victoria* 102(1): 29–39.

Koehn, J.D. & O'Connor, W. 1992, *Galaxias brevipinnis*, this is your life cycle, *Australian Society for Fish Biology Newsletter* 22(2): 38.

Koehn, J.D. & Raadik, T.A. 1991, The Tasmanian mudfish, *Galaxias cleaveri* Scott, 1934, in Victoria, *Proceedings of the Royal Society of Victoria* 103(2): 77–86.

Krumholz, L.A. 1948, Reproduction in the western mosquitofish, *Gambusia affinis affinis* (Baird and Girard), and its use in mosquito control, *Ecological Monographs*, 18: 1–43.

Kuiter, R.H. 1993, *Coastal fishes of south-eastern Australia*, Crawford House, Bathurst.

Kuiter, R.H. & Allen, G.R. 1986, A synopsis of the Australian pygmy perches (Percichthyidae) with the description of a new species, *Revue Française Aquariologie* 12(4): 109–116.

Lake, J.S. 1957, Trout populations and habitats in New South Wales, *Australian Journal of Marine and Freshwater Research* 8(4): 414–450.

Lake, J.S. 1966, Freshwater fishes of the Murray-Darling River system, *Research Bulletin New South Wales State Fisheries Department* 7: 1–48.

Lake, J.S. 1967a, Freshwater fish of the Murray-Darling River system, *New South Wales State Fisheries Bulletin* 7: 1–48.

Lake, J.S. 1967b, Principal fishes of the Murray-Darling River system, pp. 192–213 in Weatherley A.H. (ed.), *Australian inland waters and their fauna: eleven studies*, Australian National

University, Canberra.
Lake, J.S. 1967c, Rearing experiments with five species of Australian freshwater fishes. I. Inducement to spawning, *Australian Journal of Marine and Freshwater Research* 18: 137–153.
Lake, J.S. 1967d, Rearing experiments with five species of Australian freshwater fishes. II. Morphogenesis and ontogeny, *Australian Journal of Marine and Freshwater Research* 18(2): 155–173.
Lake, J.S. 1971, *Freshwater fishes and rivers of Australia*, Nelson, Melbourne.
Lake, J.S. 1978, *Australian freshwater fishes illustrated: an illustrated field guide*, Nelson, Melbourne.
Lake, J.S. & Midgley, S.H. 1970, Reproduction of freshwater Ariidae in Australia, *Australian Journal of Science* 32(11): 441.
Lake, P.S. & Fulton, W. 1983, Observations on the freshwater fish of a small Tasmanian coastal stream, *Papers and Proceedings of the Royal Society of Tasmania* 115: 163–172.
Last, P., Scott, E.O.G. & Talbot, F.H. 1983, *Fishes of Tasmania*, Tasmanian Fisheries Development Authority, Hobart.
Leggett, R. 1983, *Rhadinocentrus ornatus*, southern soft-spined sunfish, *Fishes of Sahul* 1(2): 23–24.
Leggett, R. & Merrick, J.R. 1987, *Australian native fishes for aquariums*, J.R. Merrick Publications, Artarmon.
Lethbridge, R.C. & Potter, I.C. 1981, The development of teeth and associated feeding structures during the metamorphosis of the lamprey, *Geotria australis*, *Acta Zoologica, Stockholm* 62(4): 201–214.
Lewis, F. 1942, Notes on Australian eels, *Victorian Naturalist* 59: 65–66.
Lintermans, M. 1993, Oriental weather loach *Misgurnus anguillicaudatus* in the Canberra region, *Australian Capital Territory Parks and Conservation Service Technical Report* 4: 1–22.
Lintermans, M. & Rutzou, T. 1990a, The fish fauna of the upper Cotter River catchment, *Australian Capital Territory Parks and Conservation Service Research Report* 4.
Lintermans, M. & Rutzou, T. 1990b, A new locality for the two-spined blackfish (*Gadopsis bispinosus*) outside Victoria, *Victorian Naturalist* 107(1): 26–27.
Lintermans, M., Rutzou, T. & Kubolic, K. 1990c, Introduced fish of the Canberra region: recent range expansions, *Bureau of Rural Resources Proceedings* 8.
Lintermans, M., Rutzou, T. & Kubolic, K. 1990b, The status, distribution and possible impacts of the oriental weather loach *Misgurnus anguillicaudatus* in the Ginninderra Creek catchment, *Australian Capital Territory Parks and Conservation Service Research Report* 2: 1–28.
Llewellyn, L.C. 1971, Breeding studies on the freshwater forage fish of the Murray-Darling River system, *The Fisherman (NSW)* 3(13): 1–12.
Llewellyn, L.C. 1973, Spawning, development and temperature tolerance of the spangled perch *Madigania unicolor* (Günther) from inland waters in Australia, *Australian Journal of Marine and Freshwater Research* 24(1): 73–94.
Llewellyn, L.C. 1974, Spawning, development and distribution of the southern pygmy perch, *Nannoperca australis australis* Günther, from inland waters in eastern Australia, *Australian Journal of Marine and Freshwater Research* 25(1): 121–149.
Llewellyn, L.C. 1979, Some observations on the spawning and development of the Mitchellian freshwater hardyhead *Craterocephalus fluviatilis* McCulloch from inland waters in New South Wales, *Australian Zoologist* 20: 269–288.
Lloyd, L.N. 1984, Exotic fish—useful additions or 'animal weeds', *Fishes of Sahul*, 1: 31–34, 39–42.
Lloyd, L.N. 1990, Taxonomic update: species control using taxonomy—*Gambusia affinis* no longer exists in Australia! *Australian Society for Fish Biology Newsletter* 29(2): 48.
Lloyd, L.N., Arthington, A.H. & Milton, D.A. 1986, The mosquitofish—a valuable mosquito control agent or a pest, pp. 7–25 in Kitching, R.L. (ed.) *The ecology of exotic animals and plants: some case histories*, Wiley, Brisbane.
Lloyd, L.N. & Tomasov, J.F. 1985, Taxonomic status of the Mosquitofish, *Gambusia affinis* (Poeciliidae), in Australia, *Australian Journal of Marine and Freshwater Research*, 36: 447–451.
Losse, G.F. 1968, The elopoid and clupeoid fishes of east African coastal waters, *Journal of the East African Natural History Society and National Museum* 27: 77–115.
Lynch, D.D. 1965a, Changes in Tasmanian fishery, *Australian Fisheries Newsletter* 24(4): 13–15.
Lynch, D.D. 1965b, What are whitebait? *Australian Fisheries Newsletter* 24(4): 13.
Lynch, D.D. 1973, Introduced fishes, pp. 107–112 in Banks, M.R. (ed.), *The Lake Country of Tasmania*, Royal Society of Tasmania, Hobart.
McAfee, W.R. 1966a, Eastern brook trout, pp. 240–260 in Calhoun, A. (ed.), *Inland fisheries management*, State of California Department of Fish and Game, Sacramento.
McAfee, W.R. 1966b, Rainbow trout, pp. 192–215 in Calhoun, A. (ed.), *Inland fisheries management*, State of California Department of Fish and Game, Sacramento.
McCarraher, D.B. 1986, Observations on the distribution, spawning, growth and diet of Australian bass (*Macquaria novemaculeata*) in Victorian waters, *Arthur Rylah Institute for Environmental Research Technical Report* 46: 1–9.
McCarraher, D.B. & McKenzie, J.A. 1986, Observations on the distribution, growth, spawning and diet of estuary perch (*Macquaria colonorum*) in Victorian waters, *Arthur Rylah Institute for Environmental Research Technical Report* 42: 1–21.
McDonald, C.M. 1976, *Morphological and biochemical systematics of Australian freshwater and estuarine fishes of the family Percichthyidae*, unpublished MSc thesis, Australian National University, Canberra, 185 pp.
McDowall, R.M. 1968, *Galaxias maculatus* (Jenyns), the New Zealand whitebait, *New Zealand Marine*

REFERENCES

Department Fisheries Research Bulletin 2: 1–84.

McDowall, R.M. 1971, Fishes of the family Aplochitonidae, *Journal of the Royal Society of New Zealand* 1(1): 31–52.

McDowall, R.M. 1974, Specialization in the dentition of the southern graylings—genus *Prototroctes* (Galaxioidei: Prototroctidae), *Journal of Fish Biology* 6(2): 209–213.

McDowall, R.M. 1976, Fishes of the family Prototroctidae, *Australian Journal of Marine and Freshwater Research* 27(4): 641–659.

McDowall, R.M. 1978a, A new genus and species of galaxiid fish from Australia (Salmoniformes: Galaxiidae), *Journal of the Royal Society of New Zealand* 8(1): 115–124.

McDowall, R.M. 1978b, Sexual dimorphism in an Australian galaxiid, *Australian Zoologist* 19(3): 309–314.

McDowall, R.M. 1979, Fishes of the family Retropinnidae (Pisces: Salmoniformes)—a taxonomic revision and synopsis, *Journal of the Royal Society of New Zealand* 9(1): 85–121.

McDowall, R.M. (ed.) 1980, *Freshwater fishes of south-eastern Australia (New South Wales, Victoria and Tasmania)*, first edition, Reed, Sydney.

McDowall, R.M. & Eldon, G.A. 1980, The ecology of whitebait migrations (Galaxiidae, *Galaxias* spp.), *New Zealand Ministry of Agriculture and Fisheries, Fisheries Research Bulletin* 20: 1–171.

McDowall, R.M. & Frankenberg, R.S. 1981, The galaxiid fishes of Australia, *Records of the Australian Museum* 33(10): 443–605.

McDowall, R.M. & Fulton, W. 1978a, A revision of the genus *Paragalaxias* Scott (Salmoniformes: Galaxiidae), *Australian Journal of Marine and Freshwater Research* 29(1): 93–108.

McDowall, R.M. & Fulton, W. 1978b, A further new species of *Paragalaxias* Scott (Salmoniformes: Galaxiidae) from Tasmania, with a revised key to the species, *Australian Journal of Marine and Freshwater Research* 29(4): 659–665.

McDowall, R.M., Mitchell, C.P. & Brothers, E.B. 1994, Age at migration from the sea of juvenile *Galaxias* in New Zealand (Pisces: Galaxiidae), *Bulletin of Marine Science* 54(2): 385–402.

McDowall, R.M., Robertson, D.A. & Saito, R. 1975, Occurrence of galaxiid larvae and juveniles in the sea, *New Zealand Journal of Marine and Freshwater Research* 9(1): 1–11.

McKay, N.J. 1973a, The reproductive cycle of the fire-tail gudgeon, *Hypseleotris galii*, I Seasonal histological changes in the ovary, *Australian Journal of Zoology* 21(1): 53–66.

McKay, N.J. 1973b, The reproductive cycle of the fire-tail gudgeon, *Hypseleotris galii*, II Seasonal changes in the gonads, *Australian Journal of Zoology* 21(1): 67–74.

MacKay, N.J. 1973c, Histological changes in the ovaries of the golden perch, *Plectroplites ambiguus*, associated with the reproductive cycle, *Australian Journal of Marine and Freshwater Research* 24(1): 95–101.

McKay, R.J. 1984, Introduction of exotic fishes in Australia, pp. 177–199 in Courtenay, W.R. & Stauffer, J.R. (eds), *Distribution, biology and management of exotic fishes* (430 pp.), Johns Hopkins University, Baltimore.

McKeown, K.C. 1934, Notes on the food of trout and Macquarie perch in Australia, *Records of the Australian Museum* 19: 141–152.

McKeown, K.C. 1937, The food of trout in New South Wales, 1935–36, *Records of the Australian Museum* 20(2): 38–66.

McKeown, K.C. 1955, The food of trout in New South Wales, 1938–40, *Records of the Australian Museum* 23(5): 273–279.

McMahon, B.R. & Burggren, W.W. 1987, Respiratory physiology of the intestinal breathing in the teleost fish *Misgurnus anguillicaudatus*, *Journal of Experimental Biology* 133: 371–393.

Mallen-Cooper, M. 1992, Fish migration in the Murray-Darling system and the decline of silver perch, *Australian Society for Fish Biology Newsletter* 22(3): 24.

Mallen-Cooper, M. 1993, Habitat changes and declines of freshwater fish in Australia: what is the evidence and do we need more? pp. 118–123 in Hancock, D.A. (ed.), *Proceedings of the Australian Society for Fish Biology Workshop: Sustainable fisheries through sustaining fish habitat, Victor Harbour, South Australia, 12–13 August 1992*, Australian Government Printing Services, Canberra.

Mallen-Cooper, M. 1994, How high can a fish jump? *New Scientist* 142 (1921): 32–37.

Manikiam, J.S. 1963, *Studies on the yellow-eyed mullet, Aldrichetta forsteri (Cuv. and Val.) (Mugilidae)*, unpublished MSc thesis, Victoria University of Wellington.

Marshall, J. 1988, An extension of the known range of *Rhadinocentrus ornatus*, *Fishes of Sahul* 5(1): 196–197.

Marshall, J. 1989, *Galaxias olidus* in southern Queensland, *Fishes of Sahul* 5(3): 223–225.

Mather, P.B. & Arthington, A.H. 1991, An assessment of genetic differentiation among feral Australian tilapia populations, *Australian Journal of Marine and Freshwater Research* 42: 721–728.

Merrick, J.R. & Midgley, S.H. 1985, Note on the winter diet of golden perch (*Macquaria ambigua*) in Queensland, *Proceedings of the Royal Society of Queensland* 96: 61–62.

Merrick, J.R. & Schmida, G.E. 1984, *Australian freshwater fishes: biology and management*, author, Sydney.

Mills, C.A. 1991, Reproduction and life history, pp. 483–508 in Winfield, I.J. and Nelson, J.S. (eds), *Cyprinid fishes: systematics, biology and exploitation*, Chapman & Hall, London.

Milton, D.A. & Arthington, A.H. 1983a, Reproduction and growth of *Craterocephalus marjoriae* and *C. stercusmuscarum* (Pisces: Atherinidae) in south-eastern Queensland, Australia, *Freshwater Biology* 13: 589–597.

Milton, D.A. & Arthington, A.H. 1983b, Reproductive biology of *Gambusia affinis holbrooki* Baird and

Girard, *Ziphophorus helleri* (Günther) and *X. maculatus* (Heckel) (Pisces; Poeciliidae) in Queensland, Australia, *Journal of Fish Biology*, 23: 23–41.

Milton, D.A. & Arthington, A.H. 1985, Reproductive strategy and growth of the Australian smelt, *Retropinna semoni* (Weber) (Pisces: Retropinnidae), and the olive perchlet, *Ambassis nigripinnis* (De Vis) (Pisces: Ambassidae), in Brisbane, south-eastern Queensland, *Australian Journal of Marine and Freshwater Research* 36(3): 329–341.

Milward, N.E. 1969, Development of the eggs and early larvae of the Australian smelt *Retropinna semoni* (Weber), *Research Bulletin, State Fisheries of New South Wales* 10: 1–9.

Mitchell, P.A. 1976, *A study of the behaviour and breeding biology of the southern pygmy perch, Nannoperca australis australis (Günther) (Teleostei: Nannopercidae)*, unpublished BSc Hons thesis, University of Melbourne. Parkville.

Morison, A.K. & Anderson, J.R. 1990, *Galaxias brevipinnis* in north-eastern Victoria—a new species introduced to the Murray-Darling basin, *Australian Society for Fish Biology Newsletter* 20(2): 26.

Mulley, J.C. & Shearer, K.D. 1980, Identification of natural Yanco x Boolara hybrids of the carp *Cyprinus carpio* Linnaeus, *Australian Journal of Marine and Freshwater Research* 31: 409–411.

Munro, I.S.R. 1967, *The fishes of New Guinea*, Department of Agriculture, Stock and Fisheries, Port Moresby.

Musyl, M.K. & Keenan, C.P. 1992, Population genetics and zoogeography of Australian freshwater golden perch, *Macquaria ambigua* (Richardson 1845) (Teleostei: Percichthyidae), and electrophoretic identification of a new species from the Lake Eyre basin, *Australian Journal of Marine and Freshwater Research* 43(6): 1585–1601.

Myers, G.S. 1965, *Gambusia*—the fish destroyer, *Tropical Fish Hobbyist*, 28: 315–322.

Neira, F.J., Bradley, J.S., Potter, I.C. and Hilliard, R.W. 1988, Morphological variation among widely dispersed larval populations of anadromous southern hemisphere lampreys (Geotriidae and Mordaciidae), *Zoological Journal of the Linnean Society* 92: 383–408.

Nelson, G.J. & Rothman, M.N. 1973, The species of gizzard shad (Dorosomatinae) with particular reference to the Indo-Pacific region, *Bulletin of the American Museum of Natural History* 150(2): 131–206.

Nicholls, A.G. 1958a, The population of a trout stream and the survival of released fish, *Australian Journal of Marine and Freshwater Research* 9(4): 526–536.

Nicholls, A.G. 1958b, The Tasmanian trout fishery. II. The north-west region, *Australian Journal of Marine and Freshwater Research* 9(1): 19–59.

Nicholls, A.G. 1958c, The Tasmanian trout fishery. III. The rivers of the north and east, *Australian Journal of Marine and Freshwater Research* 9(2): 167–190.

O'Connor, W.G. & Koehn, J.D. 1991, Spawning of the mountain galaxias, *Galaxias olidus* Günther in Bruces Creek, Victoria, *Proceedings of the Royal Society of Victoria* 103(2): 113–123.

Ogilby, J.D. 1903, Studies in the ichthyology of Queensland, *Proceedings of the Royal Society of Queensland* 18: 7–27.

Ovenden, J.R., Bywater, R. & White, R.W.G. 1993, Mitochondrial DNA nucleotide sequence variation in Atlantic salmon (*Salmo salar*), brown trout (*Salmo trutta*), rainbow trout (*Oncorhynchus tshawytscha*) and brook trout (*Salvelinus fontinalis*) from Tasmania, Australia, *Aquaculture* 114: 217–227.

Pandian, T.J. 1969, Feeding habits of the fish *Megalops cyprinoides* Brousonett in the Cooum Backwater, Madras, *Journal of the Bombay Natural History Society* 65: 569–580.

Parrott, A.W. 1932, Age and growth of trout from Eildon Weir, Victoria, Australia, *New Zealand Journal of Science and Technology* 14(2): 569–580.

Parrott, A.W. 1971, The age and rate of growth of quinnat salmon (*Oncorhynchus tshawytscha* (Walbaum)) in New Zealand, *New Zealand Marine Department of Fisheries Technical Report* 63: 1–66.

Persson, L. 1991, Interspecific interactions, pp. 530–551 in Winfield, I.J. & Nelson, J.S. (eds), *Cyprinid fishes: systematics, biology and exploitation*, Chapman & Hall, London.

Pollard, D.A. 1971a, The biology of a landlocked form of the normally catadromous salmoniform fish *Galaxias maculatus* (Jenyns). I. Life cycle and origin, *Australian Journal of Marine and Freshwater Research* 22(1): 91–123.

Pollard, D.A. 1971b, The biology of a landlocked form of the normally catadromous salmoniform fish *Galaxias maculatus* (Jenyns). II. Morphology and systematic relationships, *Australian Journal of Marine and Freshwater Research* 22(1): 125–137.

Pollard, D.A. 1972a, The biology of a landlocked population of the normally catadromous salmoniform fish *Galaxias maculatus* (Jenyns). III. The structure of the gonads, *Australian Journal of Marine and Freshwater Research* 23(1): 17–38.

Pollard, D.A. 1972b, The biology of a landlocked population of the normally catadromous salmoniform fish *Galaxias maculatus* (Jenyns). IV. Nutritional cycle, *Australian Journal of Marine and Freshwater Research* 21(1): 39–48.

Pollard, D.A. 1973, The biology of a landlocked form of the normally catadromous salmoniform fish *Galaxias maculatus* (Jenyns). V. Composition of the diet, *Australian Journal of Marine and Freshwater Research* 24(3): 281–295.

Pollard, D.A. 1974, *The freshwater fishes of the Alligator River, 'Uranium Province' area (Top End, Northern Territory), with particular reference to the Magela Creek catchment (East Alligator River System)*, Australian Atomic Energy Commission, Canberra.

Potter, I.C. 1970, The life cycle and ecology of Australian lampreys of the genus *Mordacia*, *Journal of Zoology, London* 161: 487–511.

REFERENCES

Potter, I.C. 1980, The Petromyzoniformes, with particular reference to paired species, *Canadian Journal of Fisheries and Aquatic Sciences* 37(11): 1595-1615.

Potter, I.C. & Hilliard, R.W. 1986, Growth and the average duration of larval life in the southern hemisphere lamprey, *Geotria australis* Gray, *Experientia* 42(10): 1170-1173.

Potter, I.C. & Hilliard, R.W. 1987, A proposal for the function and phylogenetic significance in the dentition of lampreys (Agnatha: Petromyzontiformes), *Journal of Zoology, London* 212(4): 713-717.

Potter, I.C., Hilliard, R.W. & Bird, D.J. 1980, Metamorphosis in the southern hemisphere lamprey *Geotria australis*, *Journal of Zoology, London* 190: 405-430.

Potter, I.C., Hilliard, R.W., Bird, D.J. & Macey, D.J. 1983, Quantitative data on the morphology and organ weights during the protracted spawning-run period of the southern hemisphere lamprey *Geotria australis*, *Journal of Zoology, London* 200: 1-20.

Potter, I.C., Hilliard, R.W. & Neira, F.J. 1986, The biology of Australian lampreys, pp. 207-230 in de Dekker, P. & Williams, W.D. (eds), *Limnology in Australia*, CSIRO, Melbourne, and Junk, Dordrecht.

Potter, I.C., Ivantsoff, W., Cameron, R. & Minnard, J. 1986, Life cycles and distribution of atherinids in the marine and estuarine waters of southern Australia, *Hydrobiologia* 139(3): 23-40.

Potter, I.C., Lanzing, W.J.R. & Strahan, R. 1968, Morphometric and meristic studies on populations of Australian lampreys of the genus *Mordacia*, *Journal of the Linnean Society (Zoology)* 47: 533-546.

Potter, I.C. & Robinson, E. 1991, Development of the testis during post-metamorphic life in the southern hemisphere lamprey *Geotria australis* Gray, *Acta Zoologica, Stockholm* 72: 113-119.

Potter, I.C. & Strahan, R. 1968, The taxonomy of the lampreys *Geotria* and *Mordacia* and their distribution in Australia, *Proceedings of the Linnean Society, London* 179: 229-240.

Powell, J. 1930, Notes on elvers of *Anguilla reinhardtii*, *Australian Naturalist* 8(2): 38.

Rao, A.V.P. & Ghosh, P.K. 1986, Biological control of *Oryzias melastigma* (McClelland) in less saline ponds of Bakkhali by *Megalops cyprinoides* (Broussonet), *Journal of the Marine Biological Association of India* 28(1-2): 141-143.

Reynolds, L.F. 1983, Migration patterns of five fish species in the Murray-Darling River system, *Australian Journal of Marine and Freshwater Research* 34: 857-873.

Rimmer, M.A. 1985a, Early development and buccal incubation in the fork-tailed catfish *Arius graeffei* Kner and Steindachner (Pisces: Ariidae) from the Clarence River, New South Wales, *Australian Journal of Marine and Freshwater Research* 36: 405-411.

Rimmer, M.A. 1985b, Growth, feeding and condition of the fork-tailed catfish *Arius graeffei* Kner and Steindachner (Pisces: Ariidae) from the Clarence River, New South Wales, *Australian Journal of Marine and Freshwater Research* 36: 33-39.

Rimmer, M.A. 1985c, Reproduction of the fork-tailed catfish *Arius graeffei* Kner and Steindachner (Pisces: Ariidae) from the Clarence River, New South Wales, *Australian Journal of Marine and Freshwater Research* 36: 23-32.

Rimmer, M.A. & Merrick, J.R. 1983, A review of reproduction and development in the fork-tailed catfishes (Ariidae), *Proceedings of the Linnean Society of New South Wales* 107(1): 41-50.

Rimmer, M.A. & Midgley, S.H. 1985, Techniques for hatching eggs and rearing larvae of the Australian mouthbrooding catfishes, *Arius graeffei* and *Arius leptaspis* (Ariidae), *Aquaculture* 44: 333-337.

Roberts, T.R. 1978, An ichthyological survey of the Fly River in Papua New Guinea, with descriptions of new species, *Smithsonian Contributions to Zoology* 281: 1-72.

Romanowski, N. 1986, Murray River rainbowfish (Melanotaeniidae: *Melanotaenia splendida fluviatilis*): thermal limitations and repopulation strategies for a marginal habitat, *Fishes of Sahul* 3(4): 141-144.

Rosen, D.E. & Bailey, R.M. 1963, The poeciliid fishes (Cyprinodontiformes), their structure, zoogeography and systematics, *Bulletin of the American Museum of Natural History*, 126: 1-176.

Rowland, S.J. 1983a, Spawning of the Australian freshwater fish Murray cod, *Maccullochella peeli* (Mitchell), in earthen ponds, *Journal of Fish Biology* 23: 525-534.

Rowland, S.J. 1983b, The hormone induced spawning and ovulation of the Australian freshwater fish golden perch, *Macquaria ambigua* (Richardson) (Percichthyidae), *Aquaculture* 35: 221-238.

Rowland, S.J. 1985, *Aspects of the biology and artificial breeding of the Murray cod, Maccullochella peeli, and the eastern freshwater cod, M. ikei sp. nov. (Pisces: Percichthyidae)*, unpublished PhD dissertation, Macquarie University, Sydney, 253 pp.

Rowland, S.J. 1989, Aspects of the history and fishery of the Murray cod, *Maccullochella peeli* (Mitchell) (Percichthyidae), *Proceedings of the Linnean Society of New South Wales* 111(3): 201-213.

Rowland, S.J. 1993, *Maccullochella ikei*, an endangered species of freshwater cod (Pisces: Percichthyidae) from the Clarence River System, NSW, and *M. peeli mariensis*, a new subspecies from the Mary River System, Qld, *Records of the Australian Museum* 45: 121-145.

Saeed, B., Ivantsoff, W. & Allen, G.R. 1989, Taxonomic revision of the Family Pseudomugilidae (Order Atheriniformes), *Australian Journal of Marine and Freshwater Research* (40): 719-787.

Saddlier, S.R. (1992), A research recovery plan for the variegated pigmy perch, *Nannoperca variegata*, in south-eastern Australia, unpublished report to Australian National Parks and Wildlife Service, Department of Conservation and Natural

Resources, Victoria.

Sanger, A.C. 1978, *Aspects of the ecology and evolution of the pigmy perches (Teleostei: Kuhliidae)*, unpublished BSc Hons thesis, University of Melbourne, Melbourne, 74 pp.

Sanger, A.C. 1984, Description of a new species of *Gadopsis* from Victoria, *Proceedings of the Royal Society of Victoria* 96: 93–97.

Sanger, A.C. 1990, Aspects of the life history of the two-spined blackfish, *Gadopsis bispinosus*, in King Parrot Creek, Victoria, *Proceedings of the Royal Society of Victoria* 102: 89–96.

Sanger, A.C. 1993, *The Swan galaxias recovery plan: management phase*, Inland Fisheries Commission, Hobart.

Sanger, A.C. & Fulton, W. 1991, Conservation of endangered species of Tasmanian freshwater fish, *Tasmanian Inland Fisheries Commission Occasional Report* 91-01: 1–29.

Saville-Kent, W. 1886, Notes on *Prototroctes maraena*, *Papers and Proceedings of the Royal Society of Tasmania* 1885: cv–cvi.

Schmidt, J. 1928, The freshwater eels of Australia, with some remarks on the shortfinned species of *Anguilla*, *Records of the Australian Museum* 16: 179–201.

Schuster, W.H. 1952, Fish culture in brackish water ponds of Java, *Indo-Pacific Fisheries Council Special Publication* 1: 1–143.

Scott, E.O.G. 1941, Observations on fishes of the family Galaxiidae, Part III, *Papers and Proceedings of the Royal Society of Tasmania* 1940: 55–69.

Scott, E.O.G. 1953, Observations on some Tasmanian fishes, Part 5, *Papers and Proceedings of the Royal Society of Tasmania* 87: 141–166.

Scott, E.O.G. 1971, Observations on some Tasmanian fishes, Part XVIII, *Papers and Proceedings of the Royal Society of Tasmania* 105: 119–143.

Scott, T.D., Glover, C.J.M. & Southcott, R.V. 1974, *The marine and freshwater fishes of South Australia*, Government Printer, Adelaide.

Scott, W.B. & Crossman, E.J. 1973, Freshwater fishes of Canada, *Bulletin of the Fisheries Research Board of Canada* 184: 1–996.

Semple, G.P. 1986, *Pseudomugil signifer*—maintenance, reproduction and early development of the Pacific blue-eye, *Fishes of Sahul* 3(3): 121–125.

Semple, G.P. 1991, Reproductive behaviour and early development of the honey blue-eye, *Pseudomugil mellis* Allen and Ivantsoff 1982 (Pisces: Pseudomugilidae), from the North-east Coast Division, south-eastern Queensland, *Australian Journal of Marine and Freshwater Research* 42: 277–286.

Sibbing, F.A. 1991, Food capture and oral processing, pp. 377–412 in Winfield, I.J. & Nelson, J.S. (eds), *Cyprinid fishes: systematics, biology and exploitation*, Chapman & Hall, London.

Sloane, R.D. 1983, *The truth about trout*, Tas-Trout Publications, Rosny Park.

Sloane, R.D. 1984, The upstream movements of fish in the Plenty River, Tasmania, *Papers and Proceedings of the Royal Society of Tasmania* 118: 163.

Staley, J. 1966, Brown trout, pp. 233–242 in Calhoun, A. (ed.), *Inland fisheries management*, State of California Fish and Game Department, Sacramento.

Stead, D.G. 1903, Contributions to Australian ichthyology, *Report of the Department of Fisheries, New South Wales* 2: 34–35.

Stead, D.G. 1934, Notes and exhibits, *Proceedings of the Linnean Society of New South Wales* 59: 31.

Stephenson, W. & Grant, E.M. 1957, Experiments upon impounded callop or yellowbelly (*Plectroplites ambiguus* Richardson) at Somerset Dam, *Ichthyological Notes, Brisbane* 1(3): 73–110.

Sterba, G. 1962, *Freshwater fishes of the world*, Vista Books, London, 878 pp.

Sumpton, W. & Greenwood, J. 1990, Pre- and postflood feeding ecology of four species of juvenile fish from the Logan-Albert estuarine system, Moreton Bay, Queensland, *Australian Journal of Marine and Freshwater Research* 41: 795–806.

Suzuki, R. 1983, Multiple spawning of the cyprinid loach, *Misgurnus anguillicaudatus*, *Aquaculture* 31: 233–243.

Talwar, P.K. 1990, Fishes of the Andaman and Nicobar Islands, *Journal of the Andaman Science Association* 6(2): 71–102.

Taylor, W.R. 1964, Fishes of Arnhem Land, *Report of the American Australian Scientific Expedition to Arnhem Land* 4: 45–307.

Thomson, J.M. 1953, Status of the fishery for the sea mullet (*Mugil cephalus* Linnaeus) in eastern Australia, *Australian Journal of Marine and Freshwater Research* 4: 41–81.

Thomson, J.M. 1954a, The Mugilidae of Australia and adjacent seas, *Australian Journal of Marine and Freshwater Research* 5: 70–131.

Thomson, J.M. 1954b, The organs of feeding and the food of some Australian mullet, *Australian Journal of Marine and Freshwater Research* 5: 469–485.

Thomson, J.M. 1955, The movements and migrations of mullet (*Mugil cephalus* L.), *Australian Journal of Marine and Freshwater Research* 6: 328–347.

Thomson, J.M. 1956, Fluctuations in catch of the yellow-eye mullet *Aldrichetta forsteri* (Cuvier and Valenciennes) (Mugilidae), *CSIRO Australian Division of Fisheries and Oceanography Report* 1.

Thomson, J.M. 1957a, Biological studies of economic significance of the yellow-eye mullet, *Aldrichetta forsteri* (Cuvier and Valenciennes) (Mugilidae), *Australian Journal of Marine and Freshwater Research* 8(1): 1–13.

Thomson, J.M. 1957b, Interpretation of the scales of the yellow-eye mullet *Aldrichetta forsteri* (Cuvier and Valenciennes) (Mugilidae), *Australian Journal of Marine and Freshwater Research* 8: 14–28.

Thomson, J.M. 1963, Synopsis of biological data on the grey mullet, *Mugil cephalus* Linnaeus 1758, *CSIRO Australian Division of Fisheries and Oceanography Synopsis* 1.

Thomson, J.M. 1966, The grey mullets, *Annual Review of Oceanography and Marine Biology* 4: 301–335.

Thorpe, J.E. 1977a, Morphology, physiology, behaviour and ecology of *Perca fluviatilis* L. and *P.*

REFERENCES

flavescens Mitchill, *Journal of the Fisheries Research Board of Canada* 34: 1504–1514.

Thorpe, J.E. 1977b, Synopsis of biological data on the perch, *Perca fluviatilis* Linnaeus, 1758 and *P. flavescens* Mitchill, 1814, *FAO Fish Synopsis* 113: 1–138.

Tilzey, R.D.J. 1972, The Lake Eucumbene trout fishery, *The Fisherman (NSW)* 4(2): 1–9.

Tilzey, R.D.J. 1977, Key factors in the establishment and success of trout in Australia, *Proceedings of the Ecological Society of Australia* 10: 97–105.

Tilzey, R.D.J. 1980, Introduced fish, pp. 271–279 in Williams, W.D. (ed.), *An ecological basis for water resource management* (417 pp.), Australian National University Press, Canberra.

Trewevas, E. 1983, *Tilapiine fishes of the genera Sarotheradon, Oreochromis and Danakilia*, British Museum, London.

Tsukamoto, Y. & Okiyama, M. 1993, Growth during the early life history of the Pacific tarpon, *Megalops cyprinoides*, *Japanese Journal of Ichthyology* 39(4): 379–386.

Unmack, P. 1992, Victorian pygmy perches, *Fishes of Sahul* 7(3): 321–323.

Vari, R.P. 1978, The *Terapon* perches (Percoidei, Teraponidae): a cladistic analysis and taxonomic revision, *Bulletin of the American Museum of Natural History* 159(5): 175–340.

Vronisky, B.B. 1973, Reproductive biology of the Kamchatka chinook salmon (*Oncorhynchus tshawytscha* (Walbaum)), *Journal of Ichthyology* 12(2): 259–273.

Wade, R.A. 1962, The biology of the tarpon, *Megalops atlanticus*, and the oxeye, *Megalops cyprinoides*, with emphasis on larval development, *Bulletin of Marine Science, Gulf and Caribbean* 12: 545–672.

Wager, R. 1992, The oxleyan pygmy perch: maintaining breeding populations, *Fishes of Sahul* 7(2): 310–312.

Wager, R. & Jackson, P.D. (1993), The action plan for Australian freshwater fishes, Australian Conservation Agency Endangered Species Programme Project Number 147, Canberra, 122 pp.

Watanabe, K. & Hidaka, T. 1983, Feeding behaviour of the Japanese loach *Misgurnus anguillicaudatus*, *Journal of Ethology* 1: 86–90.

Waters, J.M., Lintermans, M. & White, R.W.G. 1994, Mitochondrial DNA variation suggests river capture as a source of variance in *Gadopsis bispinosus* (Pisces: Gadopsidae), *Journal of Fish Biology* 44(3): 549–551.

Weatherley, A.H. 1958, Growth, production and survival of brown trout in a large farm dam, *Australian Journal of Marine and Freshwater Research* 9(2): 159–166.

Weatherley, A.H. 1959, Some features of the biology of the tench, *Tinca tinca* (Linnaeus) in Tasmania, *Journal of Animal Ecology* 28: 73–87.

Weatherley, A.H. 1962, Notes on distribution, taxonomy and behaviour of tench, *Tinca tinca* (Linnaeus) in Tasmania, *Annals and Magazine of Natural History* (13), 4(48): 713–719.

Weatherley, A.H. 1963, Zoogeography of *Perca fluviatilis* Linnaeus and *Perca flavescens* Mitchill, with special reference to the effects of high temperature, *Proceedings of the Zoological Society of London* 141: 557–576.

Weatherley, A.H. 1977, *Perca fluviatilis* in Australia: zoogeographical expression of a life cycle in relation to environment, *Journal of the Fisheries Research Board of Canada* 34: 1464–1466.

Weatherley, A.H. & Lake, J.S. 1967, Introduced fish species in Australian inland waters, pp. 217–239 in Weatherley, A.H. (ed.), *Australian inland waters and their fauna: eleven studies*. Australian National University, Canberra.

Wells, R.D.S. 1984, The food of the grey mullet (*Mugil cephalus*) in Lake Waahi and the Waikato River at Huntly, *New Zealand Journal of Marine and Freshwater Research* 18(1): 13–19.

Welcomme, R.L. 1988, International introductions of inland aquatic species, *FAO Fisheries Technical Paper* 294: 1–318.

Wharton, J.F.C. 1971, Spawning induction, artificial fertilisation and pond culture of the Macquarie perch (*Macquaria australasica* Cuvier, 1830), *Australian Society for Limnology Bulletin* 5: 43–65.

Wharton, J.F.C. 1973, Spawning induction, artificial fertilisation and pond culture of the Macquarie perch (*Macquaria australasica* (Cuvier, 1830)), *Australian Society for Limnology Bulletin* 5: 43–65.

Wheeler, A. 1969, *The fishes of the British Isles and north-west Europe*, Macmillan, London.

Whitley, G.P. 1956, Life history of the freshwater eel, *Australian Museum Magazine* 12(3): 89–94.

Whitley, G.P. 1957a, The freshwater fishes of Australia, 7—Herrings and smelt (cont'd), *Australasian Aqualife* 2(4): 7–10.

Whitley, G.P. 1957b, The freshwater fishes of Australia, 9—Catfishes, *Australasian Aqualife* 2(6): 6–10.

Whitley, G.P. 1957c, The freshwater fishes of Australia, 10—eels, *Australasian Aqualife* 2(7): 6–10.

Whitley, G.P. 1961, The freshwater gudgeons of temperate Australia, *Australian Museum Magazine* 13: 332–337.

Williams, D.McB. 1975, *Aspects of the ecology of three sympatric Galaxias species in an Otway stream*, unpublished BSc Hons thesis, Monash University, Melbourne, 78 pp.

Index

Abercrombie River 163
Acanthopagrus butcheri 34
Adaminaby 163
Adelaide 10, 54, 67, 82, 89, 187
adspersa, Mogurnda 201, **209–10**
affinis, Gambusia 118
affinis holbrooki, Gambusia 118
Afurcagobius tamarensis 220, **227–8**
Afurcagobius suppositus 228
agassizi, Ambassis **146–7**
Agassiz's glassfish 147
Agonostomus forsteri 196
Albury 10, 165
Aldrichetta forsteri 34, 191, **195–6**
almighty, mouth **181–2**
Ambassidae 146
Ambassis agassizii **146–7**
Ambassis castelnaui 147
Ambassis marianus 146, **148–9**
Ambassis nigripinnis 147
ambigua, Macquaria 150, **151–3**
ambiguus, Plectroplites 153
American yellow perch 184
amniculus, Craterocephalus 123, **124–5**
Amoya bifrenatus 220, **226–7**
Amoya frenatus 227
Anguilla australis **39–42**
Anguilla australis australis 42
Anguilla australis orientalis 42
Anguilla reinhardtii 39, **42–3**
anguillicaudatus, Cobitis 115
anguillicaudatus, Misgurnus **114–15**
Anguillidae 26, 39
anguilliformis, Saxilaga 65
angustifrons, Philypnodon 203
Anson River 157, 187
Aplochiton 78
Aplochitonidae 27, 78
Apogon aprion 182
Apogonidae 29, 181
aprion, Aprion 182
aprion, Glossamia **181–2**
aprion gillii, Glossamia 182
Arenigobius bifrenatus 227
Arenigobius tamarensis 228
argentea, Liza 191, **196–7**
argenteus, Neosilurus 109, **112**
argenteus, Plotosus 112
argenteus, Porochilus 112
Argentina 36, 67, 69
Ariidae 26, 107
Arius australis 108
Arius graeffei **107–8**
Arius leptaspis 108
Arius midgleyi 108
Arius thalassinus 108
Arthur River 10, 157, 187
Arthurs Lake 63, 75
Arthurs paragalaxias **75–6**
Atherina microstoma 133
Atherinasoma microstoma 123, **132–3**
Atherinidae 30, 123
atherinids 123
Atlantic salmon **85**

atlanticus, Megalops 50–1
attenuatus, Austrocobitis 69
attenuatus, Galaxias 13, 69
attenuatus scriba, Galaxias 69
atun, Leionura 34
Auckland Islands 52, 54
auratus, Carassius **99–101**
auratus, Galaxias 53, **62–3**
aureus, Oreochromis 179
australasica, Macquaria 150, **153–5**
Australian bass **155–7**
Australian Capital Territory 115, 189
Australian freshwater basses 150
Australian freshwater cods 150
Australian grayling **96–8**
Australian rainbowfish 138
Australian smelt **92–4**
Australian Society for Fish Biology 12, 35, 56, 57, 59, 64, 65, 70, 75, 97, 125, 129, 143, 160, 161, 163, 165, 172, 173, 175, 210
Australian spotted gudgeon 211
australis, Anguilla **39–42**
australis, Arius 108
australis australis, Anguilla 42
australis flindersi, Nannoperca 170
australis, Geotria **36–8**
australis, Gobiomorphus 201, **206–7**
australis, Mogurnda 207
australis, Nannoperca 168, **169–70**
australis, Neoarius 108
Austrocobitis attenuatus 69

bangos 49
barb, rosy 100
Barcoo grunter 166
barcoo, Scortum 166
Barmah Forest 115, 163
Barmah Lakes 154
barracouta 34
barramundi 134
Barron River 177
bass, Australian **155–7**
Bass Strait 10, 40, 54, 61, 64, 187
basses, Australian freshwater 150
Bathurst 10, 203
Bathurst Harbour 83
Bathurst, Lake 159
Bathygobius 225
Beechworth 163
Bega 198
Bellinger River 111
Belonidae 27
benzocaine 17
bidyan 166
Bidyanus bidyanus **164–6**
bidyanus, Terapon 166
bifrenatus, Amoya 220, **226–7**
big-headed gudgeon 203
bispinosus, Gadopsis 186, **188–90**
black bream 34, 166
black mangrove cichlid **176–7**
blackfishes, freshwater 186, 188
blackfish, river **186–8**

blackfish, two-spined **188–90**
blue cod 163
blue-eyes 141
blue-eye, common 142
blue-eye, honey **143**
blue-eye, Pacific 142
blue catfish **107–8**
blue nose cod 163
blue-spot goby **222–3**
bobby cod 167
bobby perch 167
Boiling Down Creek 124
Bolgu Island 215
bongbong, Galaxias 57
bony bream **44–6**
Bonshaw 165
Bool Lagoon 172
Boolara carp strain 102
bony-snouted gudgeon 209
Bostockia porosa 168
Bovichtidae 29, 198
Brachygalaxias pusillus 72
bream 155
bream, black 34, 166
bream, bony **44–6**
bream, silver 166
brevipinnis, Galaxias **53–5**
Brewarrina 165
bridled goby **226–7**
Brisbane 119, 121, 122, 127, 143, 179
Brisbane River 132
Broken River 163
brook char 57, **86–7**
brook trout 87
Broome 208
brown trout 34, 68, **81–4**, 172
bullhead 203
Bulloo River 10, 113, 151, 166
bullrout **144–5**
bully mullet 194
Bundaberg 143
Bundjalong National Park 175
Burdekin River 128
Burnett River 10, 128, 137, 192, 196
Burragorang, Lake 115
Burrinjuck Dam 10, 85, 115, 154
butcheri, Acanthopagrus 34
Butinae 200
Butis butis 200, **207–9**
Byfield 136

Caboolture River 40
Cairns 111, 119, 177, 179
callop 153
Campaspe River 163
Campbell Island 52, 54
Cape York 42, 211
Carassiops compressus 213
Carassiops galii 215
Carassiops klunzingeri 217
Carassius auratus **99–101**
Carassius carassius 101
Carcharhinidae 26
cardinalfish, Gill's 182

241

INDEX

cardinalfishes 181
Carnarvon 179
carps 99, 101,
carp, common 103
carp, Crucian 101
carp, European 103, 125
carp, koi 101-3
carp, Prussian 101
carp gudgeon 213
carp gudgeon, Lake's **218-19**
carp gudgeon, Midgley's **217-18**
carp gudgeon, western 200, **215-17**
carpio, Cyprinus **101-3**
castelnaui, Ambassis 167
Cataract Dam 159, 163
catfish, blue **107-8**
catfish, central Australian **112**
catfish, freshwater **109-10**
catfishes, eel-tailed 109
catfishes, fork-tailed 107
catfishes, salmon 107
Central Australian catfish **112**
Centropogon robustus 145
cephalus, Mugil 191, **193-4**
chanda perch, western 147
chanda perches 146
Chandidae 30, 146
Chanidae 27, 48
Chanos chanos **48-9**
char, brook 57, **86-7**
Charlegrark, Lake 159
Chatham Islands 40, 52, 54, 67
chequered gudgeon 210, 211
Chile 36, 67, 69, 72
Chinchilla 10, 165
chinook salmon 91
Cichlasoma nigrofasciatum 176, **178**
cichlid, black mangrove **176-7**
cichlid, convict **178**
cichlid, zebra 178
Cichlidae 31, 176
cichlids 176
cichlid, Niger 177
Clarence galaxias **57-8**
Clarence Lagoon 57, 87
Clarence River 50, 57, 108, 111, 128, 161, 182
Clarence River cod 161
cleaveri, Galaxias 52, **64-5**
cleaveri, Saxilaga 65
climbing galaxias **53-5**
Clupea cyprinoides 51
Clupeidae 27, 44
clupeids 44
Clyde River 145
cobbler, golden 108
cobbler, silver 108
Cobitidae 27, 114
Cobitis fossilis 115
Cobitis anguillicaudatus 115
Cockburn River 124
cod 160, 161
cod, blue 163
cod, bobby 167
cod, eastern **161**
cod, Mary River **160**
cod, marbled river 187
cod, Murray **158-60**
cod, trout **162-3**, 167
Coffs Harbour 136

collundera 203
colonorum, Macquaria 150, **157-8**
common blue-eye 142
common carp 103
common galaxias 69
common jollytail **67-9**
common smelt 94
common sunfish 138
compressa, Hypseleotris 201, **212-13**
compressus, Carassiops 213
conchonius, Puntius **106**
Condamine River 10, 113, 124, 165, 187
Condobolin 10, 166
conger eel 43
congolli **198-9**
convex perchlet 149
convexus, Pseudoambassis 149
convict cichlid **178**
Cooktown 197
Cooper Creek 10, 93, 100, 112
Coraki 175
Coral Sea 40
Cotter Dam 154
Cotter River 115, 189
coxii, Galaxias 55
coxii, Gobiomorphus 201, **204-5**
Cox's gudgeon **204-5**
Cox's mountain galaxias 55
Craterocephalus amniculus 123, **124-5**
Craterocephalus dalhousiensis 124, **128-9**
Craterocephalus eyresii 123, **126-7**
Craterocephalus fluviatilis 124, **125-6**, 132
Craterocephalus gloveri 124, **129-30**
Craterocephalus marjoriae 124, **127-8**
Craterocephalus stercusmuscarum 123, 132
Craterocephalus stercusmuscarum fulvus 124, **131-2**
Crescent, Lake 62
crimson spotted jewelfish 138
crimson spotted rainbowfish **138-40**
crimson-tipped flathead gudgeon 209
crimson-tipped gudgeon **207-9**
Crucian carp 101
cucumber herring 98
cucumber mullet 98
Cyprinidae 27, 99
cyprinoides, Clupea 51
cyprinoides, Megalops **50-1**
Cyprinus carpio 99, **101-3**

Daintree River 145
Dalhousie Springs 129, 130
Dalhousie Springs hardyhead **128-9**
dalhousiensis, Craterocephalus 124, **128-9**
Dandenong Creek 34
Darling River 10, 45, 56, 70, 102, 113, 124, 125, 132, 163, 209
Darling River hardyhead **124-5**
Dartmouth Dam 154, 155
Darwin 208
Dasyatidae 26
Davey River 83
Dawson River 151
Deniliquin 154
Derwent River 105, 187, 198
dewfish 111
dissimilis, Paragalaxias 53, **72-3**

dobula, Mugil 194
doody 147
Duboulay's rainbowfish **137-8**
duboulayi, Melanotaenia 134, **137-8**
Dumaresque River 165
dwarf flathead gudgeon **203-4**
dwarf galaxias 67, 72, 170

East Australian current 40
eastern cod **161**
eastern freshwater cod 161
eastern gambusia **116-18**, 200, 210
eastern little galaxias **70-2**
Edelia 168
Edelia obscura 168, 173
Eden 215
eel, conger 43
eel farming 41
eel, glass 40, 41
Eel Hole Creek 177, 178
eel, longfinned **42-3**
eel, shortfinned **39-42**
eel, spotted 43
eels, freshwater 39
eel-tailed catfishes 109
EHN 154
Eleotridinae 28, 200
eleotroides, Paragalaxias 53, **74-5**
Ellogobius olorum 222
elongatus, Myxus 191, **193**
Elopidae 27, 50
empire gudgeon 213
empirefish **212-13**
endorae, Pranesella 133
endorae, Taeniomembras 133
English perch 185
Englishman 84
Enoplosidae 28
epizootic haematopoietic necrosis 154
erebi, Nematalosa **44-6**
estuary perch **157-8**
estuary perchlet **148-9**
Eucumbene, Lake 10, 54, 115
European carp, 103, 125
European perch 185
Euston 165
Ewen Ponds 172, 187
Ewen pygmy perch **171-2**
Eyre, Lake 10, 45, 93, 112, 123, 127, 151, 166
eyresii, Craterocephalus 123, **126-7**

Falkland Islands 52, 67
fantail mullet **196**
fario, Salmo 84
Favonigobius tamarensis 228
findlayi, Galaxias 57
firetailed gudgeon **214-15**
Fitzroy River 93, 182, 217
flabby 182
flathead 199
flathead gudgeon **202-3**, 204
flathead gudgeon, crimson-tipped 209
flathead gudgeon, dwarf **203-4**
flathead galaxias 70
flathead jollytail 70
flat-tail mullet **196-7**
flavescens, Perca 184
Flinders Island 10, 40, 54, 64, 67, 72, 169

INDEX

flindersi, Nannoperca australis 170
Fluvialosa richardsoni 46
fluviatilis, Craterocephalus 124, **125–6**, 132
fluviatilis, Melanotaenia 134, **138–40**
fluviatilis, Perca **183–5**, 210
fontanus, Galaxias 52, **59–60**
fontinalis 87
fontinalis, Salvelinus 81, **86–7**
forktailed catfishes 107
forktailed catfish, freshwater 108
formalin 17
forsteri, Agonostomus 196
forsteri, Aldrichetta 34, 191, **195–6**
fossilis, Cobitis 115
fossilis, Misgurnus 115
Frankston 173
Fraser Island 10, 131, 136, 137, 143, 174, 215
Fraser Island sunfish 136
Fremantle 193, 197
freshwater basses 150
freshwater blackfishes 186
freshwater catfish **109–11**
freshwater cods 150
freshwater eels 39
freshwater forktailed catfish 108
freshwater hardyhead, western 132
freshwater herring **46–7**
freshwater jewfish 111
freshwater mullet **191–2**
freshwater perch 157
freshwater silverside 132
Frome River 127
fulvus, Craterocephalus stercusmuscarum 124, **131–2**
fuscus, Galaxias 56, 57

Gaden Hatchery 87
Gadopsidae 28, 168, 186
Gadopsis bispinosus 186, **188–90**
Gadopsis marmoratus **186–8**
gairdnerii, Salmo 14, 89
Galaxias affinis 55
Galaxias attenuatus 13, 69
Galaxias attenuatus scriba 69
Galaxias auratus 53, **62–3**
Galaxias bongbong 57
Galaxias brevipinnis **53–5**
Galaxias cleaveri 52, **64–5**
Galaxias coxii 55
Galaxias findlayi 57
Galaxias fuscus 56, 57
Galaxias fontanus 52, **59–60**
Galaxias johnstoni 53, **57–8**
Galaxias kayi 57
Galaxias maculatus 13, 53, **67–9**
Galaxias niger 55
Galaxias occellatus 62
Galaxias oconnori 57
Galaxias olidus 53, **55–7**
Galaxias ornatus 57
Galaxias parkeri 55
Galaxias parvus 52, **66–7**
Galaxias pedderensis 53, **58–9**
Galaxias planiceps 70
Galaxias rostratus 53, **69–70**
Galaxias schomburgkii 57
Galaxias scopus 62
Galaxias tanycephalus 53, **63–4**

Galaxias truttaceus 53, **60–2**
Galaxias upcheri 65
Galaxias weedoni 55
galaxias, Clarence **57–8**
galaxias, climbing **53–5**
galaxias, common 69
galaxias, Cox's mountain 55
galaxias, dwarf 67, 72, 170
galaxias, eastern little **70–2**
galaxias, flathead 70
galaxias, golden **62–3**
galaxias, inland 57
galaxias, mountain **55–7**
galaxias, mud 65
galaxias, ornate mountain 57
galaxias, Pedder **58–9**
galaxias, Pieman 55
galaxias, saddled **63–4**
galaxias, Shannon 73
galaxias, spotted **60–2**
galaxias, swamp **66–7**
galaxias, Swan **59–60**
Galaxiella pusilla 52, **70–2**
Galaxiidae 28, 52
galaxiids 52
Gale's gudgeon 215
galii, Carassiops 215
galii, Hypseleotris 201, **214–15**
galwayi, Lizagobius 222
galwayi, Mugilogobius 222
Galway's goby 222
Gambier, Mount 10, 72
Gambusia affinis 118
Gambusia affinis holbrooki 118
gambusia, eastern **116–18**, 200, 210
Gambusia holbrooki **116–18**, 200, 210
Gascoyne–Lyons River 179
Gawler River 34
Geehi River 189
George, Lake 159
Georges River 192
Geotria australis **36–8**
Geotriidae 25, 36
georgii, Valamugil 191, **196**
Geraldton 195
Gerridae 30
gillii, Glossamia 182
Gill's cardinalfish 182
Ginnindera Creek 115
Gippsland 10, 71
Gippsland Lakes 34, 102, 156
glass eel 41
glassfish, Agassiz's 147
glassfishes 146
Glenbawn Dam 163
Glencoe tandan 113
glencoensis, Neosilurus 113
Glenelg River 172, 224
Glossamia aprion **181–2**
Glossamia aprion gillii 182
gloveri, Craterocephalus 124, **129–30**
Glover's hardyhead **129–30**
gobies 220
Gobiidae 28, 220
Gobiinae 28, 220
Gobiomorphus australis 201, **206–7**
Gobiomorphus coxii 201, **204–5**
Gobius olorum 222
goby, blue-spot **222–3**

goby, bridled **226–7**
goby, Galway's 222
goby, largemouth **223–5**
goby, southern 222
goby, Swan River **221–2**, 223
goby, Tamar River **227–8**
goby, Tasmanian **225**
golden cobbler 108
golden galaxias **62–3**
golden perch **151–3**, 172, 184
goldfish **99–101**
Goobarrangandra River 189
Goodradigbee River 189
Googong Dam 154
Gordon River 10, 66
Goulburn River 154, 163
graeffei, Arius **107–8**
Grafton 175
grandiceps, Philypnodon 201, **202–3**, 204
grayling, Australian **96–8**
graylings, southern 96
greasy 188
Great Dividing Range 42, 46, 97, 137, 145, 184, 187, 146, 184, 187, 188
Great Lake 73, 74, 83
Green Lake 159
greyback 133
grey mullets 191
Grose River 97
grunters 164
grunter, Barcoo 166
grunter, spangled 167
grunter, Welch's 166
gudgeons 200
gudgeon, Australian spotted 211
gudgeon, bigheaded 203
gudgeon, bonysnouted 209
gudgeon, carp 213
gudgeon, chequered 210, 211
gudgeon, Cox's **204–5**
gudgeon, crimson-tipped flathead 209
gudgeon, dwarf flathead **203–4**
gudgeon, empire 213
gudgeon, firetailed **214–15**
gudgeon, flathead **202–3**, 204
gudgeon, Gale's 215
gudgeon, Lake's carp **218–19**
gudgeon, Midgley's carp **217–18**
gudgeon, Mulgoa 205
gudgeon northern trout **210–11**
gudgeon, purple spotted 211
gudgeon, purple striped 210
gudgeon, striped **206–7**
gudgeon, southern purple spotted **209–10**
gudgeon, trout 210, 211
gudgeon, western carp 200, **215–17**
gudgeon, Yarra 203
Gulf of Carpentaria 112, 166, 211
Gulf of Mexico 117, 119
Gulf St Vincent 10, 36
Gundagai 163
guppy 116, **118–19**

hardgut mullet 194
hardyheads 123
hardyhead, Darling River **124–5**
hardyhead, Dalhousie Springs **128–9**
hardyhead, Glover's **129–30**

243

INDEX

hardyhead, Lake Eyre **126-7**
hardyhead, Marjorie's freshwater **127-8**
hardyhead, Mary River 128
hardyhead, Mitchellian freshwater **131-2**
hardyhead, Murray **125-6**
hardyhead, smallmouthed **132-3**
hardyhead, western freshwater 132
Hawthorn, Lake 102
Hawkesbury River 10, 34, 111, 154, 156
Hazelwood Power Scheme 177, 178
helleri, Xiphophorus 116, **120-1**
Hemirhamphidae 28
Herbert River 211
herrings 44
herring, cucumber 98
herring, freshwater **46-7**
herring, Nepean 47
herring, oxeye **50-1**
herring, Yarra 98
Hexanematichthys leptaspis 108
Hiawatha, Lake 147, 175
Hobart 10, 132
holbrooki, Gambusia affinis 118
holbrooki, Gambusia **116-18**, 200, 210
honey blue-eye **143**
Hopkins River 97
hornorum, Oreochromis 179
Hume, Lake 189
Hunter River 10, 111, 195, 217
Huon River 10, 67, 187
Hypseleotris compressa 201, **212-13**
Hypseleotris galii 201, **214-15**
Hypseleotris klunzingeri 201, **215-17**
Hypseleotris species 4 201, **217-18**
Hypseleotris species 5 201, **218-19**
Hyrtl's tandan **113**
hyrtlii, Neosilurus 109, **113**

inanga 69
ikei, Maccullochella 151, **161**
immeritus, Parvigobius 222
inland galaxias 57
inland rainbow 140
irideus, Salmo 89

Jackson, Port 108
James River 77
Japanese loach 115
Japanese weatherfish 115
Japanese weatherloach 115
javanicus, Stigmatogobius 222
jewfish, freshwater 111
jewelfish 136
jewelfish, crimson spotted 138
jewel perch 167
Jindabyne, Lake 85, 87
johnstoni, Galaxias 53, **57-8**
jollytail, common **67-9**
jollytail, flathead 70
jollytail, Murray **69-70**
julianus, Paragalaxias 53, **76-7**

kayi, Galaxias 57
kenaru 111
key to atherinids 123
key to basses and cods 150
key to blackfishes 186
key to blue-eyes 141

key to cichlids 176
key to clupeids 44
key to cyprinids 99
key to eel-tailed catfishes 109
key to eleotridids 200
key to families of fishes 25
key to freshwater eels 39
key to galaxiids 52
key to glassfishes 146
key to gobies 220
key to gudgeons 200
key to hardyheads, 123
key to livebearers 116
key to mullets 191
key to pygmy perches 168
key to rainbowfishes 134
key to salmonids 81
key to shortheaded lampreys 33
key to silversides 123
key to smelts 92
key to terapontids 164
Kimberley 113
King Island 54, 67, 97, 169
king salmon 91
klunzingeri, Carassiops 217
klunzingeri, Hypseleotris 201, **215-17**
koaro 54, 55
koerin 210
koi carp 102, 103
Kosciusko, Mount 97
kroki 145
Kuhliidae 168
kurrin 210

La Trobe River 85 115, 177
Lachlan River 154, 163, 202
lake — see name of lake
Lake Eyre hardyhead **126-7**
Lake River 10, 75
Lake's carp gudgeon **218-19**
Lakes Entrance 10, 36
lamprey, non-parasitic **34-5**
lamprey, pouched **36-8**
lamprey, shortheaded 32, **33-4**
Lancefield 168
largemouth goby **223-5**
lasti, Tasmanogobius 225
latipinna, Mollienisia 120
latipinna, Poecilia 116, **119-20**
Lebistes reticulatus 119
leetus, Paradules 170
Leionura atun 34
Leiopotherapon unicolor 164, **166-7**
leptaspis, Arius 108
leptaspis, Hexanematichthys 108
leptocephalus 40, 51
line-eye 132
lineolatus, Oxyeleotris 200
little galaxias, eastern **70-2**
Little Pine River 77
livebearers 116
Liza argentea 191, **196-7**
Lizagobius galwayi 222
Lizagobius olorum 222
loach, Japanese 115
loach, mud 115
loaches 114
Logan River 127
longfinned eel **42-3**
Lord Howe Island 40, 42, 52, 67, 193

lordi, Tasmanogobius 220, **225**
Lovettia sealii 55, 68, **78-80**, 95
Lutjanidae 25, 31
Lutjanus argentimaculatus 25

Maccullochella macquariensis 13, 150, **162-3**
Maccullochella ikei 151, **161**
Maccullochella mitchelli 13, 163
Maccullochella peeli 13
Maccullochella peeli mariensis 151, **160**
Maccullochella peeli peeli 150, **158-60**
Mackay 119
Mcintyre River 124
Macleay River 137, 161
MacKenzie River 202
Macquaria ambigua 150, **151-3**
Macquaria australasica 150, **158-60**
Macquaria colonorum 150, **157-8**
Macquaria novemaculeata 150, **155-7**
Macquarie perch **153-5**, 158
Macquarie River 10, 60, 163
macquariensis, Maccullochella 13, 150, **162-3**
macrostoma, Redigobius 220, **223-5**
maculatus, Galaxias 13, 53, **67-9**
maculatus, Xiphophorus 116, **122**
Madigania unicolor 167
mangrove mullet 194
Manning River 111
Mannum 163
maraena, Prototroctes **96-8**
marbled river cod 188
marble fish 199
mariae, Tilapia **176-7**
marianus, Ambassis 146, **148-9**
mariensis, Maccullochella peeli 151, **160**
marjoriae, Craterocephalus 124, **127-8**
Marjorie's freshwater hardyhead **27-8**
marmoratus, Gadopsis **186-8**
Mary River 10, 111, 127, 128, 132, 156, 160
Mary River cod **160**
Mary River hardyhead 128
Maryborough 10, 149, 193, 206
Megalops atlanticus 50, 51
Megalops cyprinoides **50-1**
Megalops filamentosus 51
Megalops setipinnis 51
Melanotaenia duboulayi 134, **137-8**
Melanotaenia fluviatilis 134, **138-40**
Melanotaenia splendida fluviatilis 140
Melanotaeniidae 11, 29, 134
Melbourne 10, 42
mellis, Pseudomugil 141, **143**
melonfish 46
mesotes, Paragalaxias 53, **75-6**
Microperca tasmaniae 170
Microperca yarrae 173
microstoma, Atherina 133
microstoma, Atherinosoma 123, **132-3**
microstoma, Taeniomembras 133
midgleyi, Arius 108
Midgely's carp gudgeon **217-18**
Mildura 102, 163
milkfish **48-9**
millionsfish 119
minnow, trout 62
minnows 99, 127
Misgurnus anguillicaudatus **114-15**

INDEX

Misgurnus fossilis 115
mitchelli, Maccullochella 13, 163
Mitchellian hardyhead **131–2**
Mitta Mitta River 163
Mogurnda adspersa 201, **209–10**
Mogurnda australis 207
Mogurnda mogurnda 201, **210–11**
Mogurnda striata 210
Mollienisia latipinna 120
molly, sailfin **119–20**
Mongarlowe River 154
Monodactylidae 30
moonfish, silver 113
Mordacia mordax 32, **33–4**
Mordacia praecox 32, **34–5**
Mordaciidae 25, 32
mordax, Mordacia 32, **33–4**
Moreton Bay 223
Moreton Island 136, 175
Moreton Island sunfish 136
Morwell 177, 178
Moruya River 10, 33, 35
mosquitofish 118
mosquito control 117, 215
mossambicus, Oreochromis 176, **179–80**
mottled tandan 113
Mount Gambier 10, 40, 72
mountain galaxias **55–7**
mountain galaxias, ornate 57
mountain perch 155
mountain trout 62
mountain trout, spotted 62
mouth almighty **181–2**
mouthbrooders 181
mouthbrooder, Mozambique 176, **179–80**
mouthbrooder, Queensland 182
Mowbray River 147
Mozambique mouthbrooder 176, **179–80**
Mozambique tilapia 180
mud galaxias 64
mud loach 115
mud trout 65
mudfish, Tasmanian **64–5**
Mugil cephalus 191, **193–4**
Mugil dobula 194
Mugilidae 29, 191
Mugilogobius galwayi 222
Mulgoa gudgeon 205
mullets 191
mullet, bully 194
mullet, cucumber 98
mullet, fantail **196**
mullet, flat-tail **196–7**
mullet, freshwater **191–2**
mullet, hardgut 194
mullet, mangrove 194
mullet, paddy 194
mullet, river 194
mullet, sand **193**
mullet, sea **193–4**, 196
mullet, silver 196
mullet, tiger 197
mullet, yelloweyed 34, **195–6**
mullets, grey 191
Mulwala 111
Mulwala, Lake 159
Mulwarree Ponds 159
Murray cod 45, **158–60** 184

Murray–Darling River 9, 40, 45, 54, 56, 70, 93, 97, 100, 102, 104, 111, 123, 132, 139, 146, 151–4, 159, 163, 165, 166, 169, 179
Murray hardyhead **125–6**
Murray jollytail **69–70**
Murray perch 153
Murray River 9, 45, 69, 70, 102, 105, 111, 115, 125, 157, 163, 165, 169, 187, 198, 202, 203, 209
Murray River rainbow 140
Murrumbidgee River 10, 45, 70, 102, 111, 115, 125, 152, 154, 163, 169, 189, 202, 209
mykiss, Oncorhynchus 14, 81, **87–9**
Myxus elongatus 191, **193**
Myxus petardi **191–2**

Namoi River 10, 124, 131
Nannatherina 168
Nannoperca australis 168, **169–70**, 173
Nannoperca australis flindersi 170
Nannoperca obscura 168, **172–3**
Nannoperca oxleyana 168, **174–5**
Nannoperca riverinae 170
Nannoperca variegata 168, **171–2**
Nannoperca vittata 168
Nannopercidae 31, 168
Narooma 142, 149
Nematalosa erebi **44–5**
Nematalosa richmondia 46
Nemingah 124
Neoarius australis 108
neon sunfish 136
Neosilurus argenteus 109, **112**
Neosilurus glencoensis 113
Neosilurus hyrtlii 109, **113**
Nepean herring 47
New Caledonia 42, 52
New Guinea 134, 208, 213
New Zealand 11, 36, 38, 40, 52, 54, 67, 89, 103, 184, 195
Niger cichlid 177
niger, Galaxias 55
nigripinnis, Ambassis 147
nigrofasciatum, Cichlasoma 176, **178**
niloticus, Oreochromis 179
non-parasitic lamprey **34–5**
Noosa River 175
Norfolk Island 40
North Esk River 33, 187
North Stradbroke Island 136
Northern Territory 48, 107, 117
northern hemisphere smelts 78, 92, 96
northern trout gudgeon **210–11**
nosy parker 209
Notesthes robusta **144–5**
novemaculeata, Macquaria 150, **155–7**
noxious fish 179
nudiceps, Philypnodon 203
Nymboida River 111, 161

obscura, Edelia 173
obscura, Nannoperca 168, **172–3**
obscurus, Paradeles 173
occellatus, Galaxias 62
occidentalis, Anguilla australis 42
oconneri, Galaxias 57
olidus, Galaxias 53, **55–7**
olivaceous, Priops 147

olive perchlet **146–7**
olorum, Ellogobius 222
olorum, Gobius 222
olorum, Lizagobius 222
olorum, Pseudogobius **221–2**
Oncorhynchus mykiss 14, 81, **87–9**
Oncorhynchus tshawytscha 81, **89–91**
Onkaparinga River 187
Oreochromis aureus 179
Oreochromis hornorum 179
Oreochromis mossambicus 176, **179–80**
Oreochromis niloticus 179
oriental weatherloach **114–15**
ornate mountain galaxias 57
ornate rainbowfish 136
ornatus Galaxias 57
ornatus, Rhadinocentrus 118, 134, **135–6**
Osmeridae 78, 92, 96
Otway Ranges 64, 187
Ouse River 77
Ovens River 115, 163
oxeye herring **50–1**
oxleyan pygmy perch **174–5**
oxleyana, Nannoperca 168, **174–5**
Oxyeleotris lineolatus 200

Pacific blue-eye 142
paddy mullet 194
Paddys River 115
paired species 35
pallidus, Pseudambassis 147
pandaka 225
Papua New Guinea 45, 51, 108, 109, 141, 181, 182
Paradules leetus 170
Paradules obscurus 173
Paragalaxias dissimilis 53, **72–3**
Paragalaxias eleotroides 53, **74–5**
Paragalaxias julianus 53, **76–7**
Paragalaxias mesotes 53, **75–6**
Paragalaxias shannonensis 73, 75
paragalaxias, Arthur's 75–76
paragalaxias, Great Lake 73, **74–5**
paragalaxias, Shannon **72–3**
paragalaxias, western **76–7**
parrishi, Galaxias 69
Parvigobius immeritus 222
parvus, Galaxias 52, **66–7**
Patagonia 52, 67
Pedder galaxias **58–9**
Pedder, Lake 10, 58, 59, 67
pedderensis, Galaxias 53, **58–9**
Peel River 124
peeli mariesnus, Maccullochella 151,**160**
peeli peeli, Maccullochella 150, **158–60**
peladillos 78
Penstock Lagoon 73
Perca flavescens 184
Perca fluviatilis **183–5**
Percalates 157, 158
perch 153, 157, 158
perch, bobby 167
perch, English 185
perch, estuary **157–8**
perch, European 185
perch, Ewen pygmy **171–2**
perch, freshwater 157, 183
perch, golden **151–3**, 172, 184
perch, jewel 167

245

INDEX

perch, Macquarie **153-5**, 184
perch, mountain 155
perch, Murray 153
perch, oxleyan pygmy **174-5**
perch, pygmy 184
perch, redfin 154, 159, 172, **183-5**, 210
perch, silver **164-6**
perch, southern pygmy **169-70**
perch, spangled **166-7**
perch, western chanda 147
perch, white 153
perch, Yarra pygmy **172-3**
perch, yellow 184
perches, chanda 146
perches, pygmy 168
perchlet, estuary **148-9**
perchlet, olive **146-7**
perchlet, silver 149
Percichthyidae 31, 150, 168
Percidae 29, 183
petardi, Myxus **191-2**
petardi, Trachystoma 192
Philypnodon angustifrons 203
Philypnodon grandiceps 201, **202-3**, 204
Philypnodon nudiceps 203
Philypnodon sp. 201, **203-4**
Piccaninnie Ponds 172
Pieman galaxias 55
Pilbara 113
pinkear 138
pinkeye 192
planiceps, Galaxias 70
platy **122**
Platycephalidae 29, 199
Plectroplites ambiguus 153
Plotosidae 27
Poeciliidae 28, 116
Poecilia latipinna 116, **119-20**
Poecilia reticulata 116, **118-19**
Port Hacking 196
Port Jackson 108
porthole fish 136
Potamalosa richmondia 44, **46-7**
pouched lamprey **36-8**
praecox, Mordacia 32, **34-5**
Pranesella endorae 133
preservation of fish 17
Priops olivaceus 147
Pristidae 26
Prototroctes maraena **96-8**
Prototroctidae 27, 96
Prussian carp 101
Pseudambassis convexus 149
Pseudoambassis pallidus 147
Pseudambassis ramsayi 149
Pseudaphritis urvillii **198-9**
Pseudogobius olorum **221-2**
Pseudogobius sp. 9 221, **222-3**
Pseudomugilidae 141
Pseudomugil mellis 141, **143**
Pseudomugil signifer **141-2**
Pseudomugil signatus 142
Puntius conchonius **106**
purple-spotted gudgeon 211
purple-spotted gudgeon, southern **209-10**
purple-striped gudgeon 211
Purrumbete, Lake 90
pusilla, Galaxiella 52, **70-2**
puyen 69

pyberry 46
pygmy perch, Ewen **171-2**
pygmy perch, oxleyan **174-5**
pygmy perch, southern **169-70**
pygmy perch, Yarra **172-3**
pygmy perches 165, 184

Queensland mouthbrooder 182
quinnat salmon **89-91**

rainbow, inland 140
rainbow trout 54, **87-9**, 172
rainbow, Australian 138
rainbowfish, crimson spotted **138-40**
rainbowfish, Duboulayi's **137-8**
rainbowfish, Murray 140
rainbowfish, ornate 136
rainbowfish, softspined 118, **135-6**
rainbowfish, southern softspined 136
rainbowfishes 11, 134
ramsayi, Pseudambassis 149
redfin perch 154, 159, 172, **183-5**, 210
Redigobius macrostoma 220, **223-5**
reinhardtii, Anguilla 39, **42-3**
reticulata, Poecilia 116, **118-19**
reticulatus, Lebistes 119
Retropinna retropinna 94
Retropinna richardsoni 94
Retropinna semoni **92-4**
Retropinna tasmanica 79, 92, **95**
Retropinna victoriae 94
Retropinnidae 27, 92, 96
Rhadinocentrus ornatus 118, 134, **135-6**
richardsoni, Fluvialosa 46
richardsoni, Retropinna 94
Richmond River 10, 108, 157, 161, 175, 205
richmondia, Nematalosa 46
richmondia, Potamalosa 44, **46-7**
river blackfish **186-8**
river cod, marbled 188
river mullet 194
riverinae, Nannoperca 170
roach **105-6**
robusta, Notesthes **144-5**
robustus, Centropogon 145
Rockhampton 119
rostratus, Galaxias 53, **69-70**
rosy barb 106
Rubicon River 85
Rutilus rutilus 99, **105-6**
Ryans Creek 163

saddled galaxias **63-4**
sailfin molly **119-20**
salar, Salmo 81, **85**
Salmo fario 84
Salmo gairdnerii 14, 89
Salmo irideus 89
Salmo salar 81, **85**
Salmo trutta 34, **81-4**
salmon, Atlantic **85**
salmon, chinook 91
salmon farming 85, 89
salmon, king 91
salmon, quinnat **89-91**
salmon, spring 91
Salmonidae 26, 78
salmons 79
Salvelinus fontinalis 81, **86-7**

Sambell, Lake 163
sand mullet **193**
sand trout 199
sandy 199
Sarotherodon 180
Saxilaga anguilliformis 65
Saxilaga cleaveri 65
Scatophagidae 30
schomburgkii, Galaxias 57
Sciaenidae 29
scopus, Galaxias 62
Scorpaenidae 29, 144
scorpionfishes 144
Scortum barcoo 166
scriba, Galaxias attenuatus 69
sea mullet **193-4**, 196
sea trout 84
sealii, Lovettia 55, 68, **78-80**, 95
semoni, Retropinna **92-4**
Serranidae 150
Seven Creeks 163
shannonensis, Paragalaxias 73, 75
Shannon galaxias 73
Shannon Lagoon 73, 74
Shannon paragalaxias **72-3**
Shannon River 73
Shoalhaven River 10, 102, 154
shortfinned eel **39-42**
short-headed lamprey 32, **33-4**
signatus, Pseudomugil 142
signifer, Pseudomugil **141-2**
silver cobbler 108
silvereye 155
silver moonfish 113
silver mullet 196
silver perch **164-6**
silver perchlet 149
silver spray 147
silverside, freshwater 132
silversides 123
singularis, Yarra 38
sleeper 200
slimy 188
slippery 188
smallmouthed hardyhead **132-3**
smelt 127
smelt, common 94
smelt, Australian **92-4**
smelt, Tasmanian 79, **95**
smelts, southern 92
Snobs Creek 90
Snowy Mountains 87, 115
Snowy River 10, 54, 97
softspined rainbowfish 118, **135-6**
Sorell, Lake 62, 83
South Esk River 187
South Georgia 37
southern goby 222
southern graylings 96
southern purple-spotted gudgeon **209-10**
southern pygmy perch **169-70**
southern smelts 92
southern soft-spined rainbowfish 136
spangled grunter 167
spangled perch **166-7**
Sparidae 31
splendida fluviatilis, Melanotaenia 140
spotted eel 43
spotted galaxias **60-2**

246

INDEX

spotted gudgeon, Australian 211
spotted gudgeon, purple 211
spotted minnow 69
spotted mountain trout 62
spotted sunfish 138
spring salmon 91
steelhead 89
stercusmuscarum, Craterocephalus 123, 132
stercusmuscarum fulvus, Craterocephalus 124, **131–2**
Stigmatogobius javanicus 222
stinker 182
striata, Mogurnda 210
striped gudgeon **206–7**
sunfish 136
sunfish, common 138
sunfish, Fraser Island 136
sunfish, Moreton Island 136
sunfish, neon 136
sunfish, spotted 138
suppositus, Afurcagobius 228
Swamp galaxias **66–7**
Swampy Plains River 189
Swan galaxias **59–60**
Swan River 60
Swan River goby **221–2**, 223
swordtail **120–1**
Sydney 10, 46, 54, 102, 111, 159
Syngnathidae 26

Taeniomembras endorae 133
Taeniomembras microstoma 133
tailor 188
Taiwan 49
Talbingo Dam 163
Tamar River goby **227–8**
tamarensis, Afurcagobius 220, **227–8**
tamarensis, Arenigobius 228
tamarensis, Favonigobius 228
Tambo River 97
tandan 111
tandan, Glencoe 113
tandan, Hyrtl's **113**
tandan, mottled 113
tandan, white 113
tandan, yellowfin 113
Tandanus tandanus **109–11**
tanycephalus, Galaxias 53, **63–4**
tarpon 50, 51
tasmaniae, Microperca 170
Tasmanian goby 225
Tasmanian mudfish **64–5**
Tasmanian smelt 79, 94, **95**
Tasmanian whitebait 55, 62, 68, **78–80**, 95

tasmanica, Retropinna 79, 92, **95**
Tasmanogobius lasti 225
Tasmanogobius lordi 220, **225**
Taylor, Lake 159
tellagalene 193
tench **104–5**, 170
Terapon bidyanus 166
Terapontidae 31, 164
Tetraodontidae 26
thalasinus, Arius 108
Thymallus 96
tiger mullet 197
Tilapia mariae **176–7**
tilapia, Mozambique 180
Timor Sea 166
Tinana Creek 160
Tinca tinca 99, **104–5**
Tooma River 189
Torrens River 187
Torres Strait 215
Towamba River 213
Townsville 179, 193
Trachystoma petardi 192
trouts 81
trout, brook 87
trout, brown 34, 68, **81–2**, 172
trout cod **162–3**, 167
trout gudgeon 210, 211
trout gudgeon, northern **210–11**
trout minnow 62
trout, mountain 62
trout, mud 65
trout, rainbow **87–9**, 172
trout, sand 199
trout, sea 84
trout, spotted mountain 62
trutta, Salmo 34, **81–2**
truttaceus, Galaxias 53, **60–2**
tshawytscha, Oncorhynchus 81, **89–91**
Tuggerah Lakes 132
Tuggeranong Creek 115
tukari 46
Tully River 215
Tumut River 189
tupong 199
Turon River 163
Tuross River 10, 35

unicolor, Leiopotherapon 164, **166–7**
unicolor, Madigania 167
upcheri, Galaxias 65
urvillii, Pseudaphritis **198–9**

Valamugil georgii 191, **196**
Vansittart Island 40
variegata, Nannoperca 168, **171–2**

victoriae, Retropinna 94
vittata, Nannoperca 185

Wagga Wagga 10, 111
Wannon River 217
Warialda Creek 124, 125
Wangaratta 115
Warrego River 113
weatherfish 115
weatherfish, Japanese 115
weatherloach, Japanese 115
weatherloach, oriental **114–15**
weedoni, Galaxias 55
welchi, Bidyanus 166
Welch's grunter 166
Western Australia 61, 67, 100, 108, 113, 115, 117, 159, 165, 179, 182, 195, 197, 213, 222, 228
western carp gudgeon 200, **215–17**
western chanda perch 147
western freshwater hardyhead 132
western paragalaxias **76–7**
white perch 153
white eye 155
whitebait 127
whitebait farming 86
whitebait, Tasmanian 55, 62, 68, **78–80**, 95
Wilsons Promontory 64, 156, 205, 206
Wimmera River 159, 217
Wingecarribee River 115
Wollondilly River 115, 159
Woods Lake 63, 75
Woodwell 132
Wyangala Dam 154
Wye River 33

Xiphophorus helleri 116, **120–1**
Xiphophorus maculatus 116, **122**

yabby 72
Yamba 208
Yarra gudgeon 203
Yarra herring 98
Yarra pygmy perch **172–3**
Yarra River 10, 34, 105, 114, 154, 159, 169
Yarra singularis 38
yarrae, Microperca 173
Yarrawonga Weir 163
yellow perch 184
yellowbelly 153
yelloweyed mullet 34, **195–6**

zebra cichlid 178
Zostera 133